Science on the Roof of the World

When, how and why did the Himalaya become the highest mountains in the world? In 1800, Chimborazo in South America was believed to be the world's highest mountain, only succeeded by Mount Everest in 1856. *Science on the Roof of the World* tells the story of this shift, and the scientific, imaginative and political remaking needed to fit the Himalaya into a new global scientific and environmental order. Lachlan Fleetwood traces untold stories of scientific measurement and collecting, indigenous labour and expertise, and frontier-making to provide the first comprehensive account of the East India Company's imperial entanglements with the Himalaya. To make the Himalaya knowable and globally comparable, he demonstrates that it was necessary to erase both dependence on indigenous networks and scientific uncertainties, offering an innovative way of understanding science's global history, and showing how geographical features like mountains can serve as scales for new histories of empire.

Lachlan Fleetwood is a research fellow at University College Dublin.

SCIENCE IN HISTORY

Series Editors

Simon J. Schaffer, *University of Cambridge*
James A. Secord, *University of Cambridge*

Science in History is a major series of ambitious books on the history of the sciences from the mid-eighteenth century through the mid-twentieth century, highlighting work that interprets the sciences from perspectives drawn from across the discipline of history. The focus on the major epoch of global economic, industrial and social transformations is intended to encourage the use of sophisticated historical models to make sense of the ways in which the sciences have developed and changed. The series encourages the exploration of a wide range of scientific traditions and the interrelations between them. It particularly welcomes work that takes seriously the material practices of the sciences and is broad in geographical scope.

Science on the Roof of the World

Empire and the Remaking of the Himalaya

Lachlan Fleetwood
University College Dublin

CAMBRIDGE
UNIVERSITY PRESS

University Printing House, Cambridge CB2 8BS, United Kingdom

One Liberty Plaza, 20th Floor, New York, NY 10006, USA

477 Williamstown Road, Port Melbourne, VIC 3207, Australia

314–321, 3rd Floor, Plot 3, Splendor Forum, Jasola District Centre, New Delhi – 110025, India

103 Penang Road, #05–06/07, Visioncrest Commercial, Singapore 238467

Cambridge University Press is part of the University of Cambridge.

It furthers the University's mission by disseminating knowledge in the pursuit of education, learning, and research at the highest international levels of excellence.

www.cambridge.org
Information on this title: www.cambridge.org/9781009123112
DOI: 10.1017/9781009128117

First published 2022

A catalogue record for this publication is available from the British Library.

ISBN 978-1-009-12311-2 Hardback

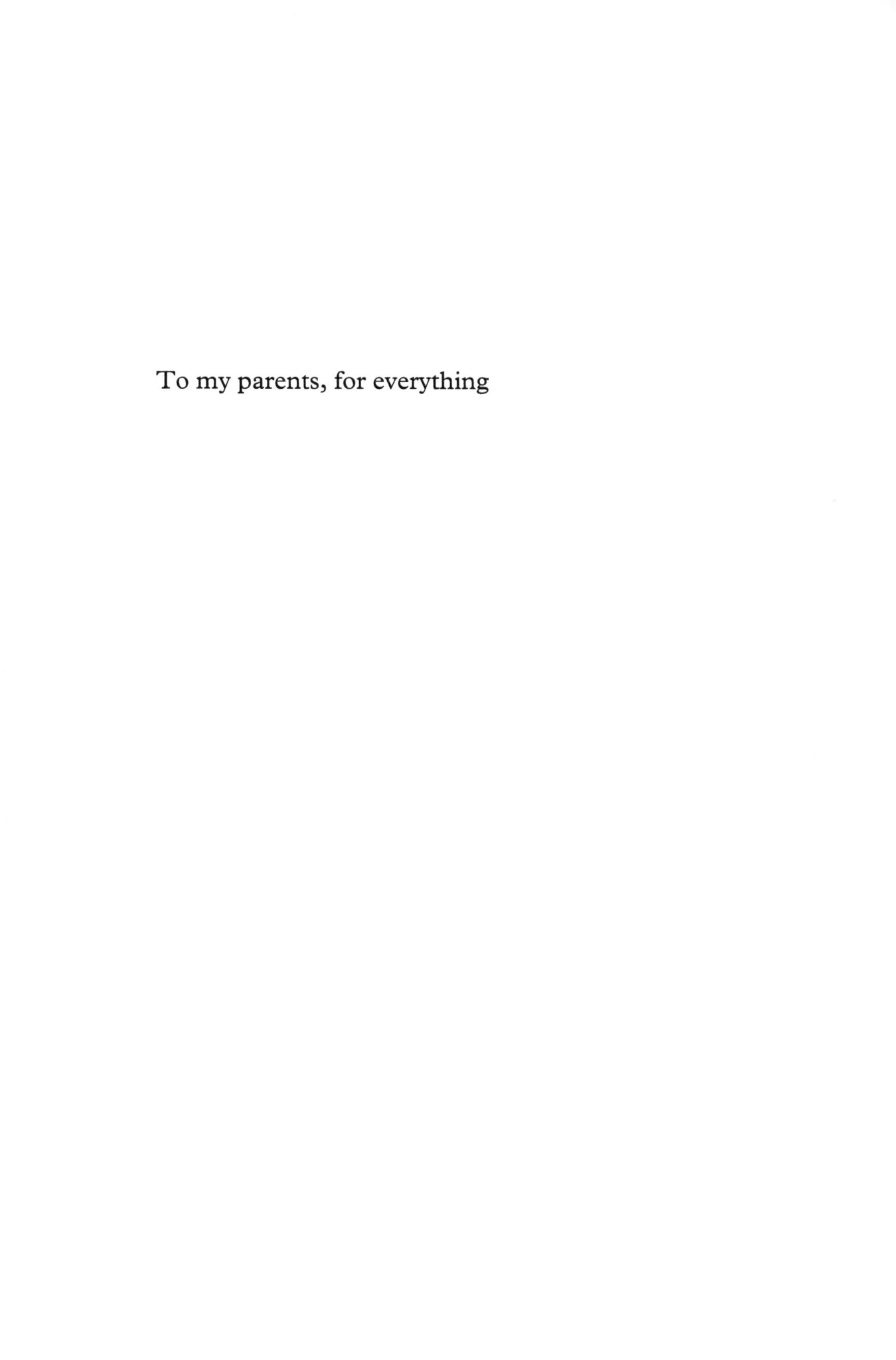

To my parents, for everything

Contents

Figures

Acknowledgements

Much as any who have been lucky enough to travel in the Himalaya, I would not have gotten very far in writing this book without depending on the knowledge, labour and expertise of many, all of whom deserve more profound thanks than an acknowledgements section alone can provide. Firstly, I owe an immeasurable debt of gratitude to my PhD supervisor, Sujit Sivasundaram, for his wonderful support throughout my doctorate at the University of Cambridge (and beyond). I thank him especially for his insightful critiques of endless drafts of my PhD thesis (from which this book arises) and his unwavering encouragement, advice and mentorship more generally. Several people also helped shape this project at pivotal moments. Special thanks, therefore, to Simon Schaffer, whose questions have resonated in all sorts of ways and whose examiners' comments have been extremely helpful in identifying ways to turn my thesis into a book. Likewise, my sincere thanks to Felix Driver for his equally invaluable comments during my PhD viva, and for his generosity and support since.

Many thanks also to Tom Simpson for his friendship, for his feedback on many drafts and for being a sounding board and collaborator in all things mountainous. Likewise, to James Poskett for his enthusiasm for the project and helpful suggestions in dealing with 'the global'. To Pratik Chakrabarti for his encouragement and for suggesting the need to think more carefully about the idea of scale. To Josh Nall for opening the Whipple Museum and its various mountain barometers and theodolites to me, and for helpful comments in the early stages. To Charlie Withers for his kind support and for various opportunities to expand my work on instruments. I am similarly grateful to various others with whom I discussed or corresponded on some aspects of this book, including Patrick Anthony, Bernard Debarbieux, Poornima Paidipaty, Michael Reidy, James Delbourgo (look out for the yeti), Bernhard Schär and Himani Upadhyaya. Likewise, to audiences in Cambridge, Manchester, London, Göttingen, Tübingen, Warsaw, Utrecht and Lausanne who listened to versions of the chapters and arguments in progress and asked great questions (even and especially the ones I could not answer).

Similarly, my gratitude to the various members of Sujit's reading group, past and present, on whom I have imposed parts (or in some cases most) of this material, namely: Tom Smith, Stephanie Mawson, Alix Chartrand, Jake Richards, James Wilson, Hatice Yildiz, Jagjeet Lally, Charu Singh, Mishka Sinha, Callie Wilkinson, Tamara Fernando, Taushif Kara, Scott Connors, Meg Foster and Marie de Rugy. Thanks for the friendship, commiserations, encouragement and numerous comments, large and small, pertinent and obscure. Similarly, my thanks to the many others who provided early career solidarity and intellectual encouragement, including Michael Sugarman, Catarina Madruga, Sebestian Kroupa, Daniel Allemann, Edwin Rose, Peter Martin, Joy Slappnig, Ed Armston-Sheret and Sarah Qidwai.

As is the case for any historical work, this book was built on the expertise and labour of numerous archivists, curators and librarians. In Cambridge, I would especially like to thank the staff at Cambridge University Library, Clare College Library, the Whipple Library and the Centre for South Asian Studies. Likewise, in London, my thanks to the teams at the British Library, Kew Gardens Library and Archives, the Natural History Museum and the Science Museum. Thanks also to the staff at the University of Nottingham Archives and the Oxford University Museum of Natural History. In India, I was made to feel welcome and was well cared for at the National Archives of India in Delhi, the Survey of India Museum in Dehradun and the Saharanpur Agri-Horticultural Institute.

I am immensely grateful to the Cambridge Commonwealth, European and International Trust for funding my PhD studies, for without their support I would never have been able to undertake the research that ultimately produced this book. In addition to the Trust, several other institutions provided financial support along the way, including Clare College, the Cambridge Faculty of History and the British Society for the History of Science. I acknowledge and sincerely thank them here. This book was revised and completed at University College Dublin (UCD), where my research has been generously funded by the Irish Research Council via a Government of Ireland Postdoctoral Fellowship (GOIPD/2020/44). At UCD, despite the many constraints of the COVID-19 pandemic, I have been very grateful for the enthusiasm and support of my postdoc mentor, Jennifer Keating.

At Cambridge University Press, I have been superbly well looked after. My thanks, in particular, to Lucy Rhymer, who read the proposal and provided wonderful support in the beginning, to Rachel Blaifeder, who was never less than fantastic in shepherding the manuscript through reviews, and to Emily Plater for her efforts corralling everything for production. Your support, help and generosity in navigating the world of academic book publishing as a first-time author has been hugely appreciated. Many thanks also to Jim Secord and Simon Schaffer for welcoming this book into the *Science in History* series. Similarly, my sincere thanks to the two anonymous reviewers at Cambridge University Press, whose generous, insightful and encouraging readers' reports helped improve the book substantially (any and all remaining shortcomings are of course my responsibility alone).

Earlier versions of two chapters of this book have previously been published in different forms: a version of Chapter 2 which focused on the period 1800–1830, as Lachlan Fleetwood, '"No Former Travellers Having Attained Such a Height on the Earth's Surface": Instruments, Inscriptions, and Bodies in the Himalaya, 1800–1830', *History of Science* 56, no. 1 (2018): 3–34; and a condensed version of Chapter 3 as Lachlan Fleetwood, 'Bodies in High Places: Exploration, Altitude Sickness, and the Problem of Bodily Comparison in the Himalaya, 1800–1850', *Itinerario* 43, no. 3 (2019): 489–515. I thank the editors of these journals, Lissa Roberts and Carolien Stolte, and the anonymous reviewers for their insightful and incisive feedback. I am likewise grateful to SAGE Publications and Cambridge University Press for allowing me to adapt and reuse material from these publications here.

Beyond these many professional and academic dues, my personal debts are even greater. Firstly, to Mirjam for her encouragement, love and friendship (and especially for tolerating the amount of my time and geographical proximity this project has taken up). Indeed, amidst all the travel, relocations and uncertainty of early career academia, your constancy and support have meant everything. Lastly, although infinitely far from least, a very many thanks to my family (Mum, Dad, Eleanor and Alice). Without your love and support, and acceptance that my interest in history would take me away from Australia to Canada, India, the United Kingdom and now Ireland, over many years, this book would never have come into being.

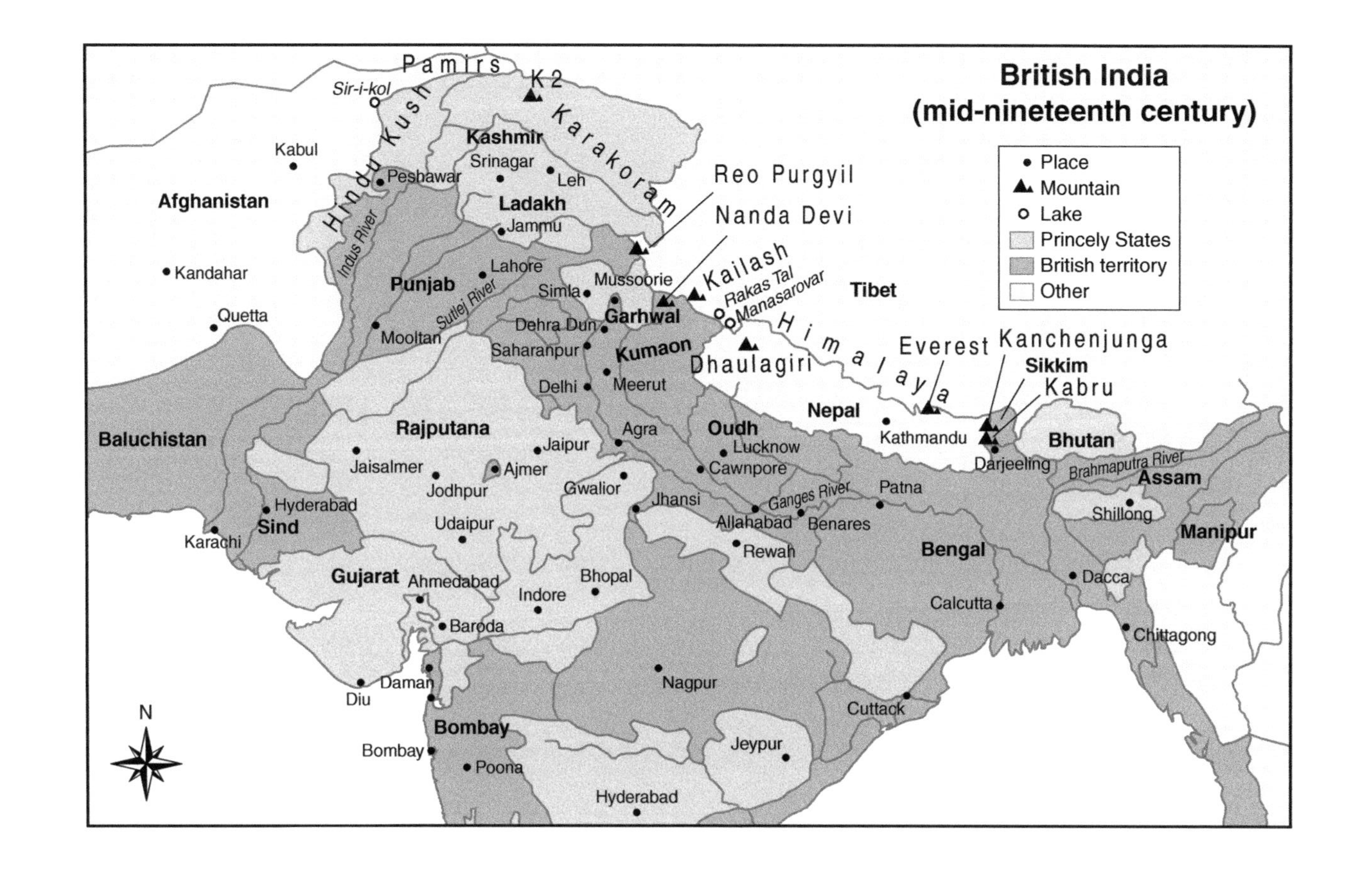
British India
(mid-nineteenth century)
Place
Mountain
Lake
Princely States
British territory
Other
Pamirs
Sir-i-kol
K2
Hindu Kush
Karakoram
Kashmir
Srinagar
Leh
Ladakh
Jammu
Kabul
Peshawar
Afghanistan
Kandahar
Quetta
Indus River
Punjab
Lahore
Sutlej River
Mooltan
Reo Purgyil
Nanda Devi
Kailash
Rakas Tal
Manasarovar
Tibet
Himalaya
Simla
Mussoorie
Garhwal
Dehra Dun
Saharanpur
Kumaon
Dhaulagiri
Delhi
Meerut
Everest
Kanchenjunga
Sikkim
Kabru
Nepal
Kathmandu
Bhutan
Darjeeling
Brahmaputra River
Assam
Shillong
Manipur
Baluchistan
Rajputana
Jaipur
Jaisalmer
Ajmer
Jodhpur
Gwalior
Agra
Oudh
Lucknow
Cawnpore
Jhansi
Ganges River
Allahabad
Benares
Patna
Hyderabad
Sind
Karachi
Udaipur
Rewah
Bengal
Dacca
Calcutta
Chittagong
Gujarat
Ahmedabad
Bhopal
Indore
Baroda
Nagpur
Daman
Diu
Cuttack
Bombay
Jeypur
Poona
Hyderabad
N

Introduction

This is a book about science, and how it changed the shape of mountains. It is also a book about how mountains changed the shape of science. In 1800, the volcano Chimborazo (in what is now Ecuador) was believed to be the highest mountain in the world. It was only around 1820 that Dhaulagiri in the Himalaya was accepted as higher, though not before a brief outrage among savants in Europe who doubted the measurements made by East India Company (EIC) surveyors. Other mountains, including Nanda Devi and Kanchenjunga, then had brief turns in the spotlight before Everest was finally confirmed as supreme in 1856. *Science on the Roof of the World* is the story of the intervening decades, and the scientific, imaginative and political remaking needed to fit the Himalaya into a new global scientific and imperial order, and to confirm Everest as the world's highest mountain.

Early efforts to map the Himalaya meant accounting for what were, at the turn of the nineteenth century, unprecedented if not unimaginable heights. Spanning some 2,400 kilometres in a roughly crescent- shaped band across Asia, the Himalaya are one of the most striking and celebrated geographical features of our planet. In the early nineteenth century, however, they remained little known to European science and geography, and far from a cohesive place. Within the vertiginous mountain locales and sweeping valleys of the Himalaya, geographies – both real and imagined – thus played central roles in the experiences of European surveyors and naturalists in a region of increasingly elevated political importance. They also shaped explorers' interactions with the Himalayan peoples, who assisted and resisted their attempts to map the Himalaya as an imperial and scientific frontier. In illuminating these uncertain and contested spaces, *Science on the Roof of the World* traces interweaving and largely untold stories of scientific measurement and collecting, indigenous labour and expertise and frontier making and breaking, through the mountains and across the first half of the nineteenth century. In turn, these moments in the mountains become part of a broader story; of how uplands became aberrant, science became three-dimensional and the globe became vertical.

While framed around the particularities of space, topography, terrain and landscape, this book is also about the Himalaya at a particular time. As much as for scientific edification, mapping and measuring greater and greater heights in the Himalaya in the first half of the nineteenth century was motivated by growing fears around the lack of information about the northern frontiers of EIC's territory in India. This was coupled with mounting concerns about the relative permeability rather than impenetrability of the range, which meant that the mountains increasingly took on the role of a threatening 'blank space' at the edges of the empire. Indeed, despite the meanderings of several Jesuits, and the trade missions of George Bogle (1746–1781) and Samuel Turner (1759–1802) to Tibet in the late eighteenth century, European knowledge of the Himalaya in 1800 remained sparse and fragmentary.[1] As Peter Bishop has eloquently argued, 'the immense *verticality* of the mountains, with their steep contrast between perpetually silent, snow-clad peaks and dark, densely vegetated valleys, echoed the intense *horizontal* mystery of the frontier' in this period, but 'as yet . . . the "frontier" lacked imaginative coherence'.[2] This is thus the story of uneven, incomplete and contested attempts to impose scientific, imperial and imaginative coherence on the Himalaya; a story bound up with both an increasingly expansionist British Empire in South Asia and an emerging global environmental order around altitude.

A central premise of this book is that the methods for surviving in these mountain environments mirrored the methods for making knowledge about them.[3] Attempts by European surveyors and naturalists to address the scientific, political and imaginative incoherence of the Himalaya in the first half of the nineteenth century is particularly reflected in the reconfiguration of practices and theories. Here I am especially interested in the moments that instruments, collections and bodies broke down, as these are revealing of the social relationships that underpinned the knowledge being produced.[4] Unlike many recent studies of imperial knowledge production which limit themselves to a particular branch of science – for

[1] For a useful conceptualisation of 'blank space', see Dane Kennedy, *The Last Blank Spaces: Exploring Africa and Australia* (Cambridge, MA: Harvard University Press, 2013). For Bogle and Turner, see Peter Bishop, *The Myth of Shangri-La: Tibet, Travel Writing, and the Western Creation of Sacred Landscape* (Berkeley: University of California Press, 1989); Kate Teltscher, *The High Road to China: George Bogle, the Panchen Lama and the First British Expedition to Tibet* (London: Bloomsbury, 2006); Gordon Stewart, *Journeys to Empire: Enlightenment, Imperialism and the British Encounter with Tibet* (Cambridge: Cambridge University Press, 2009).

[2] Bishop, *The Myth of Shangri-La*, 88–89.

[3] See Charlotte Bigg, David Aubin and Philipp Felsch, 'Introduction: The Laboratory of Nature – Science in the Mountains', *Science in Context* 22, no. 3 (2009): 314.

[4] Simon Schaffer, 'Easily Cracked: Scientific Instruments in States of Disrepair', *Isis* 102, no. 4 (2011): 706.

example, botany or medicine or astronomy – it is necessary here to address a broad range of sciences in order to understand the remaking of the Himalaya in the first half of the nineteenth century. As a result, this book ranges across different aspects of science and scientific practice, from instrumental measurement to altitude physiology, from fossils to botanic gardens and from glaciology to plant geography. While, on the one hand, a study of eclectic knowledge-making in the context of empire, this is thus necessarily also a story of the development of various modern scientific disciplines – from geology to biogeography – in which mountains were becoming central. Tracing these interlinked stories ultimately points to larger shifts in understandings of altitude in science, as the Himalaya were shaped in this period through the lens of global comparison, especially with the Alps and the Andes, and remade as commensurable within in a framework of empire. In turn, examining these reconfigurations offers new ways of comprehending the entangled relationship between scientific practice, environmental imaginaries and the reordering of global space in nineteenth-century imperial geography.

Even as it attends to these broader developments, *Science on the Roof of the World* also maintains a central focus on the roles of the Himalayan people who served as brokers, guides, porters and translators in European scientific and expeditionary practice. These demonstrate the overwhelming extent to which measuring the mountains depended on pre-existing local routes, expertise and labour. Even while ostensibly exploring the mountains, surveyors were almost never stepping off paths that had existed for millennia prior to their scientific interest. As remains true in the context of Himalayan mountaineering and trekking even today, foreign travellers would not have gotten very far or very high if they had not been able to rely on Himalayan peoples to identify the correct routes, transport their instruments and supplies, and share (or sometimes assume) the not insignificant risks of mountain travel.[5] Indeed, this book features an eclectic cast of long- overlooked brokers and guides, including Pati Ram and Bhauna Hatwal Khasiah, and technicians like Hari Singh and Murdan Ali, as well as a large number of (usually unnamed and unidentifiable) 'Bhotiyas', 'Lepchas' and 'Tartars', who served as interpreters, porters and informants (see for example

[5] For the continuation of this, and especially the later mythologisation of the Sherpa, see Peter H. Hansen, 'Partners: Guides and Sherpas in the Alps and Himalayas, 1850s–1950s', in *Voyages and Visions: Towards a Cultural History of Travel*, ed. Jaś Elsner and Joan-Pau Rubiés (London: Reaktion Books, 1999), 210–31; Sherry B. Ortner, *Life and Death on Mt. Everest: Sherpas and Himalayan Mountaineering* (Princeton: Princeton University Press, 2001).

Figure 0.1 'The Snowy Range from Tyne or Marma', published in George Francis White's *Views in India* (1838).[7] This is a typically romanticised image of Himalayan exploration, centred on a European traveller gazing heroically out to the mountains through a telescope (which implies his technological mastery). Meanwhile, the Bhotiya porters in the foreground are depicted lounging about, neatly fitting the trope of the 'lazy native'. However, their presence ultimately reminds us that surveyors and naturalists were always dependent on the expertise and labour of Himalayan peoples in the mountains.

Figure 0.1).[6] In examining everyday interactions within expedition parties, I aim to reveal the many ways these relationships were central to the remaking of the mountains. More broadly, these interactions matter because they shaped the subsequent relationships of Himalayan peoples

[6] These terms for ethnicity are problematic, frequently homogenising different groups of people. Alternatives are nevertheless often difficult to read out of the colonial documents, even as their imprecision was acknowledged by some contemporary travellers. For example, Joseph Hooker noted that 'the inhabitants of these frontier districts belong to two very different tribes, but all are alike called Bhoteeas (from Bhote, the proper name of Tibet)'. Joseph Dalton Hooker, *Himalayan Journals; Or, Notes of a Naturalist in Bengal, the Sikkim and Nepal Himalayas, the Khasia Mountains* (London: John Murray, 1854), Vol 1, 215. For more on these issues, see Felix Driver, 'Hidden Histories Made Visible? Reflections on a Geographical Exhibition', *Transactions of the Institute of British Geographers* 38, no. 3 (2013): 425; Christoph Bergmann, *The Himalayan Border Region: Trade, Identity and Mobility in Kumaon, India* (Dordrecht: Springer, 2016), 7–10.

[7] George Francis White, *Views in India: Chiefly among the Himalaya Mountains*, ed. Emma Roberts (London: Fisher, Son, and Co., 1838), 38.

to empire, and later to postcolonial South Asian states. In tracing the reconfiguration of practices and relationships, this book thus explains the many ways that the mountains came to be constituted as politically, culturally and environmentally marginal spaces in this period by imperial agents. In this context, I show how uplands were ultimately consolidated and confirmed as peripheral places in relation to the lowlands, which simultaneously came to be seen as both normal and normative – something that remains largely true today.

In revealing these histories, *Science on the Roof of the World* develops two main arguments through the mountains and across the first half of the nineteenth century. Firstly, it presents a close study of the sheer laboriousness of doing science in the Himalaya and the inherent dependency of surveyors and naturalists on pre-existing networks of labour and expertise, arguing for the necessity of further decentring the spaces of science. Secondly, it details the role of global comparison in the making of the mountains, and especially highlights the need to trace both connection and disconnection in understanding the rise of a vertically oriented view of the world. It is worth now considering these two sets of arguments separately and in turn, before reflecting on the way that together they help us to better understand the trajectories and implications of imperial science and geography in the nineteenth century.

Science at the Edge of Empire

As historical geographers and historians of science have come to insist, places 'are very far from being just matters of physical location and symbolic meaning. They are also constitutive of social exchange, enabling or constraining activities that carried out within their confines'.[8] My argument in this book follows from the 'spatial turn', a broadening body of scholarship that recognises the situatedness rather than placelessness of scientific practice, and increasingly acknowledges the analytical benefits to mapping out the geographies of science.[9] In what follows, the high spaces of the Himalaya – the mountainous topography, the social and cultural geography, the human and non-human dimensions – are cast in protagonistic roles. In so doing, I argue for further decentring the spaces of science. Scholars have already convincingly broken down classic diffusionist

[8] David N. Livingstone and Charles W. J. Withers, *Geographies of Nineteenth-Century Science* (Chicago: University of Chicago Press, 2011), 7.

[9] For an overview, see David N. Livingstone, *Putting Science in Its Place: Geographies of Scientific Knowledge* (Chicago: University of Chicago Press, 2003).

models of the spread of scientific knowledge from Europe to colonial peripheries, and effected a reorientation to India-centred perspectives on scientific practice.[10] The dismantling of these older paradigms, as brought about through the notion of 'circulation', and by scholarship that emphasises networks and webs, has been an important step.[11] However, in this book, I seek to facilitate a further move away from viewing peoples and practices in science outside of Europe through, to use Pratik Chakrabarti's term, 'centres in the periphery'.[12] Indeed, more needs to be done to effect an additional step, and to go beyond the overwhelming scholarly focus on major South Asian scientific 'centres', especially Calcutta, Madras, Benares and Bombay.[13] As French naturalist and traveller Victor Jacquemont (1801–1832) wrote in the early 1830s: '[at] the distance at which I am not only from Europe, but also from Calcutta and Bombay, nothing is so much a matter of chance as the arrival of my letters' (let alone journals or books with up-to-date scientific information).[14] As he continued, with perhaps a touch of bombast: 'I am waiting with great impatience for news ... [but] it is fourteen thousand miles from Calcutta to London, and fifteen hundred from hence to Calcutta' and moreover 'the post in India goes on foot, and tigers sometimes eat the letter-carriers'.[15] A central pillar of this book is

[10] Kapil Raj, *Relocating Modern Science: Circulation and the Construction of Knowledge in South Asia and Europe, 1650–1900* (New York: Palgrave Macmillan, 2007). For an example of a non-European scientific 'centre', see Savithri Preetha Nair, 'Native Collecting and Natural Knowledge (1798–1832): Raja Serfoji II of Tanjore as a "Centre of Calculation"', *Journal of the Royal Asiatic Society* 15, no. 3 (2005): 279–302. See also Deepak Kumar, 'Scientific Institutions as Sites for Dissemination and Contestation: Emergence of Colonial Calcutta as a Science City', in *Sites of Modernity: Asian Cities in the Transitory Moments of Trade, Colonialism, and Nationalism*, ed. Wasana Wongsurawat (Berlin: Springer, 2016), 33–46.

[11] The literature is vast, but see Bruno Latour, *Science in Action: How to Follow Scientists and Engineers through Society* (Cambridge, MA: Harvard University Press, 1987); Tony Ballantyne, *Orientalism and Race: Aryanism in the British Empire* (Basingstoke: Palgrave Macmillan, 2002); James A. Secord, 'Knowledge in Transit', *Isis* 95, no. 4 (2004): 654–72; Raj, *Relocating Modern Science*. For an example of classic centre/periphery approaches, see Robert Stafford, *Scientist of Empire: Sir Roderick Murchison, Scientific Exploration and Victorian Imperialism* (Cambridge: Cambridge University Press, 1989).

[12] Pratik Chakrabarti, *Western Science in Modern India: Metropolitan Methods, Colonial Practices* (Delhi: Permanent Black, 2004), 48, 94. For a similar argument, see Ballantyne, *Orientalism and Race*, 15.

[13] For the related idea of a 'double displacement' from London to Calcutta to Darjeeling, and how naturalists could sometimes use this to their advantage, see Thomas Trautmann, 'Foreword', in *The Origins of Himalayan Studies: Brian Houghton Hodgson in Nepal and Darjeeling 1820–1858*, ed. David Waterhouse (London: Routledge Curzon, 2004), xviii.

[14] Victor Jacquemont, *Letters from India*, trans. anon (London: Edward Churlton, 1834), Vol 1, 294.

[15] Jacquemont, *Letters from India*, Vol 1, 301.

thus investigating not only the unevenness with which information was carried out of the Himalaya, but also the unevenness with which up-to-date information was available to the eclectic array of naturalists, surveyors and travellers tasked with mapping and measuring what were only just becoming recognised were the world's highest mountains.

Historians have not been insensitive to the tendency to seek out new scientific 'centres' (not least for archival reasons).[16] Moreover, work on Calcutta – especially institutions like the Asiatic Society and the Botanic Garden – has been essential, not only to a more representative history of science in South Asia, but also to the particular stories told in this book. Indeed, both of these institutions played critical roles in circulating – if haphazardly – material and people between Europe and the mountains, and ultimately in the remaking of the Himalaya as a globally commensurable space. Likewise, Calcutta was itself a space of overlapping knowledge traditions, centres and peripheries, and incomplete attempts to create bounded spaces for science. Here the Asiatic Society, especially through its publications *Asiatic Researches* and the *Journal of the Asiatic Society of Bengal*, facilitated, alongside the natural historical investigations traced in this book, significant studies into literature, linguistics and ethnography, all of which also represent important strands of European imaginative investment in the Himalaya in this period. A different book might even have chosen to tell the story of the remaking of the Himalaya from these perspectives. In looking to illuminate some of the lesser-studied spaces and natural historical practices in the high mountains, however, I consider this scholarship on Calcutta as complementary, but ultimately only an essential step in a larger move. If what happened in the mountains is in some ways only half of the story, it is the half that has yet to be fully told. What follows is thus a sustained account of operating in displaced, disconnected and unevenly resourced locations, and the idiosyncrasies insisted on by those doing science there. Ultimately, this focus on decentred locations allows for new insights into labour dependency, scientific practice and imperial insecurity. At the same time, I seek to move beyond simply delineating the relationship between science, adventure and authority in the 'field'.[17] Instead, I aim to demonstrate the potential

[16] Chakrabarti, *Western Science in Modern India*, 38–39; David Arnold, *The Tropics and the Traveling Gaze: India, Landscape, and Science, 1800–1856* (Seattle: University of Washington Press, 2006), 169.

[17] See for example Bruce Hevly, 'The Heroic Science of Glacier Motion', *Osiris* 11 (1996): 66–86; Dorinda Outram, 'On Being Perseus: New Knowledge, Dislocation, and Enlightenment Exploration', in *Geography and Enlightenment*, ed. David N. Livingstone and Charles W. J. Withers (Chicago: University of Chicago Press, 1999), 281–94. Debate has centred around the tensions between 'laboratory' and 'field' sciences (even

of a mountain-oriented approach to understanding these spaces as both imperial and scientific frontiers. We are increasingly reminded that frontiers mattered, often in unexpected ways, and that the 'process of empire' is especially visible at the edges.[18] This book thus extends historical work on margins and peripheries, emphasising the way frontiers circumscribed the limits of knowledge and mastery in the early nineteenth century, and became conducive to particular forms of scientific practice and imperial ambition.

Scholars are nevertheless right to insist that European surveyors and naturalists, however peripherally they were located, were motivated by the understanding that they were participating in wider imperial and scientific projects, and that 'somewhere in their minds, the centrality of Europe remained overwhelming' as 'the final site of fame, recognition, and support', even if these would more often than not prove elusive.[19] This book features an eclectic mix of surgeons and naturalists, including William Griffith and George Govan, adventurers such as William Moorcroft, 'Mrs' Hervey and James Baillie Fraser, and Bengal Infantry officers seconded as surveyors and administrators, including William Webb, James Herbert and Alexander Gerard.[20] In many cases, rather than institutionally or state-sponsored explorers, these were EIC employees eclectically grafting their scientific proclivities onto regular duties.[21] Few gained financially (itself still a suspect reason for pursuing science), and indeed most were unable to recoup their costs, even via less tangible rewards like social advancement and esteem, which remained strong but often elusive motivations for many on the edges of empire. The way these surveyors and travellers positioned themselves – from necessity and also choice – vis-à-vis both Calcutta *and*

while acknowledging that this dichotomy was always somewhat ambiguous). For an overview, see Robert E. Kohler and Jeremy Vetter, 'The Field', in *A Companion to the History of Science*, ed. Bernard Lightman (Chichester: John Wiley, 2016), 282–95. See also especially Vanessa Heggie, *Higher and Colder: A History of Extreme Physiology and Exploration* (Chicago: University of Chicago Press, 2019).

[18] David Ludden, 'The Process of Empire: Frontiers and Borderlands', in *Tributary Empires in Global History*, ed. Peter Fibiger Bang and Christopher Bayly (Basingstoke: Palgrave Macmillan, 2011), 135–36.

[19] Chakrabarti, *Western Science in Modern India*, 81.

[20] For a narrative account of some of these expeditions, see John Keay, *When Men and Mountains Meet: The Explorers of the Western Himalayas, 1820–75* (London: John Murray, 1977). For earlier histories, see Clements Markham, *A Memoir on the Indian Surveys* (London: W. H. Allen, 1871); Kenneth Mason, *Abode of Snow: A History of Himalayan Exploration and Mountaineering from Earliest Times to the Ascent of Everest* (London: Rupert Hart-Davis, 1955).

[21] See David Arnold, *Science, Technology and Medicine in Colonial India* (Cambridge: Cambridge University Press, 2004), 24–25; Chakrabarti, *Western Science in Modern India*, 34, 86–87.

London thus becomes an important facet of the story. Disproportionately, they made their careers mostly or wholly in Asia rather than Europe. Some were born there, many learned their science there and a significant number died there – hoping but perhaps not fully expecting to see England, Scotland or France again. This study does feature notable exceptions like Hugh Falconer (1808–65) and John Forbes Royle (1798–1858), who transitioned to metropolitan circles with some success (the latter, for example, curating the Indian displays at the Great Exhibition of 1851), or those such as Victor Jacquemont, who achieved relatively widespread – if posthumous – fame in Europe (albeit in this case a notoriety largely resulting from a series of published letters full of injudicious and incendiary comments about various EIC officials). Even in these cases, however, it was their experiences and the scientific practices and theories they developed in the Himalaya that they were able to leverage for their success, and with which their intellectual reputations were bound up. We therefore need to take seriously that for many of the actors in this study, it was not just 'alternative centres' like Calcutta, but the displaced spaces of the Himalaya that were overwhelmingly the tableaux on which their scientific orientations played out.

This book thus seeks to explain the many ways that, in locations isolated even from Calcutta, let alone London, naturalists and surveyors adapted – or devised new – instruments, practices and theories using the knowledge, techniques and resources they had available to them, and the varying degrees to which they were successful. Central to this was the materiality of the mountains, and the ways that both human and non-human worlds were affected by altitude. Instruments and bodies functioned differently – often poorly – in the mountains, and fossils, plants and animals were arranged in particular ways. These difficulties were further compounded by the way surveyors' and naturalists' abilities to operate in Himalayan environments were overwhelmingly dependent on networks of labour and expertise in which their agency was often limited and sometimes resisted. At the same time, while the challenges travellers dealt with in the high mountains were real, it is important to recognise that they often sought to emphasise these logistical difficulties and physiological hardships. In what follows, I pay particular attention to moments when travellers exaggerated the idiosyncrasy of their surroundings, either to excuse their failings or to leverage their ability to overcome these difficulties and produce authoritative knowledge. Indeed, accounts of scientific practice in the Himalaya in this period are rife with explanations of, and a sometimes almost desperate insistence on, challenges that could supposedly only be resolved by those who had first-hand experience of the high mountains (even as similar claims were made by imperial agents engaging with other 'extreme' environments, whether oceans, jungles or deserts). As this book argues, it is precisely this tension – between the necessary insistence on the peculiarity of remote locations and the universalising needs of early nineteenth-century imperial geography – that makes the story of the appropriation and remaking of the Himalaya such an instructive one, opening up the possibilities for examining circulation and connection, and global and globalising sciences anew.

Mountain Science as Global Science

These overlapping stories of scientific practice and indigenous labour ultimately contribute to a larger story. Indeed, they underpin my second main argument around the importance of global comparison in the rise of verticality as a framework for understanding both human and non-human worlds. As much as it is a close study of everyday scientific practice and brokered exploration in the high spaces of the Himalaya, this is thus also a global history of mountain sciences. Measuring altitude with any real degree of accuracy had never really been necessary or even desirable before about the mid-eighteenth century. As the nineteenth century dawned, however, accurate measurements of elevation were becoming a critical variable in many of the sciences of the period, including plant geography, physiology and geology. The height of a mountain became an essential facet of its identity, and a principal factor in its significance to an emerging global pantheon.[22] Altitude was made a key characteristic, a point of data without which a given specimen – whether plant or rock or physiological reaction – had only a limited use in advancing the cause of natural history.[23] By the mid-century, altitude was pervasive. As the Bengal Infantry Officer Henry Strachey (1816–1912) put it when describing Western Tibet in 1853: 'having sketched the general plan of *Nari* in

[22] See Jon Mathieu, *The Third Dimension: A Comparative History of Mountains in the Modern Era*, trans. Katherine Brun (Cambridge: White Horse Press, 2011), 27–29; Bernard Debarbieux and Gilles Rudaz, *The Mountain: A Political History from the Enlightenment to the Present*, trans. Jane Marie Todd (Chicago: University of Chicago Press, 2015), 22–26; Veronica Della Dora, *Mountain: Nature and Culture* (London: Reaktion Books, 2016), 21–23. For the earlier history, see Sylvain Jouty, 'Naissance de l'altitude', *Compar(a)ison: An International Journal of Comparative Literature* 1 (1998): 17–32.

[23] Marie-Noëlle Bourguet, 'Landscape with Numbers: Natural History, Travel and Instruments in the Late Eighteenth and Early Nineteenth Centuries', in *Instruments, Travel and Science: Itineraries of Precision from the Seventeenth to the Twentieth Century*, ed. Marie-Noëlle Bourguet, Christian Licoppe and H. Otto Sibum (London: Routledge, 2002), 116.

horizontal extension, I must now explain its vertical relief, as this is developed on such a gigantic scale that no true idea of the country could be formed without equal attention to *the third co-ordinate*'.[24]

Science on the Roof of the World is thus the story of a growing recognition – and insistence – that mountain environments were commensurable, and that natural phenomena needed to be understood and mapped in three dimensions. In other words, accounting for the Himalaya in this period ultimately meant accounting for a globe that was not only round, but also vertical. These changing perceptions gave shape to what I call the 'vertical globe', and point to the increasing importance of verticality as a framework in organising and understanding scientific phenomena of all kinds. In what follows, the idea of the vertical globe encompasses a range of attempts to organise space in three dimensions, as well as the wider impacts of these reconfigurations on natural history, imperial geography, environmental imaginaries and frontier politics. Indeed, the practices and imaginative reorientations developed to account for the vertical globe ultimately highlight the imperial utility that underpinned nineteenth-century scientific applications of 'the globe' more widely. This, in turn, gives greater urgency to historicising and denaturalising these global and globalising imperial visions, and recognising the many vestiges of them which remain with us today, and continue to shape and distort our geographical imaginations.

These reorientations to the vertical had their origins in the second half of the eighteenth century when, as Marie-Noëlle Bourguet argues, 'the powerful image of the mountain as an ideal microcosm' became 'a common theme in the literature of natural history, and a scheme that framed travellers' perceptions'.[25] Writing about biogeography, Janet Browne similarly illustrates how in this period mountains came to be characterised as 'hemisphere[s] in miniature, where the floral zones of earth were found repeated on a vertical scale'.[26] As these perceptions were increasingly adopted in the first half of the nineteenth century, however, they had to be adapted as improved methods for accurately measuring elevation, and innovative techniques for mapping data were brought to bear on new parts of the globe. As in the usurpation of Chimborazo by Dhaulagiri, Kanchenjunga and finally Everest in the mid-1850s, the addition of the Himalaya to the vertical globe in the first half of the nineteenth century transformed it both rapidly and radically.

[24] Henry Strachey, 'Physical Geography of Western Tibet', *Journal of the Royal Geographical Society* 23 (1853): 16–17.
[25] Bourguet, 'Landscape with Numbers', 103.
[26] Janet Browne, *The Secular Ark: Studies in the History of Biogeography* (New Haven: Yale University Press, 1983), 46.

Taking a mountain-centric approach thus allows for the examination of the Himalaya as both a space and a subject of science. Altitude is treated here as both an object of study and a factor with particular implications for scientific and cross-cultural knowledge-making – simultaneously 'the *where* and the *what*' of scientific practice.[27] In this period, attempts to develop a coherent geography and natural history of the Himalaya unfolded alongside, advanced and diverged from major themes in the historiography of nineteenth-century science, including a new emphasis on precise instrumental measurement, an accelerating shift from amateur to professional practice, the geological time revolution and increasingly pressing questions around human origins. However, previous studies of science and surveying in the Himalaya have overwhelmingly gravitated towards mid-century figures, especially the Schlagintweit brothers and the 'Pundits', both of whom have been the subject of major recent studies.[28] The activities of English botanist and later director of Kew Gardens Joseph Dalton Hooker (1817–1911) and Kathmandu and Darjeeling-based naturalist and ethnologist Brian Houghton Hodgson (c.1801–1894) in the late 1840s have also received significant attention.[29] While particularly concerned with Hooker and Hodgson, David Arnold provides perhaps the most eloquent exception. Indeed, he argues that especially from the 1830s onwards, far 'from being

[27] Bigg, Aubin and Felsch, 'Introduction', 314.

[28] For the Schlagintweit brothers, see Gabriel Finkelstein, 'Conquerors of the Künlün? The Schlagintweit Mission to High Asia, 1854–57', *History of Science* 38, no. 2 (2000): 179–218; Moritz Von Brescius, *German Science in the Age of Empire: Enterprise, Opportunity and the Schlagintweit Brothers* (Cambridge: Cambridge University Press, 2019); Felix Driver, 'Intermediaries and the Archive of Exploration', in *Indigenous Intermediaries: New Perspectives on Exploration Archives*, ed. Shino Konishi, Maria Nugent and Tiffany Shellam (Canberra: Australian National University Press, 2015), 11–30. For the 'Pundits', see Derek Waller, *The Pundits: British Exploration of Tibet and Central Asia* (Lexington: University Press of Kentucky, 1990); Kapil Raj, 'When Human Travellers Become Instruments: The Indo-British Exploration of Central Asia in the Nineteenth Century', in *Instruments, Travel and Science: Itineraries of Precision from the Seventeenth to the Twentieth Century*, ed. Christian Licoppe, Marie-Noëlle Bourguet and H. Otto Sibum (London: Routledge, 2002), 156–88; Lowri Jones, 'Local Knowledge and Indigenous Agency in the History of Exploration' (PhD Dissertation, Royal Holloway, University of London, 2010); Tapsi Mathur, 'How Professionals Became Natives: Geography and Trans-Frontier Exploration in Colonial India' (PhD Dissertation, University of Michigan, 2018).

[29] For Hooker, see especially Jim Endersby, *Imperial Nature: Joseph Hooker and the Practices of Victorian Science* (Chicago: University of Chicago Press, 2008); Arnold, *The Tropics and the Traveling Gaze*; David Arnold, 'Envisioning the Tropics: Joseph Hooker in India and the Himalayas, 1848–1850', in *Tropical Visions in an Age of Empire*, ed. Felix Driver and Luciana Martins (Chicago: University of Chicago Press, 2005), 211–34. For Hodgson, see the essays collected in David Waterhouse, ed., *The Origins of Himalayan Studies: Brian Houghton Hodgson in Nepal and Darjeeling 1820–1858* (London: Routledge Curzon, 2004).

a remote, almost spectral, appendage to India, the Himalayas began to appear as a crossroads, a point at which, ethnographically as well as botanically and zoologically, China, Europe, and the Malay world met and mingled in bizarre and unexpected ways'.[30] This is an important insight, but this book nevertheless emphasises the need to pay as much attention to the period before 1830 as after it, when new instruments and practices were first brought to bear, the true scale of the mountains became apparent, and their basic formulations as scientific and imperial frontiers were imagined and laid down. While Hooker and Hodgson both make appearances in what follows, they are thus treated as part of a longer and larger project – albeit a haphazard one only intermittently endorsed by the EIC – of grappling with altitude and the makeup of the world's tallest mountains, rather than as exceptional. The first half of the nineteenth century is thus revealed as not simply a precursor or period of 'pioneering', but as foundational in terms of imperial knowledge of the Himalayan mountains and their significance in a world of vertical science (even if many of the subjects of this book were far less able and successful in publicising their contributions than their successors).

In particular, the early nineteenth century matters because accounting for the Himalaya was an inherently comparative process. Travellers were constantly measuring their expectations and experiences against other mountain ranges, especially the Alps and the Andes.[31] Here they were occasionally drawing from personal experience, but more usually from written descriptions. These comparisons could be inevitable and automatic, as well as deliberate strategies to aid explanation. For example, as EIC surveyor and administrator Richard Strachey (1817–1908) wrote, 'it will assist with the apprehension of the magnitude of these Himalayan masses if, before proceeding further, I compare a portion of them with the [Swiss Alps] with which so many of my European readers will be familiar'.[32] Such comparative lenses were so widely adopted in this period that a 'Mont Blanc' came to be a de facto unit of measurement, and a scale to which all other heights were referred. This of course did several kinds of work: on the one hand assisting in appropriating the mountains into a framework of European geography, but also helping explorers render explicable to themselves and to audiences at home the sheer scale of the

[30] David Arnold, 'Hodgson, Hooker and the Himalayan Frontier, 1848–1850', in *The Origins of Himalayan Studies: Brian Houghton Hodgson in Nepal and Darjeeling 1820–1858*, ed. David Waterhouse (London: Routledge Curzon, 2004), 196–97.

[31] Peter Bishop notes that as well as access to the Himalaya, these comparisons depended on wide familiarity with the Alps, itself a late eighteenth-century phenomenon. Bishop, *The Myth of Shangri-La*, 45.

[32] Richard Strachey, 'Physical Geography of the Himalaya (Unbound Proof Copy), c.1851', British Library, Mss. Eur. F127/200, f29.

Himalaya. Indeed, as Richard Strachey continued: 'it is a remark that I have constantly made ... that what in Europe would be called a mountain, [in the Andes or the Himalaya] we look upon as a mere spur of hill, and that the whole scale of our nomenclature is shifted in a similar manner'.[33] Across the first half of the nineteenth century, measuring Himalayan mountains thus required determining where they should be placed on an imagined vertical globe that was already crowded with the likes of Teneriffe, Chimborazo and Mont Blanc.

Making this accounting possible nevertheless first (or perhaps simultaneously) required the development of the Himalaya as a coherent territory that could be deployed as a unit of long-distance comparison. Indeed, without laborious measurements, made possible by imperial frameworks and resources, meaningful global comparisons were impossible. At the same time, these comparisons did not automatically bring clarity, and the recognition of the commensurability of mountain environments in South America and Europe could serve to increase, as much as to alleviate, the uncertainties around scientific phenomena like altitude sickness or the line of perpetual snow.[34] I am thus especially interested in moments when theories and understandings of mountains derived from norms established in the Alps and the Andes failed to transfer easily or broke down in the Himalaya. These are reflective of the inherent unevenness of global comparisons, and who had access to what information and when. More broadly, these moments echo growing calls to pay attention to not only circulation and connection in global history, but also disconnection, failure and resistance.[35]

In researching this book, I have frequently been struck by the way that scholarship on mountains is characterised by its fragmentation across multiple disciplines, including history, the history of science, geography, anthropology and literary and cultural studies.[36] One consistent strand across these disciplines has followed mountaineering, which, never far from the popular imagination, has also recently been the subject of sophisticated scholarship, for example on mountains and modernity.[37] To be clear though, while this is

[33] Strachey, 'Physical Geography of the Himalaya (Unbound Proof Copy), c.1851', f29.

[34] See Mathieu, *The Third Dimension*, 10. Comparisons could also be problematic because in South America the uplands were often centres of political power and population rather than vice versa. See Jon Mathieu, 'Long-Term History of Mountains: Southeast Asia and South America Compared', *Environmental History* 18, no. 3 (2013): 562–63.

[35] Sujit Sivasundaram, 'Focus: Global Histories of Science: Introduction', *Isis* 101, no. 1 (2010): 96.

[36] See for example the essays collected in Christos Kakalis and Emily Goetsch, eds., *Mountains, Mobilities and Movement* (London: Palgrave Macmillan, 2018).

[37] Peter H. Hansen, *The Summits of Modern Man: Mountaineering after the Enlightenment* (Cambridge, MA: Harvard University Press, 2013); Thomas Simpson, 'Modern

a history of mountains, it is not a history of mountaineering. Even if already being enthusiastically developed in the Alps, mountaineering as a sporting pursuit, shorn of scientific pretensions, was still decades away from reaching the Himalaya when this study takes its leave. Likewise, key tropes of imperial mountaineering (and the later history of Victorian exploration more widely) – including relentless publicisation, ostentatious egomania and sometimes patriotic nationalism – are not usually especially helpful in understanding the activities and motivations of most of the actors traced in this book. That said, some recent themes the history of mountaineering has pursued – for example the visibility of social relationships around class and gender in extreme environments – can be usefully co-opted into describing the sociability of Himalayan exploration.[38] The relationship between mountains, science and scientific practice has also received intermittent – though growing – attention, even if scholars have tended to gravitate to the later nineteenth and early twentieth centuries (and to activities in Europe and the Americas).[39] Notable exceptions have been made for savants like Genevan naturalist Horace-Bénédict de Saussure, who made the third ascent of Mont Blanc, and most extensively for Prussian polymath Alexander von Humboldt, whose influence and Andean legacy loomed large over many of the EIC naturalists and surveyors featured in this book, much as it does over histories of science, empire and geography in this period more widely.

In tracing the origins of the vertical globe through its inherently comparative dimensions, this book thus seeks to reconstruct and historicise the formation of what were both sciences of the globe *and* global sciences in practice. While many recent histories of global science have shied away from addressing larger, ostensibly universal framings, often preferring microstudies – and indeed my chapters often do the same – this is nevertheless a global history with *the globe* in it (or at least the parts of the globe that are significantly above sea level).[40] This book is thus

Mountains from the Enlightenment to the Anthropocene', *The Historical Journal* 62, no. 2 (2019): 553–81.

[38] See for example Michael S. Reidy, 'Mountaineering, Masculinity, and the Male Body in Mid-Victorian Britain', *Osiris* 30, no. 1 (2015): 160–61.

[39] See especially the essays in the special issue of *Science in Context* following Bigg, Aubin and Felsch, 'Introduction'. Francophone scholars have tended to pay more attention to the longer history: Jean-Claude Pont and Jan Lacki, eds., *Une cordée originale: histoire des relations entre science et montagne* (Chêne-Bourg: Georg, 2000); Aymon Baud, Philippe Forêt and Svetlana Gorshenina, *La Haute-Asie telle qu'ils l'ont vue: explorateurs et scientifiques de 1820 à 1940* (Geneva: Olizane, 2003). See also Veronica Della Dora and Denis Cosgrove, for whom 'high places' are defined as having either high altitude or high latitude: Denis E. Cosgrove and Veronica Della Dora, *High Places: Cultural Geographies of Mountains, Ice and Science* (London: I. B. Tauris, 2009).

[40] For the longer history of European notions of the globe, see Denis E. Cosgrove, *Apollo's Eye: A Cartographic Genealogy of the Earth in the Western Imagination* (Baltimore: Johns Hopkins University Press, 2001).

explicitly intended as a contribution to new methods for new global histories of science. While increasingly popular, the turn to 'the global' in the history of science has not been without caveats, especially around sources and the possibilities for escaping older and inherently Eurocentric framings without losing conceptual coherence.[41] In what follows, my approach is thus to interrogate the imperial utility of ideas of 'the globe' itself, as well as to address notions of 'the global' on multiple levels. This includes a material dimension in the imperial circulation and movement of enormous quantities of things – for example, specimens, inscriptions, drawings and personnel. However, I also particularly examine the imaginative dimension, considering the way surveyors and naturalists envisioned and positioned their science in relation to mountains in other parts of the world.[42]

Here 'the global' is not 'the universal', but rather a means of grasping mountains within a framework of commensurability reflecting the worldviews of its makers, and that itself circulated imperfectly as it was made and deployed for a range of sometimes contradictory ends. Nor, as it has sometimes been framed, is global science here a shorthand for either non-Western science or Western science outside of Europe; rather, it is treated as a category in the making, both a means and an end. In taking this approach, I ultimately argue that imagining and representing the world in three dimensions in this period, with heights precisely quantified by new instruments and techniques, had long-term implications for the political and economic positioning of mountain peoples and mountain environments. In the first half of the nineteenth century, the way naturalists, surveyors and administrators thought about and imagined upland environments and peoples as aberrations from horizontal norms, and Himalayan spaces as deviations from exemplar mountains in the Alps and the Andes, all mattered. The application of these visions ultimately did a particularly insidious kind of imperial work, paving the way for the scientific, aesthetic and political appropriation of the mountains, if – as this book ultimately shows – unevenly and incompletely.

[41] See Lissa Roberts, 'Situating Science in Global History: Local Exchanges and Networks of Circulation', *Itinerario* 33, no. 1 (2009): 9–30; Sujit Sivasundaram, 'Sciences and the Global: On Methods, Questions, and Theory', *Isis* 101, no. 1 (2010): 146–58; Marwa Elshakry, 'When Science Became Western: Historiographical Reflections', *Isis* 101, no. 1 (2010): 98–109; Fa-ti Fan, 'The Global Turn in the History of Science', *East Asian Science, Technology and Society: An International Journal* 6 (2012): 249–58.

[42] Here my approach follows James Poskett, *Materials of the Mind: Phrenology, Race, and the Global History of Science, 1815–1920* (Chicago: University of Chicago Press, 2019), 3–8.

Imaginative Geographies and Indigenous Topographies

Science on the Roof of the World is about a place both extraordinary and ordinary. It is about a region central to imperial imaginations, but peripheral to empires: a place at once always there, and yet one that was constantly reconfigured and recreated. In this book, I treat the 'Himalaya' in an inclusive sense, drawing in the Hindu Kush, Tien Shan, Pamirs and Karakoram, as well as sub-Himalayan ranges, including the Pir Panjal and the Siwaliks.[43] This reflects the understanding of many contemporaries, and in 1851, for example, Richard Strachey suggested 'the name Himalaya which is only strictly applied by the natives of India to the ranges they see covered with perpetual snow, has however naturally been long used by Europeans to designate the whole mountainous region'.[44] Such subdivisions are also to some extent arbitrary – the Trans-Himalaya are, after all, the result of the same geophysical processes, being formed only relatively recently in geological time by the collision between the Indian and Eurasian tectonic plates (see Figure 0.2).[45]

As Figure 0.2 also shows, the Trans-Himalayan belt spans what is now Pakistan, China (including the Xinjiang and Tibet Autonomous Regions), India, Nepal and Bhutan, as well as touching on Afghanistan, Tajikistan and Myanmar. In following surveyors, naturalists and guides across the mountains, we will sometimes encounter moments that contributed to the genesis of these modern state borders. Though worth keeping in mind given that they are the settings for ongoing and intractable conflicts and the legacies of empire, these borders are nevertheless not always a useful way to engage with the region even today, let alone in the nineteenth century.[46] Ultimately, the edges of the Himalaya remained much fuzzier than those of the states that came to ostensibly contain them, and people and goods regularly moved into and out of the mountains – spaces in which imperial and state power was anyway often difficult to bring to bear. Similarly, even generalised observations about the Himalayan mountains – for example, that they tend to be sparsely inhabited or feature harsh environments – are not uniformly true. Studying 'the Himalaya'

[43] For a succinct overview of Himalayan sub-classifications, see Maurice Isserman and Stewart Weaver, *Fallen Giants: A History of Himalayan Mountaineering from the Age of Empire to the Age of Extremes* (New Haven: Yale University Press, 2008), 1–8.

[44] Strachey, 'Physical Geography of the Himalaya (Unbound Proof Copy), c.1851', f2.

[45] See Mike Searle, *Colliding Continents: A Geological Exploration of the Himalaya, Karakoram, and Tibet* (Oxford: Oxford University Press, 2013).

[46] See Dan Smyer Yü, 'Introduction: Trans-Himalayas as Multistate Margins', in *Trans-Himalayan Borderlands: Livelihoods, Territorialities, Modernities*, ed. Dan Smyer Yü and Jean Michaud (Amsterdam: Amsterdam University Press, 2017), 16.

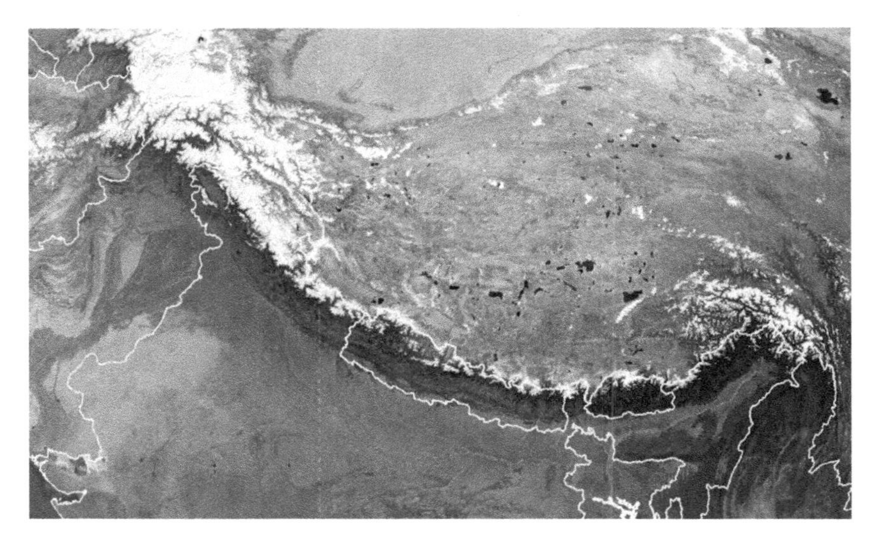

Figure 0.2 Satellite image showing the contiguous nature of the Trans-Himalayan belt. The borders of the modern nation-states that encompass the Himalaya are traced in white. This also shows the divide between the lowland monsoonal plains to the south of the range, and the high-altitude deserts that characterise the north.
Source: NASA Landsat 7.

holistically might thus inadvertently perpetuate lowlanders' perspectives, and reflect vestiges of older romantic, orientalist and imperial fascination with Asia's notoriously 'mysterious' and 'exotic' mountains – though these framings, too, have their histories.[47] Indeed, as I argue in this book, these homogenising tropes had particular imperial utility, and were ultimately essential to the process of appropriating the Himalaya and making them globally commensurable (even as they simultaneously compromised attempts to govern and control the uplands). In what follows, it is thus worth remembering that there is not just one Himalaya but many, and that experiences and understandings of the mountains could and did vary greatly among those who interacted with them.

While the Himalaya began to take on new imaginative, political and scientific importance for Europeans in the first half of the nineteenth century, they had long held particular semiotic significances for South Asians, whether residents of the mountains, or from the subcontinent and

[47] See especially Bishop, *The Myth of Shangri-La*; Tom Neuhaus, *Tibet in the Western Imagination* (New York: Palgrave Macmillan, 2012).

beyond.[48] The word 'Himalaya' has its etymology in Sanskrit, with 'Himā' meaning snow, and 'alāya' house or abode. The entire construction is thus often translated as the 'Abode of Snow', an intermittently apt descriptor but only one name among many.[49] The Himalaya hold key places in Buddhist, Jain, Sikh and Hindu cosmology. Himalayan peaks and lakes feature extensively in South Asian religious tradition and are important sites of pilgrimage. Mount Kailash, for example, serves as the resting place of the gods in the Hindu epics *Mahabharata* and *Ramayana* and is also the setting of the Buddhist classic *The Hundred Thousand Songs of Milarepa*.[50] The Himalaya also feature in the *Puranas*, especially in the *Skanda Purana*, in which the narrator warns the reader that 'in a hundred ages of the gods I could not tell thee of the glories of Himachal, where Shiva lived and where the Ganges falls from the foot of Vishnu like the slender thread of the Lotus flower'.[51] As this indicates, the Himalaya have long been recognised as the source of the rivers – specifically, the Ganges, Sutlej, Indus and Brahmaputra – which are crucial to agricultural production and hence life on the subcontinent.[52] Both historically and in the present, these texts have also frequently been used to support the idea of the Himalaya as a 'natural' barrier. Indeed, references to the idea of a 'Hindustan' separated and delineated by the Himalaya appear at least as far back as the first century BCE, where the *Vishnu Purana* refers to them as the 'shield of India'.[53]

These long-standing framings of the Himalaya, by both South Asian uplanders and lowlanders, were very apparent to European visitors, even if attempts to record them were riddled with misunderstandings and mistranslations. In the case of rivers, for example, Irish-born officer and botanist Edward Madden (1805–1856) wrote that: 'we may consider the

[48] Arnold, *The Tropics and the Traveling Gaze*, 102. In this book, I sometimes use the collective terms 'Asian' and 'South Asian' despite their relative anachronism. This is because as generic terms they are sometimes more widely applicable than 'Indian', as well as having the advantage of being hierarchically equal and equivalent to 'European' (even if only by being similarly non-specific).

[49] Isserman and Weaver, *Fallen Giants*, 1–2. This etymology also helps explain why I prefer the singular 'Himalaya' when referring to the region in this book, rather than the plural 'Himalayas' as is equally common. The pluralised version is itself arguably a legacy of colonialism: see Maharaj K. Pandit, *Life in the Himalaya: An Ecosystem at Risk* (Cambridge, MA: Harvard University Press, 2017), 3.

[50] For an overview, see Edwin Bernbaum, *Sacred Mountains of the World* (Berkeley: University of California Press, 1998).

[51] Quoted in Isserman and Weaver, *Fallen Giants*, 2.

[52] See Aniket Alam, *Becoming India: Western Himalayas under British Rule* (New Delhi: Foundation Books, 2008), 1–2.

[53] Nayanika Mathur, 'Naturalizing the Himalaya-as-Border in Uttarakhand', in *Borderland Lives in Northern South Asia*, ed. David N. Gellner (Durham: Duke University Press, 2013), 72.

Himalaya as nature's vast reservoir for the irrigation of empires ... it is probable, that a portion of the Hindoo veneration for the range is owing to its containing the springs of so many of the rivers which fertilize their country'.[54] In considering European scientific and imaginative appropriations of the Himalaya, it is thus important to remember that imperial places 'did not emerge straightforwardly out of a blank slate, an unknown or unnamed landscape'.[55] While not my central focus here, the many ways that South Asian conceptions of the Himalaya, in their myriad and dynamic permutations, continued to exist alongside, interact with, compete against, ignore, resist and infiltrate European attempts to categorise the mountains in this period are essential facets of the story. These, in turn, meant that global visions based on the Alps and the Andes, and the commensurability of mountain environments, could only ever be unevenly and imperfectly applied, even if these applications might have pervasive consequences.

The role of imaginative geographies in the histories of exploration, science and empire owes much to the formative work of Paul Carter, who took as his starting point that travellers 'invented places, rather than found them', and that in naming and historicising places into European knowledge frameworks, these inventions were inherently insidious.[56] These ideas map particularly well onto the Himalaya in the early nineteenth century, a time in which European knowledge production was wrapped up in attempts to bring the mountains within the purview of empire while simultaneously framing them as frontiers. In terms of mountain imaginaries, early nineteenth-century renderings also drew especially on romanticism, the picturesque and the sublime. Here Marjorie Hope Nicolson's classic trajectory traces the shift from fear and avoidance of mountains to fascination and aesthetic appreciation in the eighteenth and nineteenth centuries – or from 'mountain gloom' to 'mountain glory'.[57] European naturalists, surveyors and travellers brought these aesthetic traditions with them to the Himalaya, and the material practice of science is thus impossible to separate from the

[54] Edward Madden, 'Diary of an Excursion to the Shatool and Boorun Passes over the Himalaya, in September, 1845', *Journal of the Asiatic Society of Bengal* 15 (1846): 112.

[55] Sujit Sivasundaram, 'Islanded: Natural History in the British Colonization of Ceylon', in *Geographies of Nineteenth Century Science*, ed. David N. Livingstone and Charles W. J. Withers (Chicago: University of Chicago Press, 2011), 143.

[56] Paul Carter, *The Road to Botany Bay: An Exploration of Landscape and History* (Minneapolis: University of Minnesota Press, 1987), 51. For this process in India, see Arnold, *The Tropics and the Traveling Gaze*, 5.

[57] Marjorie Hope Nicolson, *Mountain Gloom and Mountain Glory: The Development of the Aesthetics of the Infinite* (Ithaca: Cornell University Press, 1959). See also Robert Macfarlane, *Mountains of the Mind: A History of a Fascination* (London: Granta Books, 2003).

discursive appropriation of the mountains in this period. This tension between the 'real' physical or material reality of mountains and their representational or imaginative existences is one that thus reappears throughout this story, but is one that was echoed by the contemporary travellers as much as it is by modern scholars.

Mountains as a Scale for Global History

'The Himalaya' is of course not the only possible term with which to engage Asia's most famous mountains. The designation 'High Asia' or '*Haute-Asie*' – as especially used in Francophone scholarship on the region – might at times be better.[58] It has the advantage of highlighting the key shared feature of what is otherwise a hugely diverse area, namely: significant elevation above sea level, the implications of which – both for scientific practice, and as an object of study in and of itself – lie at the heart of this history. Perhaps even more precise is the recently proposed term 'Northern South Asia', which while lacking simplicity helpfully reflects the way that this book largely considers exploration and interactions with the Himalaya originating upwards and from the south.[59] Nevertheless, imperial and scientific imaginations were guided not by these scholarly neologisms, but by imaginative geographies of 'the Himalaya' as the highest mountains on the globe. This is also why I have borrowed, for the book's title, another phrase that contemporaries came to apply to the region, namely: the 'Roof of the World'. An evocative expression, and one that readily reflects the new imperial imaginaries that emerged around the mountains in this period, using it is nevertheless not without caveats. In particular, in its original usage, it is rather more regionally specific, properly referring to the Pamirs (largely in what is today Tajikistan, and extending into Kyrgyzstan, Afghanistan and China) rather than the broader span of mountains that this book considers, or indeed the narrower Himalayan massif at all. Here travellers suggested the expression originated as a Persian phrase, as well as being in use in local languages (Wakhi and Tajik). Indeed, while on a mission to trace the source of the Oxus River in 1838, Scottish naval officer and surveyor John Wood (1812–1871) recorded that he eventually found himself standing, 'to use a native expression, upon the *Bam-i-Dúniah*, or "*Roof of the World*"'.[60] These etymological and geographical associations were later

[58] See for example, Baud, Forêt and Gorshenina, *La Haute-Asie telle qu'ils l'ont vue*.
[59] See David N. Gellner, ed., *Borderland Lives in Northern South Asia* (Durham: Duke University Press, 2013).
[60] John Wood, *Narrative of a Journey to the Source of the River Oxus* (London: John Murray, 1841), 331–32, 354, 359. See also Thomas Edward Gordon, *The Roof of the World: Being*

stripped away, however, and the expression came to be frequently applied to High Asia more generally (and by the turn of the twentieth century, especially to the Tibetan plateau).[61] As an evocative yet ultimately problematic phrase adorning a book about the imperial remaking of Asia's mountains, *Science on the Roof of the World* is thus a suitably ironic title, and one which reflects the story of European epistemological and imaginative appropriation told between its covers.

Writing about the Himalaya nevertheless requires special sensitivity to the way that such a wide lens can have a flattening effect, potentially ignoring diversity and difference and obscuring the importance of local experiences. Such a framework runs the risk of geographical determinism and essentialising Himalayan peoples who are better characterised by their heterogeneity.[62] However, much as with anthropological studies of 'Zomia' discussed later, the advantage is an alternative arrangement of the world as a way of escaping a focus on a particular region, state or group of people.[63] Taking a wider view of the Himalaya also has the advantage of counteracting the tendency, especially in some Western scholarship, of using the Himalaya as a shorthand for Tibet.[64] The fleetingness of Tibet's appearances in this book are thus in part a deliberate move to shine a light on different parts of the mountains, though in practical terms are also a consequence of focusing on a time period in which Europeans were almost wholly excluded from the mountain kingdom.

This is not to say that all parts of the Himalaya will be looked at in equal detail or even at all. This history – even if moving away from some of the most well-known stories – still follows the contours of political events and the itineraries of imperial travellers, inevitably privileging certain places. Indeed, given the limitations on access to Nepal and Tibet, European scientific interactions with the mountains in the first half of the nineteenth century took place in only piecemeal fashion. They tended to cluster initially on Kumaon, Garhwal and Ladakh, as well as later on Kashmir, Xinjiang, Afghanistan and Sikkim. Recognising the unevenness with which these imperial explorations unfolded nevertheless matters, because the more accessible parts of the mountains ultimately provided the

the Narrative of a Journey over the High Plateau of Tibet to the Russian Frontier and the Oxus Sources on Pamir (Edinburgh: Edmonston and Douglas, 1876).

[61] See Bishop, *The Myth of Shangri-La*, 122–23, 149.

[62] Chetan Singh, ed., *Recognizing Diversity: Society and Culture in the Himalaya* (Oxford: Oxford University Press, 2011).

[63] For an overview, see Jean Michaud, 'Editorial – Zomia and Beyond', *Journal of Global History* 5, no. 2 (2010): 200–2.

[64] For scholarship that slips into this register, see for example Alastair Lamb, *British India and Tibet, 1766–1910* (London: Routledge & Kegan Paul, 1986); Bishop, *The Myth of Shangri-La*.

imaginative toolbox for representations of the whole. It is also worth noting that while explorations were unfolding at the same time, experiences in the Eastern Himalaya (especially Assam and Myanmar) are not addressed in the same way or to the same extent here. These regions demand sometimes quite different frameworks for understanding both their topography and politics, and although important imperial frontiers, they were less central to contemporary scientific enquiries into altitude and are thus largely beyond the scope of this volume.[65]

Much as transnational geographic regions have aroused the interest of scholars, geographical features themselves are increasingly being employed as productive scales for global and transnational histories. Scholars have adopted oceans and seas in particular, and successfully used these to disrupt older nationalist historiographies.[66] Other geographical features, whether rivers, islands or beaches, have allowed scholars to tell important and often previously invisible transnational, transimperial and translocal stories.[67] Sujit Sivasundaram, for example, argues that in the case of Sri Lanka, 'an island-centred approach may reveal better the local entanglements of natural knowledge'.[68] This book contends that a mountain-centred approach may do the same. Meanwhile, as Ruth Rogaski demonstrates, 'putting the mountain at the centre of inquiry not only allows us to understand the interaction between humans and their environment, but also facilitates examination of human-to-human interactions'.[69] Mountains have nevertheless been less quickly co-opted into this trend of framing research around geographical features than oceans, islands and rivers. This is perhaps because whereas watery bodies seem to emphasise movement and flow, mountains might appear to have the opposite effect. However, despite tropes casting the Himalaya as 'impenetrable' or the ultimate barrier to travel and communication, the range was – and long had been – highly porous (see Figure 0.3). Extensive trade networks operated within and through the Himalaya for millennia prior to European

[65] For cross-frontier comparison in the later period, see especially Thomas Simpson, *The Frontier in British India: Space, Science, and Power in the Nineteenth Century* (Cambridge: Cambridge University Press, 2021).

[66] See for example David Armitage, Alison Bashford and Sujit Sivasundaram, *Oceanic Histories* (Cambridge: Cambridge University Press, 2017).

[67] Though traceable back to Fernand Braudel and earlier, recent scholarship framing itself around geographical features owes much to Greg Dening, *Islands and Beaches: Discourse on a Silent Land: Marquesas, 1774–1880* (Honolulu: University Press of Hawaii, 1980). See also Nicholas Thomas, *Islanders: The Pacific in the Age of Empire* (New Haven: Yale University Press, 2010); Sujit Sivasundaram, *Islanded: Britain, Sri Lanka, and the Bounds of an Indian Ocean Colony* (Chicago: University of Chicago Press, 2013).

[68] Sivasundaram, 'Islanded: Natural History in the British Colonization of Ceylon', 130.

[69] Ruth Rogaski, 'Knowing a Sentient Mountain: Space, Science, and the Sacred in Ascents of Mount Paektu/Changbai', *Modern Asian Studies* 52, no. 2 (2018): 721.

Figure 0.3 'Specimen of a Mountain Road in the Himalayahs' (1853). Watercolour by Conway Shipley, from the volume 'India, Tibet and Kashmir'. Himalayan exploration overwhelmingly followed long-standing routes, relied on existing infrastructure such as bridges and roads, and depended on local knowledge of conditions and techniques for safe travel in the mountains. At the same time, this sort of exaggerated depiction of a mountain road dramatises the 'extremes' and 'dangers' of such travels for audiences at home.
Source: With kind permission of the Central Asia Library of The Henry S. Hall, Jr. American Alpine Club Library.

interest.[70] Much as European appropriation of knowledge of the sub-continent and the Indian Ocean followed networks of exchange mapped onto existing trade networks, imperial science and exploration in the Himalaya advanced by following the routes and advice of traders, migrants and pilgrims.[71]

Especially in their western reaches, the Himalaya had long been a global space of contact, exchange and movement – both vertical and horizontal – as the so-called 'crossroads' of Asia. This is not to say that Himalayan travel and trade did not present some substantial and particular difficulties – material as well as cultural and political – with valleys and passes dictating routes, and altitude placing limitations on the sorts of goods that could be moved, and on the speed of travel. Movement could also be highly seasonal, regulated by the opening and closing of key high passes. Ultimately though, we need to move beyond semiotic assumptions about mountains as barriers and limiters, which are insufficient to explain the roles of the Himalayan mountains as borderlands, and as spaces of trade, migration and cross-cultural exchange. Mountains, I argue, are thus eminently suitable for co-opting into the increasingly productive trend of using geographical features as sites and as scales for conducting world history, and for telling global stories.

Exploration and Indigenous Labour

In this book, broader questions about empire, geography, labour and scientific practice are all tied together by what is necessarily also a history of exploration. Such histories largely fell out of fashion in academic circles in the 1960s, though heroic – and occasionally anti-heroic – depictions of individual explorers have remained staples of popular writing.[72] The re-emergence of scholarly interest in exploration in the 1990s, especially in the Arctic, Africa and Australia, has been articulated around the necessity of placing individual explorers firmly within their social, cultural and political contexts, as well as the

[70] Michaud, 'Editorial – Zomia and Beyond', 194–95.

[71] Lamb, *British India and Tibet, 1766–1910*; Alam, *Becoming India*; Raj, *Relocating Modern Science*. For pursuit of the shawl wool trade, a key (if largely unfulfilled) economic motivator, see Christian Jahoda, *Socio-Economic Organisation in a Border Area of Tibetan Culture: Tabo, Spiti Valley, Himachal Pradesh, India* (Vienna: Austrian Academy of Sciences Press, 2015).

[72] See Dane Kennedy, 'British Exploration in the Nineteenth Century: A Historiographical Survey', *History Compass* 5, no. 6 (2007): 1879–1900; Dane Kennedy, ed., *Reinterpreting Exploration: The West in the World* (Oxford: Oxford University Press, 2014); Peter R. Martin and Edward Armston-Sheret, 'Off the Beaten Track? Critical Approaches to Exploration Studies', *Geography Compass* 14, no. 1 (2020): e12476, https://doi.org/10.1111/gec3.12476.

production and reception of texts, and the role of institutions like the Royal Geographical Society (founded in 1830). Felix Driver, one of the leading figures in rehabilitating exploration scholarship, sums this up in his assertion that exploration needs to be 'understood as a set of cultural practices'.[73] Another key theme to come out of this scholarship is the relative vulnerability of European explorers, and their reliance on local assistance and cooperation. Scholars have demonstrated that although agents of a frequently brutal imperialism, explorers' ability to act out their supposed superiority was often limited in remote regions.[74] Showing that explorers could be confused and that exploration was contingent nevertheless brings with it the risk of diminishing the imperial implications of these journeys for the peoples who assisted European travellers, surveyors and explorers in delimiting the borders of empire. Michael Robinson goes further, arguing that there is an additional caveat in that 'ironically, developing a global framework for exploration means acknowledging the asymmetries of science and exploration as Western cultural practices'.[75]

It is thus important to disaggregate those involved in exploration in this period. Alexander Gerard and Joseph Hooker, for example, both made significant contributions to the scientific and imaginative constitution of the Himalaya. They did so, however, with very different resources – both material and social – and it would be wrong to suggest they were part of a coherent project or motivated by the same expectations and ends. Meanwhile, tracing their stories is not without distinct methodological challenges, and the archival record for some individuals – especially Bengal Infantrymen like Gerard – is often sporadic. In particular, personal correspondences are few and far between. Many of the activities of the surveyors and surgeons must thus be pieced together and read out of the archives of those better known, and more socially elevated, such as English botanist (and Joseph's father) William Hooker at Kew. As a result, this book draws especially on articles published in nascent India-based scientific journals, including the *Asiatic Researches, Gleanings in Science* and the *Journal of the Asiatic Society of Bengal*, all of which also contributed to the production of 'centres in the periphery' like Calcutta.[76]

[73] Felix Driver, *Geography Militant: Cultures of Exploration and Empire* (Malden: Blackwell, 2001), 8.

[74] Nigel Leask, *Curiosity and the Aesthetics of Travel Writing, 1770–1840: From an Antique Land* (Oxford: Oxford University Press, 2002), 16–17; Kennedy, *The Last Blank Spaces*, 230.

[75] Michael Robinson, 'Science and Exploration', in *Reinterpreting Exploration: The West in the World*, ed. Dane Kennedy (Oxford: Oxford University Press, 2014), 32.

[76] Many of the Indian articles (often abridged) later also made their way into the *Asiatic Journal, Edinburgh Philosophical Journal* and *Edinburgh Journal of Science*, or were translated for the likes of the *Bulletin de la Société de géographie*.

Similarly, I make extensive use of official correspondences, survey field-books and maps in order to ascertain collective experiences, even while acknowledging that what survives for many of the individual surveyors is haphazard and incomplete. I also draw on the surviving material sources – especially plant and mineral specimens, and scientific instruments – which provide further insights into the fragility and limits of practice, attempts to bolster authenticity and credibility, and the centrality and role of indigenous labour. Finally, the eclectic range of visual representations of the mountains that were produced in this period provide evidence of the initially unimaginable scale of the Himalaya, and the challenges involved in making verticality legible. As Ann Colley has argued, what sometimes 'gets lost in this inevitable concentration upon conquest and control, or upon Britain's "imperial project" in the Himalaya, is the aesthetic sensibility that accompanied it' and 'forgotten is the reality that these explorers were frequently overwhelmed by the sheer sublimity of their surroundings'.[77] As I will show, even while revealing limits, these various visual representations were nevertheless carefully designed to fit with and reinforce ethnographic expectations and facilitate the 'othering' of both landscapes and peoples for audiences a world away in Europe.

If tracing the activities of itinerant EIC surveyors and naturalists can be challenging, this is even more so for the Himalayan peoples involved in these expeditions. In recent decades, scholars of science and exploration have increasingly recognised the critical roles of 'intermediaries' and 'local informants' in imperial knowledge-making, administration and science, though this project remains far from complete and is often characterised by methodological angst.[78] In examining the interactions between naturalists and their guides in this book, a particularly helpful concept is that of the broker or 'go-between'. This is a useful analytical lens, because the 'go-between' was not a 'simple agent of cross-cultural diffusion, but someone who articulates relationships between disparate worlds or cultures by being able to translate between them'.[79] In considering the production of knowledge in cross-cultural contexts, notions of

[77] Ann C. Colley, *Victorians in the Mountains: Sinking the Sublime* (Farnham: Ashgate, 2010), 221. On visual representations of the Himalaya in the later period, see also Simpson, *The Frontier in British India*.

[78] The literature is now expansive, but see especially Simon Schaffer et al., eds., *The Brokered World: Go-Betweens and Global Intelligence, 1770–1820* (Sagamore Beach: Science History Publications, 2009); Driver, 'Hidden Histories Made Visible?'; Shino Konishi, Maria Nugent and Tiffany Shellam, eds., *Indigenous Intermediaries: New Perspectives on Exploration Archives* (Canberra: Australian National University Press, 2015); Tiffany Shellam et al., eds., *Brokers and Boundaries: Colonial Exploration in Indigenous Territory* (Canberra: Australian National University Press, 2016).

[79] Schaffer et al., *The Brokered World*, xiv.

hybridity and joint- or co-production have also had considerable appeal. More attention is now being given to the slippages between knowledge traditions, and the way boundaries – if there are boundaries – might be characterised as fluid and open to active renegotiation.[80] Here, there has been recognition that while Europeans might actively appropriate knowledge, operating in these spaces also resulted in accidental transfer, which undermined naturalists attempts to categorise and control.[81] In spite – or perhaps *because* – of the urgency of questions around indigenous knowledge, significant debates and caveats continue in the scholarship. Scholars have especially noted the dangers of this approach in potentially minimising the asymmetrical nature of power relations in these negotiations and the inherent violence – not only physical but also epistemological – in the imposition of European scientific norms.[82] Another key concern is that studying 'indigenous' knowledge may serve to either unhelpfully reify it or to create an artificial and problematic dichotomy in which it is cast as opposite to – and even if not inferior, at least essentially different from – 'Western' knowledge.[83]

The reconstruction of the roles of Himalayan peoples in exploration and science is also challenging in practical terms, because individual intermediaries are often rendered invisible or homogenised into obscurity, becoming ghosts in the narrative and silences in the archive.[84] This is sometimes also perpetuated in modern scholarship, whether on the Americas or Africa or Asia, which might feature faceless 'guides' and 'assistants' defined by function and lacking personality or idiosyncrasy.[85] These issues are further compounded by the way that the written records for early nineteenth-century exploration and science in the Himalaya are overwhelmingly European and colonial. Indeed, it is travel narratives, journals, articles, fieldbooks and correspondences,

[80] See, among others, Mary Louise Pratt, *Imperial Eyes: Travel Writing and Transculturation*, 2nd ed. (London: Routledge, 2008); Richard Grove, *Green Imperialism: Colonial Expansion, Tropical Island Edens and the Origins of Environmentalism, 1600–1860* (Cambridge: Cambridge University Press, 1995); Anna Winterbottom, *Hybrid Knowledge in the Early East India Company World* (Basingstoke: Palgrave Macmillan, 2016).

[81] See especially Fa-ti Fan, *British Naturalists in Qing China* (Cambridge, MA: Harvard University Press, 2004), 153; Fa-ti Fan, 'Science in Cultural Borderlands: Methodological Reflections on the Study of Science, European Imperialism, and Cultural Encounter', *East Asian Science, Technology and Society* 1, no. 2 (2007): 213–31.

[82] See Arnold, *The Tropics and the Traveling Gaze*, 8.

[83] Sivasundaram, *Islanded: Britain, Sri Lanka, and the Bounds of an Indian Ocean Colony*, 174.

[84] See Michel-Rolph Trouillot, *Silencing the Past: Power and the Production of History* (Boston: Beacon Press, 1995).

[85] Heather F. Roller, 'River Guides, Geographical Informants, and Colonial Field Agents in the Portuguese Amazon', *Colonial Latin American Review* 21, no. 1 (2012): 1.

written predominantly by Europeans – and only in few exceptional cases by South Asians – that form the central body of sources for this history. These narratives thus need to be examined as textual productions that present particular experiences of the spaces of the high Himalaya, even while these experiences underwent significant curation – both in terms of the self-fashioning of their authors, as well as massaging by metropolitan publishers and editors.[86] Writing about South Asian actors in the remaking of the mountains thus inevitably involves addressing ongoing methodological questions arising from the use of colonial archives. In reconstructing non-European perspectives from European sources, it is nevertheless possible to employ techniques like reading 'against the grain' and looking for 'indigenous countersigns', as well as the use of alternate sources such as specimens or images to reveal traces of labour and expertise.[87] Even while acknowledging the limits of viewing the world through these colonially tinged lenses, we thus can – and must – do more.

In addressing these caveats, my approach is to pay close attention to the practical, everyday aspects of doing science in remote locations. Here I follow insights, especially from the history of astronomy, where Alex Soojung-Kim Pang has argued that when examining expeditionary science it is important to 'recapture the emotional texture of science along with the messy details of its practice', and Joydeep Sen has emphasised the value of focusing on 'the practical engagement between Europeans and Indians' in order to get at the 'experiential texture' of doing science.[88] Likewise, I take seriously Moritz von Brescius's assertion that we need to 'overcome the myth of the western solitary traveller by taking a new and multi-perspective look at the inner life of expeditions' and ultimately treat them as 'moving colon[ies]', characterised by hierarchies of interests and authority that were far more ambiguous than the European naturalists would have liked.[89] In drawing on these insights, I thus aim to trace the realities and travails of the practice of science in spaces characterised by the crushing effects of altitude sickness, sensory derangement and cross-cultural tensions, and consider the ways these affect how we should understand the knowledge produced in and of the Himalaya. In other

[86] See Innes M. Keighren, Charles W. J. Withers and Bill Bell, eds., *Travels into Print: Exploration, Writing, and Publishing with John Murray, 1773–1859* (Chicago: University of Chicago Press, 2015).

[87] Ann Laura Stoler, *Along the Archival Grain: Epistemic Anxieties and Colonial Common Sense* (Princeton: Princeton University Press, 2010); Bronwen Douglas, *Science, Voyages, and Encounters in Oceania, 1511–1850* (London: Palgrave Macmillan, 2014).

[88] Alex Soojung-Kim Pang, *Empire and the Sun: Victorian Solar Eclipse Expeditions* (Stanford: Stanford University Press, 2002), 5; Joydeep Sen, *Astronomy in India, 1784–1876* (London: Pickering & Chatto, 2014), 9.

[89] Brescius, *German Science in the Age of Empire*, 4, 161–216.

words, I examine European engagement with local knowledge in a dangerous and unfamiliar environment, less from an epistemological or ontological standpoint, and more from one grounded in everyday practice and the complexities of expedition sociability. Here objects and specimens, be they instruments, tents or even firewood, also have interesting stories to tell. In examining these, I ultimately aim to show that Himalayan peoples and knowledge systems were not incidental to the way surveyors and naturalists pursued altitude sciences in this period, and demonstrate the extent to which the agency of European explorers might be resisted and subverted in ill-understood and challenging high spaces and frontiers.

Outline of the Book

Science on the Roof of the World unfolds across six chapters that are organised in thematic rather than chronological fashion. Each chapter deals with a different branch of scientific practice, though these inevitably sometimes overlap. Chapter 1 on 'Measuring Mountains' considers debates over the recognition of the Himalaya as the highest mountains in the world, and their simultaneous constitution as an often-insecure imperial frontier. In discussing issues of social status and reliance on indigenous labour, this chapter sets the scene for the mapping of all kinds of natural historical and ethnographic knowledge on a globe both vertical and round. Chapter 2 considers 'Unstable Instruments' and the practices developed to determine altitude accurately, even while highlighting the problems of breakage and sensory derangement. In examining the tools necessary to make altitude a basis for imposing commensurability, this chapter thus explains how the challenges insisted on by those operating in remote locations complicated necessary global comparisons. Chapter 3 examines 'Suffering Bodies' and early understandings of medical topographies of high mountains, particularly by considering altitude sickness and expedition sociability. It focuses on bodily comparisons between European travellers and their Bhotiya and Tartar guides, demonstrating that these are often revealing of dependency. Chapter 4 turns to 'Frozen Relics', considering especially how frontiers circumscribed the limits of knowledge, and the uneven ways that pre-existing networks moved material from the uplands to the lowlands. The chapter focuses on the materiality of specimens and the role of labour, as well as new understandings of time in explaining the upheavement of mountains and the retreat of glaciers (in the latter case drawing on the memories of Himalayan peoples). It also considers overlapping cosmological understandings of the mountains, particularly by looking at fossils that were both ritual objects and scientific specimens. Chapter 5 focuses on 'Higher Gardens', shifting the lens from itinerant expeditions to institutions by examining the EIC gardens at Saharanpur and Mussoorie. In focusing on the ambiguous position of these gardens – straddling the uplands and lowlands – this chapter stresses the inherent haziness in attempts to graduate the vertical globe. It also demonstrates the central role of South Asian gardeners and collectors in the remaking of the mountains. Chapter 6 considers 'Vertical Limits', and traces debates around the elevational limits of flora and fauna, as well as the vertical dynamics of the human and non-human worlds of the Himalaya more broadly. In so doing, it emphasises the insidiousness of borrowing horizontal norms to explain the vertical, and the way that theories developed in the Alps and the Andes, such as the line of perpetual snow, might fail to account for the Himalaya. Finally, in the conclusion, I expand the lens by examining comparative tableaux of mountains in European atlases. In so doing, I ultimately consider the way that universality and global commensurability were imperfectly imposed onto mountain environments in this period by obscuring both uncertain scientific practices and the roles of Himalayan peoples.

1 Measuring Mountains

In 1802, when Alexander von Humboldt climbed to more than 19,000 feet on the side of the volcano Chimborazo, it was the highest mountain in the world. Or at least, he believed that it was.[1] In the first decades of the nineteenth century, as the shape of the vertical globe changed radically, so too did the title of world's highest mountain, finally settling on Mount Everest only in 1856. This shift was neither straightforward nor uncontroversial at the time. While it is now taken for granted that Everest is the 'third pole' or pinnacle of the world, this is actually very hard to know without the assistance of precise instrumental measurements. Indeed, it is telling that Chimborazo had only secured its supremacy in the eighteenth century over the Peak of Tenerife (Teide), which had previously been considered by many European commentators as a good candidate for the world's highest mountain (this is partly explained by its 'prominence', rising dramatically out of the sea). Moreover, from the perspective of many in the early modern period, the exact height of a mountain was neither especially important nor interesting.[2] Other factors, including aesthetics, location or association with historical or religious events, all mattered more in assigning significance. Gaining momentum in the latter half of the eighteenth century and reaching the Himalaya in the 1800s, the reorganisation of the vertical globe based on altitude thus relied on both imperial expansion and novel technologies to bring about not only new possibilities for measuring mountains accurately, but also new philosophical, imperial and imaginative interest in the precise ranking of mountains according to their height.

[1] The summit of Chimborazo, at 20,548 feet, nevertheless remained out of Humboldt's reach. An argument is sometimes made that Chimborazo could still be considered the world's highest mountain because its position near the equator means its summit is further from the centre of the earth than Everest's. See Mathieu, *The Third Dimension*, 30.

[2] The currently accepted height for Teide is 12,198 feet, significantly less than half that of Everest, and indicative of the difficulty of judging height merely by eye. For growing interest in altitude as far back as the seventeenth century, see, however, Jouty, 'Naissance de l'altitude'.

This chapter introduces the key scientific, imaginative and imperial developments in and beyond the early nineteenth-century Himalaya that saw Everest finally installed as the highest mountain in the world. In so doing, it demonstrates how global comparisons were wrapped up in imperial efforts to measure the Himalaya and map their natural history from the outset. Indeed, essential to establishing Everest's supremacy were reliable measurements of other mountains around the world, something that required sufficient imperial penetration of the world's high places. By looking at when and how European surveyors started measuring the Himalaya accurately, this chapter thus also considers why this occurred later than in the Alps and the Andes.[3] Here, a key part of the delay in recognising the supremacy of the Himalaya stems from the often insecure nature of the mountain frontiers in the early nineteenth century. Specifically, a lack of in situ observations contributed to significant uncertainties – especially around refraction – when measuring mountains from the plains looking up to the peaks. However, while the East India Company (EIC) was consolidating its power in the lowlands of South Asia, it remained significantly less capable of bringing this growing imperial confidence to bear in the uplands. Similarly, in operating in uncertain environments and contested borderlands, surveyors always relied on Bhotiyas, Tartars and Lepchas to identify safe routes, procure sufficient supplies, provide information about seasonal possibilities and navigate complex local and regional politics. This chapter thus also considers fraught attempts to map the new frontiers that were opened up from the 1810s onwards, especially after the Anglo-Gurkha War of 1814–1816, and how they were bound up in the redrawing of the vertical globe. In so doing, I ultimately demonstrate how insecurity around the frontiers simultaneously propelled Himalayan exploration and limited the ability of EIC surveyors to operate in the mountains.[4]

In thinking about the constitution of the Himalaya as both the highest mountains in the world and an insecure frontier, this emerges as not only an imperial story, but also one bound up in the particular tribulations of the British Empire. The majority of the surveyors and naturalists who measured the mountains and mapped their natural history in the first half of the nineteenth century were British (a notably high proportion Scottish, who might extend the vertical globe into the realm of cultural comparison, for example, imputing the supposed characteristics, both good and bad, of Scottish highlanders onto Himalayan peoples). However, this was not exclusively so, and Frenchman Victor Jacquemont, for example, also

[3] On this point, see also Mathieu, *The Third Dimension*, 30.
[4] For the role of the frontier more widely, see Simpson, *The Frontier in British India*.

plays a notable role in the story. Although there were occasional outbursts of xenophobia towards foreigners masquerading as fear of scientific discoveries being usurped, or of finding military service with Indian rulers, these nevertheless rarely spilled over from rhetoric into real hostility towards the presence of 'foreign' scientific travellers in the 'blank spaces' of the high mountains.[5] The British Empire in its guise as the EIC is nevertheless omnipresent in at least one important respect, in that surveyors and naturalists shared in common complex relationships with the Company hierarchy – whether as their employers or as gatekeepers in granting permission to travel.[6] The EIC did at times limit 'foreign' access (if inconsistently) and most notoriously turned down Alexander von Humboldt's multiple requests to visit.[7] This was later lamented by some EIC employees in the Himalaya, and as Richard Strachey put it: 'men of science will still long have to regret that this illustrious traveller was prevented from visiting the east; Englishmen alone need remember that he was prevented by them'.[8] The greatest counterfactual to the story of the Himalaya's emergence as the highest mountains in the world remains how different it might have looked if the EIC had not said no to Humboldt.

It is thus no accident that this chapter begins with Humboldt, whose biography is never far from the story of measuring mountains. Here my argument especially addresses recent work on verticality by the likes of Michael Reidy, Jon Mathieu and Bernard Debarbieux, all of whom deal, to a greater or lesser extent, with Humboldt. Reidy, for example, takes a broad view, demonstrating how Humboldt was imbricated in 'a larger vertical consciousness that engulfed science in the early nineteenth century' (indeed, Reidy and others have been just as interested in verticality below the ground and under the sea as in relation to mountains).[9]

[5] For the argument that British chauvinism played a larger role, especially around jealously over French scientific patronage, see Andrew Grout, 'Geology and India, 1770–1851: A Study in the Methods and Motivations of a Colonial Science' (PhD Dissertation, School of Oriental and African Studies, 1995), 123–33.

[6] For more on transnational scientific careers in the EIC, see David Arnold, 'Globalization and Contingent Colonialism: Towards a Transnational History of "British" India', *Journal of Colonialism and Colonial History* 16, no. 2 (2015), https://doi.org/10.1353/cch.2015.0019; Brescius, *German Science in the Age of Empire*.

[7] Initially sounding out the idea in 1808, he formally requested access while visiting London in 1814 and again in 1817. Denied his Himalayan dream, Humboldt instead travelled through parts of Central Asia (with Russian permission) in 1829. See Jean Théodoridès, 'Humboldt and England', *The British Journal for the History of Science* 3, no. 1 (1966): 39–55.

[8] Richard Strachey, 'On the Snow-Line in the Himalaya', *Journal of the Asiatic Society of Bengal* 18, part 1 (1849): 287.

[9] Michael S. Reidy, *Tides of History: Ocean Science and Her Majesty's Navy* (Chicago: University of Chicago Press, 2009), 280. See also Michael S. Reidy, 'From the Oceans to the Mountains: Spatial Science in an Age of Empire', in *Knowing Global Environments:*

Meanwhile, Bernard Debarbieux argues that for Humboldt mountains 'provided comparable settings useful to his project of building a global knowledge' and similarly Jon Mathieu suggests that Humboldt was the central figure in 'the historical genesis of a "mountain world" stretching all around the globe'.[10] I take such contentions seriously, but in examining EIC surveyors and naturalists' sometimes eclectic attempts to measure the Himalaya, look to go beyond a standard historical narrative that features a trajectory from Alexander von Humboldt to Joseph Dalton Hooker and Charles Darwin, without considering the essential transformations of the intervening decades.[11]

In contrast, this book argues that the seemingly magnetic (and often archivally driven) pull towards the biographies of savants or 'great men' of science – in both popular and scholarly works – has notable limits, not only for understanding the imperial remaking of the Himalaya, but also for the histories of science, exploration and geography in this period more widely. Among other things, treating these iconic figures as part of wider processes is essential to illuminating the networks of indigenous labour and expertise on which the measurement of mountains relied, and ultimately to understanding the imperial utility that underpinned the imposition of ostensibly global categories. Of course, this also makes beginning the chapter with Humboldt on Chimborazo rather ironic, even if this is also where many EIC surveyors began when imagining their own activities in scientifically mapping and measuring the Himalaya. The writings of European naturalists who worked in the Himalaya, such as John Forbes Royle, Victor Jacquemont and Thomas Thomson (1817–1878) and, indeed, Joseph Hooker himself towards the mid-century, are peppered

New Historical Perspectives on the Field Sciences, ed. Jeremy Vetter (New Brunswick: Rutgers University Press, 2011), 17–38; Michael S. Reidy, 'The Most Recent Orogeny: Verticality and Why Mountains Matter', *Historical Studies in the Natural Sciences* 47, no. 4 (2017): 578–87. On the idea of verticality in other contexts, see Bruce Braun, 'Producing Vertical Territory: Geology and Governmentality in Late Victorian Canada', *Ecumene* 7, no. 1 (2000): 7–46; Eyal Weizman, *Hollow Land: Israel's Architecture of Occupation* (London: Verso, 2007); Stuart Elden, 'Secure the Volume: Vertical Geopolitics and the Depth of Power', *Political Geography* 34 (2013): 35–51. See also the *Centaurus* special issue introduced by Wilko Graf von Hardenberg and Martin Mahony, 'Introduction – Up, Down, Round and Round: Verticalities in the History of Science', *Centaurus* 62, no. 4 (2020): 595–611.

[10] Bernard Debarbieux, 'The Various Figures of Mountains in Humboldt's Science and Rhetoric [Figures et Unité de l'idée de montagne chez Alexandre von Humboldt]', *Cybergeo: European Journal of Geography* [En ligne] (2012), http://cybergeo.revues.org/25488; Mathieu, *The Third Dimension*, 22.

[11] Michael Reidy, for example, seems to take at face value some of the claims to novelty made by Joseph Hooker with regard to his activities in the 'untouched' Himalaya. See Reidy, 'From the Oceans to the Mountains', 28, 30. In fact, Hooker had to settle for Sikkim because it remained relatively unvisited by naturalists compared to other parts of the mountains like Kumaon, Garhwal and Kashmir.

with deferential references to 'the great Humboldt', 'the high authority of Humboldt' and, most frequently, 'the illustrious Humboldt'.[12] Such reflexivity appears throughout the accounts of surveyors in Asia, much as in other parts of the world. The name of Humboldt is ultimately inescapable, even if exactly what Humboldt meant and to whom is sometimes harder to quantify.

Humboldt also looms large in the way questions around the global commensurability of mountain environments saw the development of innovative charts and diagrams for representing and visualising altitudinal relationships, with the most famous – if not entirely original – example being his sea-to-summit profile of Chimborazo.[13] As a result, the knowledge production examined here in relation to altitude might be considered as what scholars often refer to as 'Humboldtian science'. This term was inaugurated by Susan Faye Canon in the 1970s and has since been adopted widely, usually to indicate an emphasis on aesthetics and precise measurement, as well as an impulse to study global phenomena comparatively.[14] However, in its now common and sometimes uncritical usage, it has also become a shorthand or trope that can obscure a series of heterogeneous practices that need to be disaggregated and interrogated on their own terms. Indeed, the scholarly conceit of 'Humboldtian science' is not necessarily the best lens to examine the scientific practices of EIC employees and itinerant travellers in the Himalaya.[15]

[12] John Forbes Royle, *Illustrations of the Botany and Other Branches of the Natural History of the Himalayan Mountains* (London: W. H. Allen, 1839), Vol 1, 3, 87, 113, 147, 191, 281, 283, 284, 375, 394; Thomas Thomson, *Western Himalaya and Tibet: A Narrative of a Journey through the Mountains of Northern India during the Years 1847–8* (London: Reeve, 1852), 474; Hooker, *Himalayan Journals*, Vol 2, 401; Jacquemont, *Letters from India*, Vol 1, 49, 59, 165, 237, 239, 311, 332.

[13] This was published as the centrepiece of Alexander von Humboldt and Aimé Bonpland, *Essai Sur La Géographie Des Plantes* (Paris: F. Schoell, 1807).

[14] Susan Faye Cannon, *Science in Culture: The Early Victorian Period* (New York: Science History Publications, 1978). See also Malcolm Nicolson, 'Alexander von Humboldt, Humboldtian Science and the Origins of the Study of Vegetation', *History of Science* 25, no. 2 (1987): 167–94; Malcolm Nicolson, 'Alexander von Humboldt and the Geography of Vegetation', in *Romanticism and the Sciences*, ed. Andrew Cunningham and Nicholas Jardine (Cambridge: Cambridge University Press, 1990), 169–85; Michael Dettelbach, 'The Face of Nature: Precise Measurement, Mapping, and Sensibility in the Work of Alexander von Humboldt', *Studies in History and Philosophy of Biological & Biomedical Science* 30, no. 4 (1999): 473–504; Michael Dettelbach, 'Global Physics and Aesthetic Empire: Humboldt's Physical Portraits of the Tropics', in *Visions of Empire: Voyages, Botany and Representations of Nature*, ed. David Philip Miller and Peter Hanns Reill (Cambridge: Cambridge University Press, 2002), 258–92. For a recent call for rethinking 'Humboldtian Science', see however Patrick Anthony, 'Mining as the Working World of Alexander von Humboldt's Plant Geography and Vertical Cartography', *Isis* 109, no. 1 (2018): 28–55.

[15] For a prescient if largely unheeded earlier critique, see Jane R. Camerini, 'Heinrich Berghaus's Map of Human Diseases', *Medical History* 44, no. S20 (2000): 189. For

Instead, this book pivots away from much of the recent scholarship by tracing eclectic attempts to apply Humboldt's science by those in displaced locations lacking many of Humboldt's resources. Put differently, rather than the extent to which the activities of European naturalists measuring the Himalaya were 'Humboldtian', I am more interested in the way Himalayan travellers read and positioned themselves in relation to Humboldt. Here it becomes apparent that the pervasiveness of Humboldt's ideas, especially those developed in relation to the Andes, could cause interpretive problems in early encounters with the Himalaya. Indeed, the 'high authority' of Humboldt sometimes delayed recognition of problems with theories, and estimates of scale, when it came to explaining Asia's mountains. In particular, Humboldt's early theories with regard to elevation, latitude and the line of perpetual snow failed, more or less utterly, to translate to the Himalaya. This was a particular impediment to recognising the Himalaya as the highest mountains in the world, as it was eventually proved to occur far higher in Asia than theories based on the Alps and the Andes suggested that it could be (an oversight that meant, for example, that reported measurements taken as high as 16,000 in places without snow were treated with scepticism and even scorn).

In considering the identification of the Himalaya as the world's highest mountains and their simultaneous constitution as an imperial frontier, this chapter thus sets the scene for the story of the remaking of the Himalaya through imperial global comparison. It begins by considering the initial controversy over the realisation that the Himalaya were the highest mountains in the world and reflects on the role of social status in the making of geographical knowledge. I then consider the emergence of the Himalaya as a key frontier for the EIC's burgeoning empire in South Asia and the way exploration and science were shaped by political realities which limited imperial agency in the mountains. The third and final section of the chapter lays the groundwork for another of the key threads that run throughout this book, namely: the dependency of European surveyors and naturalists on the labour and expertise of Himalayan peoples, and the consequences of these imperial activities for those whose homes were surveyed and subsumed into both an uneven imperial framework and an emerging global environmental order. Here the many and dynamic roles played by Himalayan peoples in EIC exploration and the ways they might have resisted and subverted surveyors' agency in the mountains all tie into the story of the ascendancy of the Himalaya over

a related point about the limits of John Forbes Royle's 'Humboldtian' aspirations, see also Arnold, *Science, Technology and Medicine in Colonial India*, 52.

the Andes. Ultimately, these overlapping factors – political, technological and social – in the acceptance of the Himalaya as the world's highest mountains demonstrate the uneven way that knowledge moved around the globe in the period, and the growing centrality of global comparison in understanding environmental and scientific phenomena in the nineteenth century.

The Highest Mountains in the World

Everyone knows that the Himalaya are the highest mountains in the world, but two hundred years ago this was not the case. Indeed, the mere suggestion that the Himalaya might overtop the Andes was initially controversial.[16] This is not to say that the existence of the Himalaya, or at least a significant range of mountains in Asia, had not long been known to Europeans. They had been written about at least as far back as Herodotus and Ptolemy. However, as James Bell noted in a general geography book of 1832, although '*Imaus* and *Emodus* were well known to the ancients, as ranges clothed in perpetual snow . . . they had not the most distant idea of their real height'. Hastening to defend European scientific honour, he went on to add that 'the Hindoos were equally ignorant of their elevation'.[17] Here, however, he rather misses the point. Elevation above sea level – today the metric by which many mountains become important – is not the only way to ascribe significance. The most important mountains in South Asian cosmology, for example, are not necessarily the highest in geographical terms. Indeed, the mythical Meru (as seen in Figure 1.1), sometimes considered the centre or axis of the world, and the real Kailash, an important site of pilgrimage, remains more important in Buddhist, Hindu and Jain tradition than locally important but otherwise relatively minor peaks like Everest, and especially the almost entirely obscure K2; that is, mountains whose main claim to fame is their great height above the level of the sea.[18]

[16] Reginald Henry Phillimore, *Historical Records of the Survey of India* (Dehradun: Survey of India, 1954), Vol 3, 29–48; John Keay, *The Great Arc: The Dramatic Tale of How India Was Mapped and Everest Was Named* (London: HarperCollins, 2000). See also Mathieu, *The Third Dimension*, 27–30.

[17] James Bell, *A System of Geography, Popular and Scientific: Or a Physical, Political, and Statistical Account of the World and Its Various Divisions* (Glasgow: A. Fullarton, 1832), Vol 4, 458. This line was lifted from 'Himalaya Mountains and Lake Manasawara', *The London Quarterly Review* 17 (1817), 430–31.

[18] Indeed, K2 retains its original survey designation after European surveyors were unable to identify an indigenous name, indicative of its limited significance other than in terms of its great altitude. See Keay, *The Great Arc*, 158–72.

Figure 1.1 A nineteenth-century Tibetan painting of Mount Meru. The sacred mountain is depicted here as the axis or centre of the universe. Reproduced with kind permission of the Rubin Museum of Art C2006.66.558 (HAR 1038).

In encountering and sometimes engaging with these cosmologies, European naturalists mapping the mountains nevertheless often noted the link between sacredness and height (indeed, the sacred is associated with high places in a variety of traditions globally). Bengal Infantry officer and surveyor James Dowling Herbert (1791–1833) nevertheless complained about the difficulty identifying the sacred mountain Kailash because his Kinnaura guides, 'as well as the people of the plains' seemed to 'call every high place by the term Kailas', and he found it difficult to impose his understanding of what constituted a mountain onto the indigenous topography.[19] Elsewhere, Alexander Cunningham wrote of Kailash that:

> I have ventured to call it the *Kailás*, or *Gangri* range, because those names are equally celebrated by the Hindus and Tibetans. *Kailás*, or 'Ice-mountain', is the Indian Olympus, the abode of Siva and the celestials. *Gang-ri*, or 'Ice-mountain', is called *Ri-gyal*, or King of Mountains, by the Tibetans, who look upon Ti-se, or the Kailás Peak, as the highest mountain in the world.[20]

Here as well as drawing in comparisons to European – in this case Greek – mythology, Cunningham suggests that the meaning of 'highest' is far from assured. In what follows, it is thus worth keeping in mind the sometimes garbled ways that surveyors and naturalists drew on existing understanding of high places in South Asian cosmology, even while placing themselves and their new instrumental epistemologies higher than those of their guides and informants.

If the story told in this book has an underlying animating thread, it is that height – and more specifically, height above sea level – only became an essential characteristic of mountains in both European science and imaginative geographies relatively recently. Indeed, scholars have suggested that it was only 'once a method for measuring the height of mountains had been agreed upon at the end of the eighteenth century, a mountain's precise height, just as much as its name, became part of its identification'.[21] However, even as it emerged as a question of both scientific and philosophical interest in the late eighteenth century, the technology to measure mountains accurately remained limited. For example, James Rennell, the prominent eighteenth-century cartographer and first Surveyor General of Bengal, merely noted in his 1788 edition of the *Memoir of a Map of Hindoostan* that the Himalaya must be 'among the highest of the mountains of the old hemisphere'. He went on to explain this vagary as follows: 'I was not able to determine their height; but it may in some measure be guessed, by the circumstance of their rising

[19] James Herbert, 'An Account of a Tour Made to Lay Down the Course and Levels of the River Setlej', *Asiatic Researches* 15 (1825): 351. See also Alex McKay, *Kailas Histories: Renunciate Traditions and the Construction of Himalayan Sacred Geography* (Leiden: Brill, 2015), 174–76.

[20] Alexander Cunningham, *Ladak, Physical, Statistical, and Historical; with Notices of the Surrounding Countries* (London: W. H. Allen, 1854), 43.

[21] Marie-Noëlle Bourguet, Christian Licoppe and H. Otto Sibum, *Instruments, Travel and Science: Itineraries of Precision from the Seventeenth to the Twentieth Century* (London: Routledge, 2002), 12.

considerably above the horizon when viewed from the plains of Bengal, at a distance of 150 miles'.[22] While their visibility from great distances and the occurrence of perpetual snow indicated that the Himalaya were obviously significant mountains, among eighteenth-century commentators it was thus still entirely reasonable to consider that they might not match even the Alps, let alone the Andes.

However, in 1802, the same year that Humboldt was toiling upwards on Chimborazo, EIC officer Charles Crawford (1760–1836) began making measurements of angles looking up to the mountains rising above the Kathmandu Valley. While limited in scope, these measurements, made while attached to the short-lived first British Residency in Nepal (1802–3), represented the first serious attempt to scientifically measure the height of the Himalaya. Crawford's results suggested that the mountains were not only high but possibly stupendously so. However, although Crawford's measurements had the potential to drastically reconfigure the shape of the vertical globe, all of his notes and drawings were lost before they could make much of an impact (indeed, commentators who followed acknowledged Crawford's contributions, even as they mourned the unavailability of his specific calculations).[23] A few years later, in 1807, Robert Hyde Colebrooke, then Surveyor General of Bengal, measured some more angles looking up from the plains, and his calculations too suggested that the Himalaya might well supersede the Andes. He was initially cautious with these findings, acknowledging uncertainties around measuring from a considerable distance away from the mountains, and particularly around the potential of terrestrial refraction (i.e. the way light bends when passing through the atmosphere) to introduce errors and make distant objects appear higher than they actually were.[24] Before he could investigate further, however, Colebrooke fell seriously ill – a not uncommon fate among the EIC surveyors and naturalists in this period – and had to give up refining his measurements. Instead, he sent an assistant, Bengal Infantry officer William Spencer Webb (1784–1865), to continue the survey, and who between 1808 and 1810 fixed Dhaulagiri, a prominent mountain in Nepal, from four different stations. Using trigonometry, Webb went on to calculate that the summit of this peak was a shocking 26,862 feet above sea level, or more than 5,000 feet higher than Humboldt's

[22] James Rennell, *Memoir of a Map of Hindoostan; or the Mogul Empire* (London: M. Brown, 1788), 256.

[23] Henry Thomas Colebrooke, 'On the Height of the Himálaya Mountains', *Asiatick Researches* 12 (1816): 256.

[24] See Colebrooke, 'On the Height of the Himálaya Mountains', 254–56.

Chimborazo (famously, Webb's measurement later proved to be correct to within 70 feet of the currently accepted height).[25]

A lowly lieutenant in the Bengal army, Webb duly submitted his calculations to the EIC, where they were soon picked up by the orientalist scholar and mathematician Henry Thomas Colebrooke (cousin of Surveyor General Robert). Colebrooke, who had developed what would amount to a lifelong fascination with the Himalaya in the 1790s, was initially careful, noting in the 1810 volume of the Calcutta-based *Asiatic Researches* that 'whether the altitude of the highest peaks of *Himálaya* be quite so great as Lieut. WEBB infers from observation, I will not venture to affirm' especially as 'owing, to disappointment in the supply of instruments, no barometrical observation could be made to confirm or check the conclusions of a trigonometrical calculation'.[26] Indeed, this lack of in situ observations – and difficulties with the instruments needed to make these once politics allowed, as Chapter 2 details – was an ongoing refrain. Colebrooke thus went on to suggest that 'without however supposing the *Himálaya* to exceed the *Andes*, there is still room to argue, that an extensive range of mountains, which rears, high above the line of perpetual snow, in an almost tropical latitude . . . is neither surpassed nor rivalled by any other chain of mountains but the *Cordilleras* of the *Andes*'.[27] By 1816, however, this equivocation was gone, and Colebrooke published a second essay in the *Asiatic Researches* with the slightly understated title of 'On the height of the Himálaya Mountains'. In this, Henry Colebrooke stated that he had initially 'thought it right to speak thus guardedly; not having been then enabled to examine the particulars of the altitudes taken'.[28] However, after receiving further observations from Webb and comparing them to the indications of Crawford and the preliminary work of Robert Colebrooke, he considered 'the evidence to be now sufficient to authorize an unreserved declaration of the opinion, that the *Himálaya* is the loftiest range of Alpine mountains which has been yet noticed, its most elevated peaks greatly exceeding the highest of the *Andes*'.[29] Colebrooke was clearly aware of the potentially explosive nature of this claim and carefully

[25] Colebrooke, 'On the Height of the Himálaya Mountains', 264–66. See also Felix Vincent Raper, 'Narrative of a Survey for the Purpose of Discovering the Sources of the Ganges', *Asiatic Researches* 11 (1810): 446–563. The currently accepted height of Dhaulagiri is 26,795 feet.

[26] Henry Thomas Colebrooke, 'On the Sources of the Ganges, in the Himádri or Emodus', *Asiatic Researches* 11 (1810): 445. For more on Colebrooke and the Himalaya, see Rosane Rocher and Ludo Rocher, *The Making of Western Indology: Henry Thomas Colebrooke and the East India Company* (Abingdon: Routledge, 2014), 108–11, 152.

[27] Colebrooke, 'On the Sources of the Ganges, in the Himádri or Emodus', 445.

[28] Colebrooke, 'On the Height of the Himálaya Mountains', 251–52.

[29] Colebrooke, 'On the Height of the Himálaya Mountains', 252.

laid out what he considered as sufficient evidence. This was coupled with a pre-emptive discussion and careful refutation of possible objections around terrestrial refraction (which always had to be compensated for in calculations and which was complicated by uncertainties around the rarefied atmosphere in high mountains), as well as generalised assumptions about the level at which perpetual snow could exist (which were based on experiences in the Alps and the Andes but failed to translate to the Himalaya). Bases thus covered, Henry Colebrooke went on to back up Webb, concluding that 'we may safely then pronounce, that the elevation of *Dhawalagiri*, the white mountain of the *Indian* Alps, exceeds 26862 feet above the level of the sea'.[30] Although Colebrooke felt no need to spell it out in the essay, the implication of this was clear: Chimborazo was finished, and the world had a new highest mountain.

In Europe, the initial response to these claims was scepticism, escalating to umbrage in some quarters. Most notably, anonymous writers in the *Quarterly Review* of 1817 published a condescending critique of Colebrooke's essay, and particularly of Webb's measurements. Here the authors wrote that 'with unfeigned respect for the talent and erudition of Mr. Colebrooke, whose name is a host in Oriental literature, we cannot help thinking that he has come to this conclusion rather hastily'.[31] Here the reviewers focused their attack not on the method per se, but on the quality of the data: 'we have not one word to offer against his calculations nor his formula . . . all we mean to protest against, is the insufficiency of his facts to authorize the conclusion which he has drawn from them'.[32] While the respected savant Colebrooke was attacked with some care, the lowly officer Webb was more robustly taken to task, the authors noting for example that one reported mountain height seemed to have been 'stretched out . . . [and] all the others seem to have grown in the same proportion'.[33] From the comfort of their London armchairs, the reviewers could thus not 'resist the conclusion that the elevation of the Himalaya range has been greatly exaggerated'.[34] Their conclusion, that 'the height of the Himalaya mountains has not yet been determined with

[30] Colebrooke, 'On the Height of the Himálaya Mountains', 266. Colebrooke made further defences of these measurements in following years. See Henry Thomas Colebrooke, 'On the Limit of Constant Congelation in the Himalaya Mountains', *Quarterly Journal of Literature, Science and the Arts* 7 (1819): 38–43; Henry Thomas Colebrooke, 'Height of the Himalaya Mountains', *Journal of Science and the Arts* 6 (1819): 51–65; Henry Thomas Colebrooke, 'On the Height of the Dhawalagiri, the White Mountain of Himalaya', *Quarterly Journal of Science, Literature and the Arts* 11 (1821): 240–46.
[31] 'Himalaya Mountains and Lake Manasawara', 431.
[32] 'Himalaya Mountains and Lake Manasawara', 431.
[33] 'Himalaya Mountains and Lake Manasawara', 441.
[34] 'Himalaya Mountains and Lake Manasawara', 439.

sufficient accuracy, to assert their superiority over the Cordilleras of the Andes', while condescendingly put, was nevertheless not entirely unreasonable, especially given the lack of barometrical measurements made in the mountains themselves.[35] In taking this position, they were not alone, and in an unevenly updated 1822 edition of *The Hundred Wonders of the World*, for example, the author could still assert that 'the Andes, in South America, are the loftiest, the most extensive, and, therefore, the most wonderful' mountains, while the Himalaya 'will be found to fall considerably short of the height attributed to them by Mr. Colebrooke'.[36] However, resistance quickly crumbled as more commentators weighed in, and more and more damning measurements from the Himalaya piled up. Smug in the comfort of hindsight, James Bell was soon able to mock these defenders of the Andes, framing the story thus: 'Colonel Crawford sounded the first alarm, by the actual measurement of several peaks in the vicinity of Nepaul', here sarcastically indicating the value of measurement in situ as opposed to theoretical speculation. As he went on, 'the second and third alarms were given by Colebrook [sic] and Webb. The matter now became serious; the theory was in danger, and it was felt a matter of incumbent duty to defend it against such audacious statements'.[37]

In this story – as indeed that of the barometers and other instruments discussed in Chapter 2 – the social status of the surveyors involved was always an underlying issue. That their standing as Company employees rather than gentlemanly savants made their task more difficult was not lost on those surveying in the Himalaya. As James Herbert wrote in a letter he submitted to his superiors in Calcutta 1819: 'I would first observe, that this survey involves as a principal point the determination of the height of the Himmaleh now acknowledged to be the highest range of mountains in the world by all except' those 'at home [who] think science confined to Europe & that it is impossible for an officer in the Company's service to measure the height of a mountain'.[38] As he continued, 'to satisfy such prejudiced ... judges it is evident that something more is required than the more routine work which it is the lot of most surveyors to furnish'. Here he indicated he 'need only refer for the truth of this assertion to the very unkind remarks passed on Captain Webb by the Quarterly Reviewers in return for his polite communication of some of the

[35] 'Himalaya Mountains and Lake Manasawara', 441.
[36] C.C. Clarke, *The Hundred Wonders of the World, and of the Three Kingdoms of Nature*, 15th ed. (London: Richard Phillips, 1822), 1, 88. Though see also p. 3, where Clarke implies the Himalaya are in fact higher.
[37] Bell, *A System of Geography, Popular and Scientific*, Vol 4, 458.
[38] Herbert to Colin Mackenzie, 1819, NAI/SOI/DDn. 152, f130.

heights he had determined'.[39] Webb himself made an effort to seem sanguine about this, noting in a letter written in 1819 while returning from a survey of the Kedarnath Temple in Garhwal that he had received the 1817 volume of the *Quarterly Review* 'not very flattering nor encouraging to my labours'.[40] The *Quarterly Review* itself proved far more generous in an 1820 follow-up, which, ostensibly a review of Humboldt, effectively served as a *mea culpa* for the high dudgeon of 1817.[41] As they put it, at Kedarnath in 1819 'Captain Webb received (and we notice the circumstance with some emotion of pride and pleasure) a copy of our Journal' in 'which it may be recollected that we freely stated all the difficulties we felt in reconciling the enormous elevation of the Himalaya mountains'. Here they noted, 'taking for our guide the theory which in Europe has been found to correspond with sufficient accuracy to ascertained facts' around the limit of the snowline, 'drew conclusions ... as erroneous, it now appears, as those of the Baron de Humboldt; so little applicable is that theory to the upper regions of India and Tartary'.[42] The apology is thus slightly undercut by the rhetorical invocation of the greatest authority on mountains of the age: we might have been wrong, they admit, but then so was Humboldt.

In a way, this retraction only served to confirm Herbert's assertion that 'it is quite clear from the spirit in which these remarks have been written that no determination of heights will ever satisfy the curious in Europe that is not accompanied with ample details as to the original observations as well as a full exposition of the methods of calculation'.[43] While in this book I have mostly resisted expanding the idea of verticality from science and topography and applying it to social relationships, this is at times too tempting. Indeed, verticality and the metaphors of 'people in high places' or 'high circles' map rather well onto the social hierarchies of surveyors and naturalists vis-à-vis the EIC and the empire, and guides and porters vis-à-vis their employers. Such metaphors also help explain the economics of knowledge production in remote locations. The historian David Ludden suggests that resource limits were endemic in peripheries because of 'the low rank of frontier sites and officials. Imperial priorities gravitate toward higher interests and higher purposes. (Again, divine authority

[39] Herbert to Colin Mackenzie, 1819, NAI/SOI/DDn. 152, f130–31.
[40] William Webb, 'Extract of a Letter from Captain William Spencer Webb, 29th March, 1819', *Quarterly Journal of Science, Literature and the Arts* 9 (1820): 63.
[41] 'Passage of the Himalaya Mountains', *Quarterly Review* 22 (1820): 415–30.
[42] 'Passage of the Himalaya Mountains', 417–18.
[43] Herbert to Colin Mackenzie, 1819, NAI/SOI/DDn. 152, f130–31. For just such an exposition, see John Hodgson and James Herbert, 'An Account of Trigonometrical and Astronomical Operations for Determining the Heights and Positions of the Principal Peaks of the Himalaya Mountains', *Asiatic Researches* 14 (1822): 187–372.

comes to mind.)'[44] This meant that 'imperial investments must travel imperial space that is horizontal (geographical distance) and vertical (status ranks)'.[45] Tensions could thus emerge between geographical and social verticality, as they did when surveyors began reporting that the Himalaya might, in fact, be higher than the Andes.

Another notable example of this comes from John Anthony Hodgson (1777–1848), Herbert's compatriot and erstwhile Surveyor General, who similarly complained that 'people in Europe are unwilling to believe that the Himalaya are higher than the Andes', suggesting that 'perhaps in England they think that officers of the army are unequal to the task, but really it is not mysterious; good instruments, time, care, perseverance, and a moderate skill in calculation, are all that is required'.[46] Of course, here Herbert and Hodgson are also taking advantage of their privileged positions in the high spaces of the mountains to produce credible knowledge of those places, and insisting on their ability to contribute to major scientific debates (in turn, reflecting a desire for metropolitan recognition, which remained a significant motivation for their contributions). However, the fact remained that unlike the Andes, whose measurements were underwritten by the great Humboldt, the Himalaya were largely put on the map by EIC employees seconded to surveys from other tasks and unable to draw on gentlemanly credibility. Hodgson reflected on this too in his tirade against metropolitan condescension, invoking Humboldt when noting that the Himalaya clearly surpassed the Andes 'by some thousands of feet, which that prince of travellers and excellent observer, M. F. Humboldt, will prove if he comes, and I wish he were here to do so'.[47] As is so often the case in this story, global comparisons thus emerge as an essential facet, evident here in the way those operating in the Himalaya felt the need to position themselves in relation to Humboldt in order to bolster their claim to producing reliable knowledge.

As for Humboldt himself, he appears to have accepted the supremacy of the Himalaya – and the methods of the Bengal Infantry surveyors – as early as 1816, and even seemed amused that 'this announcement was received in England with great incredulity'.[48] He did note the ongoing lack of essential in situ measurements with barometers, and continued to

[44] Ludden, 'The Process of Empire', 138–39.
[45] Ludden, 'The Process of Empire', 139.
[46] [John Hodgson], 'Letter from the Himmalaya', *Gleanings in Science* 2 (1830): 49.
[47] [John Hodgson], 'Letter from the Himmalaya', 49.
[48] Alexander von Humboldt, *Views of Nature: Or Contemplations on the Sublime Phenomena of Creation*, trans. Elise Charlotte Otté and Henry George Bohn (London: Henry G. Bohn, 1850), 70. For Humboldt's earlier validation of Webb's measurements and acceptance of the supremacy of the Himalaya, see Alexander von Humboldt, 'Sur l'elévation des montagnes de l'Inde', *Annales de chimie et de physique* 3 (1816): 297–317; Alexander

grapple with assumptions about the line of perpetual snow, which was far higher in the Himalaya than his theory suggested it should be, but was ultimately willing to accept that these early measurements were sufficient proof. Perhaps ever so slightly miffed by the blow to his beloved Andes, he did, however, later console himself with their ostensible quality, writing in his major late-career work *Cosmos* that 'although the mountains of India greatly surpass the Cordilleras of South America, by their astonishing elevation, (which after being long contested has at last been confirmed by accurate measurements,) they cannot' however 'from their geographical position, present the same inexhaustible variety of phenomena by which the latter are characterised. The impression produced by the grander aspects of nature does not depend exclusively on height'.[49] Here he was not alone, and Victor Jacquemont thought that whatever the Himalaya had going for them in scale, they lacked in aesthetics: 'I have had many fatigues and privations to suffer; but I think myself sufficiently rewarded by the interesting nature of all that I have seen' among the Himalaya, but 'it is a purely scientific interest: the landscape is poor and monotonous. In the highest mountains in the world there is necessarily grandeur; but it is grandeur without beauty' (he preferred the Alps).[50]

Once accepted, as indeed it broadly was by the early 1820s, the sheer magnitude of Chimborazo's demotion also soon became apparent. In one table of newly measured mountains in the Himalaya, James Herbert noted, for example, that there were 'twenty-eight as high, or higher than Chimborazo'.[51] (And to add insult to injury, Chimborazo would itself turn out to not even be the highest of the Andes, with several peaks, including Aconcagua, Illimani and Sorata, soon proving loftier.) However, even when the Himalaya had clearly displaced the Andes, doubts remained about the fixing of the highest peak and the true successor to Chimborazo as the world's highest mountain (both Dhaulagiri, Nanda Devi was considered a serious contender in the 1820s and 1830s). Measuring in the mountains in situ with barometers, essential to assuaging doubts around refraction and for fixing the elevation of trigonometrical stations allowing even greater precision, remained a highly fraught process (a story taken up in Chapter 2). This all meant that while Dhaulagiri initially took the crown, there was a definite sense that this

von Humboldt, 'Sur la limité inférieure des neiges perpétuelles dans les montagnes de l'Himalaya et les régions Équatoriales', *Annales de chimie et de physique* 14 (1820): 5–56.

[49] Alexander von Humboldt, *Cosmos: A Sketch of a Physical Description of the Universe*, trans. Elise Charlotte Otté (London: Henry G. Bohn, 1848), 8.

[50] Jacquemont, *Letters from India*, Vol 1, 241–42.

[51] James Herbert, 'Report upon the Mineralogical Survey of the Himalayan Mountains', *Journal of the Asiatic Society of Bengal* 11, part 2 (1842): xxvi.

might only be temporary, and that even more enormous mountains could still lurk out there somewhere, waiting to be found.[52] Joseph Dalton Hooker, for example, perhaps wary at the rate, which the title of world's highest mountain was claimed and surrendered, chose to refer to Kanchenjunga (which took the throne in 1848) as 'the loftiest hitherto measured mountain'.[53] This framing left plenty of scope for it to be superseded, as it duly was by Everest only a few years later. From this mid-century moment, however (and the odd speculation in the 1860s about the as- yet- unmeasured K2 notwithstanding), the highest place on earth had been settled. Across the half-century span of this book – bracketed by the consolidation of Company rule in South Asia and the eve of its demise – the vertical globe thus took on the shape by which it is still recognisable today.

An Insecure Imperial Frontier

In a commentary published in 1820, Humboldt noted that in extracting knowledge from the Himalaya, 'the obstacles offered by the suspicious policy of the Tartar and Tibetan chiefs dependent on the Chinese government are even greater than the difficulties which arise from the geological constitution of these mountains crisscrossed by crevasses, and topped by inaccessible peaks'.[54] The emergence of the Himalaya as an important imperial frontier is inextricable from the story of measuring them as the world's highest mountains, and compounded both the topographical challenges and scientific uncertainties. Indeed, to their immense and ongoing irritation, European naturalists were almost entirely blocked from entering Tibet across the period of this study (and indeed, long afterwards as well). Likewise, the continuing autonomy of powerful rulers and states like Ranjit Singh in the Punjab, and ongoing resistance from the Kingdom of Nepal, all circumscribed surveyors' ambitions for measuring the heights. This latter opposition continued even after the Anglo-Gurkha War of 1814–1816, which, although ostensibly a British victory, saw the Gurkhas maintain considerable autonomy (for example in limiting the British Resident to the Kathmandu Valley).

[52] In the 1810s and 1820s, there was also ongoing speculation as to whether there might be a range of mountains behind the Himalaya that was higher again. See for example, Humboldt, 'Sur l'Elévation Des Montagnes de l'Inde', 304.

[53] Hooker, *Himalayan Journals*, Vol 2, 386.

[54] Humboldt, 'Sur La Limité Inférieure Des Neiges Perpétuelles', 7. ['Les entraves qu'offre la politique méfiante des chefs tartares et thibétains dépendans du gouvernement chinois, sont plus grandes encore que les difficultés qui naissent de la constitution géologique de ces montagnes sillonnées par des crevasses, et surmontées de pics inaccessibles'.]

The Anglo-Gurkha War is nevertheless significant to this story as it resulted in the British acquisition of the provinces of Kumaon and Garhwal. This meant relatively easier and more reliable access to a much higher elevational cross-sectional profile of the Himalaya than was available before, even if Company authority in these newly acquired territories remained uneven and contested (and institutional interest changeable and sometimes ambivalent).[55] Throughout the period of this study, access to the mountains was thus dictated by fluctuating alliances and conflicts, which could result in gains like Kumaon and Garhwal, or Kashmir following the Anglo-Sikh wars beginning in 1845, as well as significant reversals such as in Afghanistan in 1839–1842. When considering the confirmation of the Himalaya as the highest in the world, it is therefore important to remember that to a significant extent it was negotiations with indigenous states, regional rulers and even local villagers that dictated the possibilities for travel and science in the mountains. In turn, these political realities and insecurities shaped the contours of scientific exploration, and meant that places like Kumaon and Garhwal came to play outsized roles in imperial visions of the Himalaya more widely.

These political limits were a frequent point of discussion and contention among surveyors. While operating on the edge of the Tibetan plateau in 1821, for example, the EIC officer and surveyor Alexander Gerard (1792–1839) became concerned that he was about to be stopped by Tartar border guards at the behest of the Qing Empire. As he later wrote, 'upon the surrounding heights near the Pass are many shughars or piles of stones sacred to the gods, and which at a distance exactly resembled men'.[56] Gerard continued that 'the instant my people observed them, they said they were the Tartars waiting for me; I thought the same, as they had a very suspicious appearance from below', and 'I could not divest myself of the belief (although the guides assured me that they were shughars) till I looked through the glass'.[57] Fears assuaged by the deployment of his telescope and 'seeing clearly that the supposed Tartars were stones', Gerard went on to admit some hope of penetrating further into what was becoming one of the most pressing 'blank spaces' on European maps.[58] These were quickly dashed,

[55] Lamb, *British India and Tibet, 1766–1910*; Arnold, *The Tropics and the Traveling Gaze*, 99–100; Bernardo A. Michael, *Statemaking and Territory in South Asia: Lessons from the Anglo-Gorkha War (1814–1816)* (London: Anthem Press, 2012); Bergmann, *The Himalayan Border Region*.

[56] William Lloyd and Alexander Gerard, *Narrative of a Journey from Caunpoor to the Boorendo Pass, in the Himalaya Mountains* (London: J. Madden, 1840), Vol 2, 120.

[57] Lloyd and Gerard, *Narrative of a Journey*, Vol 2, 120–21.

[58] Lloyd and Gerard, *Narrative of a Journey*, Vol 2, 121.

however, as he crossed the altitude sickness-inducing high pass only to meet a group of Tartars – real, this time, rather than illusory – who had some time ago learned that he was coming, and were waiting to politely but firmly send him back to the lowlands. This episode reveals much about the nature of Himalayan surveying in the first decades of the nineteenth century: the limits of existing knowledge, the dependence on local guides and pre-existing networks of labour and the growing insecurity of the EIC about the mountains to the north of their burgeoning Indian Empire. Meanwhile, the *shughars* or cairns – piles of stone and cloth which served as both waymarkers and shrines to the mountain spirits – are a reminder that whatever the state of imperial knowledge in this period, these had long been lived and inscribed landscapes.

As well as checks by regional political entities, increasing imperial ambitions on the subcontinent brought new concerns about Chinese and Russian activities to the north and northwest of the Himalaya, respectively. Initially, the sheer verticality of the mountains had seemed reassuring.[59] Early European travellers echoed the supposed impenetrability of Himalaya, with Englishman Samuel Turner, who led an EIC diplomatic mission in 1783, remarking of Bhutan that 'these rugged and impracticable ways, certainly lessen the importance of those military posts, we so lately passed ... the Booteeas cannot possibly have a better security, than in such a chain of inaccessible mountains, and in the barrenness of their frontier'.[60] However, such comforting rhetoric quickly became unsustainable and as Bengal Secretary John Garstin wrote in 1812: 'there are many passes into the hills from which ... the inhabitants of the Mountains might make excursions into the Plains, carrying destruction in their train, and return with impunity from our want of knowledge of the roads leading to their fastness'.[61] The surveyors seconded to the mountains were clearly aware of the potential military implications of their journeys, many having served within the 'enormous dells and craggy heights' while personally involved in the hostilities with Nepal, and had experienced first-hand the porosity of the 'frontier which was penetrated at different points by the invading columns'.[62]

The untangling of Himalayan heights and passes also became intimately linked with the so-called Great Game and fears about Russian

[59] For the idea of the Himalaya as a 'natural frontier' later in the century, see Kyle Gardner, 'Moving Watersheds, Borderless Maps, and Imperial Geography in India's Northwestern Himalaya', *The Historical Journal* 62, no. 1 (2019): 149–70.

[60] Samuel Turner, *Account of an Embassy to the Court of the Teshoo Lama in Tibet, Containing a Narrative of a Journey through Bootan and Part of Tibet* (London: W. Bulmer, 1800), 190.

[61] John Garstin to Charles Wright Gardiner, 7 March 1812, NAI/SOI/DDn. 128, f74.

[62] Lloyd and Gerard, *Narrative of a Journey*, Vol 1, 110.

aspirations for India. William Webb, for example, was rather put out to learn in 1816 that a friend had obtained from St Petersburg a publication describing his having crossed the Himalaya, 'regarded as inaccessible' but 'by which a route can be opened through Tartary to Russia'.[63] Historical writing on the 'Great Game' is expansive and of variable quality. It has also seen considerable revision in recent decades, with scholars pointing out the way that classic stories of Anglo-Russian rivalry tend to marginalise Central Asian peoples and agency, take exaggerated imperial rhetoric for reality and conflate threat of invasion with the threat of loss of authority and prestige within India itself.[64] Compounding environmental, social and political factors nevertheless meant that India's frontiers were increasingly seen as insecure, both on the ground and especially – perhaps even mostly – in imperial imaginations. In this book, the idea of insecurity is thus intended to signal a relative lack of knowledge and the threatening nature of this lacunae, something that evolved and fluctuated, but was never really overcome during the course of the nineteenth century.[65] In this context, Chris Bayly has argued that intermittent 'information famines' and 'information panics' were central to imperial expansion in South Asia, even as they could also put the entire apparatus of colonial power at risk.[66] Here insecurity might become an impetus for

[63] Webb to Colin Mackenzie, 2 December 1817, NAI/SOI/DDn. 150, f27. ['regardées comme inaccessibles' but 'par lesquelles on peut ouvrir une route par la Tartarie jusqu'en Russie'.]

[64] See Benjamin Hopkins, *The Making of Modern Afghanistan* (London: Palgrave Macmillan, 2008); Alexander Morrison, 'Introduction: Killing the Cotton Canard and Getting Rid of the Great Game: Rewriting the Russian Conquest of Central Asia, 1814–1895', *Central Asian Survey* 33, no. 2 (2014): 131–42; Ian W. Campbell, '"Our Friendly Rivals": Rethinking the Great Game in Ya'qub Beg's Kashgaria, 1867–77', *Central Asian Survey* 33, no. 2 (2014): 199–214. For the sort of classic 'Great Game' scholarship these authors are revising, see Peter Hopkirk, *The Great Game: On Secret Service in High Asia* (London: John Murray, 1990).

[65] For the continuation of this later in the century, see Thomas Simpson, '"Clean Out of the Map": Knowing and Doubting Space at India's High Imperial Frontiers', *History of Science* 55, no. 1 (2017): 3–36; Kyle Gardner, *The Frontier Complex: Geopolitics and the Making of the India-China Border, 1846–1962* (Cambridge: Cambridge University Press, 2021). For insecurity, anxiety and 'information panics' in British India more generally, see Jon Wilson, *The Domination of Strangers: Modern Governance in Eastern India, 1780–1835* (Basingstoke: Palgrave Macmillan, 2008); Mark Condos, *The Insecurity State: Punjab and the Making of Colonial Power in British India* (Cambridge: Cambridge University Press, 2017). But for a critique of these analyses, see Joshua Ehrlich, 'Anxiety, Chaos, and the Raj', *The Historical Journal* 63, no. 3 (2020): 777–87.

[66] Christopher Bayly, *Empire & Information: Intelligence Gathering and Social Communication in India, 1780–1870* (Cambridge: Cambridge University Press, 1997), 97–113, 144–50, 165–78. For the argument that frontier officials sometimes tried to deliberately perpetuate these information famines, see Thomas Simpson, 'Forgetting Like a State in Colonial North-East India', in *Mountstuart Elphinstone in South Asia: Pioneer of British Colonial Rule*, ed. Shah Mahmoud Hanifi (Oxford: Oxford University Press, 2019), 223–48.

imperial knowledge-making, even if generating more knowledge – for instance the existence of passes by which the mountains might be crossed – could actually increase rather than alleviate this insecurity. Likewise, insecurity about the mountains was only exacerbated by surveyors' reliance – acknowledged or otherwise – on existing information orders and networks for extracting and circulating information, networks that might actively resist their aims, or which their prejudices might limit their access and receptiveness to.

Measuring in the mountains could likewise be a political act in and of itself. As in the case of Alexander Gerard, once gaining more reliable access in the years following the Anglo-Gurkha War, surveyors had to be particularly wary about using their instruments near the edge of Tibet, lest they arouse the suspicion of the watching agents of the Qing Empire. Here it is worth briefly digressing that this book only examines exploration of the Himalaya that originated from the south, a lens that serves to highlight the uncertainties of the contemporary actors as to what was occurring in the north and northwest. Chinese interactions with the Himalaya were nevertheless not wholly dissimilar to those playing out from the British Indian side, as administrators dealt with geographical uncertainties, unruly informants, the complexities of border-making and a growing awareness of a foreign power flexing on the opposite side of the mountains.[67] Exploration from the northwest, especially by Russians like Ivan Vitkevich (Jan Prosper Witkiewicz), also sometimes features similar themes. Indeed Vitkevich, though mostly engaged in political efforts rather than natural history, acted as translator for Alexander von Humboldt during part of his travels in Central Asia.[68] However, the results of these studies were largely inaccessible and unknown to EIC surveyors and naturalists operating in the mountains in the first half of the century, and thus remain on the fringes of the story, much as the Russians themselves.

Measuring the Himalaya was thus often politically fraught, even as surveyors felt compelled to make as many observations as possible. This reflects the inherent tension faced by imperial surveyors between 'boundary making … and boundary crossing'.[69] As Alexander Gerard wrote while near Shipke, 'we did not think it advisable to use the theodolite in

[67] See Matthew W. Mosca, *From Frontier Policy to Foreign Policy: The Question of India and the Transformation of Geopolitics in Qing China* (Stanford: Stanford University Press, 2013); Ulrike Hillemann, *Asian Empire and British Knowledge: China and the Networks of British Imperial Expansion* (Basingstoke: Palgrave Macmillan, 2009).

[68] See, for example, Alexander Morrison, 'Twin Imperial Disasters: The Invasions of Khiva and Afghanistan in the Russian and British Official Mind, 1839–1842', *Modern Asian Studies* 48, no. 1 (2014): 253–300; Svetlana Gorshenina, *Explorateurs en Asie centrale: voyageurs et aventuriers de Marco Polo à Ella Maillart* (Geneva: Editions Olizane, 2003).

[69] D. Graham Burnett, *Masters of All They Surveyed: Exploration, Geography, and a British El Dorado* (Chicago: University of Chicago Press, 2000), 255.

the presence of the inhabitants, knowing their extreme jealousy'.[70] He was duly chastened and sent back, recording this in his journal as a carefully choreographed performance of the limits of the empire. Indeed, having already lowered his ambitions to a rough route survey by compass and wheel – rather than a more precise trigonometrical survey using the theodolite – Gerard was ultimately unsuccessful even in this more modest aim. This sort of self-policing of instrumental practice near the borders is indicative of the limitations that remained on British power and imperial mastery in the Himalaya throughout the first half of the nineteenth century, even while they were moving towards political dominance in the lowlands. These political realities are also reflected in a map Alexander Gerard produced from his surveys, which abruptly ends in 'blank space' (the termination is marked with a note, 'stopped here by the Chinese', as can be seen in Figure 1.2).

East India Company surveyors made numerous attempts to visit Tibet, though never to any avail, as Tartars with strict orders were always waiting to politely send them back the way they came. As Alexander Gerard wrote: 'I reached the limits of their country in four different quarters, but was not allowed to advance a step farther'.[71] These meetings – including the one following his paranoid encounter with the *shughars* – were intricate acts of diplomacy, and Gerard described the Tartar *Lafa* he negotiated with as 'inflexible in his determination' though he 'met me with an air of openness and good humour seldom equalled'.[72] All of Gerard's attempts at negotiation were for naught, and neither letters of introduction nor the offer of substantial bribes were effective. The reality was that if he had decided to resort to violence, the Tartars would not have been able to stop him. He records the *Lafa* acknowledging this: 'we will post ourselves on the road, but you have a sufficient number of people to force the passage, for we will not fight; we, however, trust you will not attempt it without permission'.[73] Especially after a duplicitous and unsanctioned use of disguises by English adventurer William Moorcroft (1767–1825) to reach Lake Manasarovar in 1812 – which Gerard cited as a precedent for ongoing tensions – further diplomatic incidents were not something the EIC was in a position to entertain, and the Bengal Infantry

[70] Alexander Gerard, *Account of Koonawur, in the Himalaya*, ed. George Lloyd (London: James Madden, 1841), 284.
[71] Gerard, *Account of Koonawur, in the Himalaya*, 104–5. See also Dorothy Woodman, *Himalayan Frontiers: A Political Review of British, Chinese, Indian and Russian Rivalries* (London: The Cresset Press, 1969), 30–34.
[72] Gerard, *Account of Koonawur, in the Himalaya*, 104–6. *Lafa* was, according to Gerard, a generic term for chief.
[73] Gerard, *Account of Koonawur, in the Himalaya*, 104.

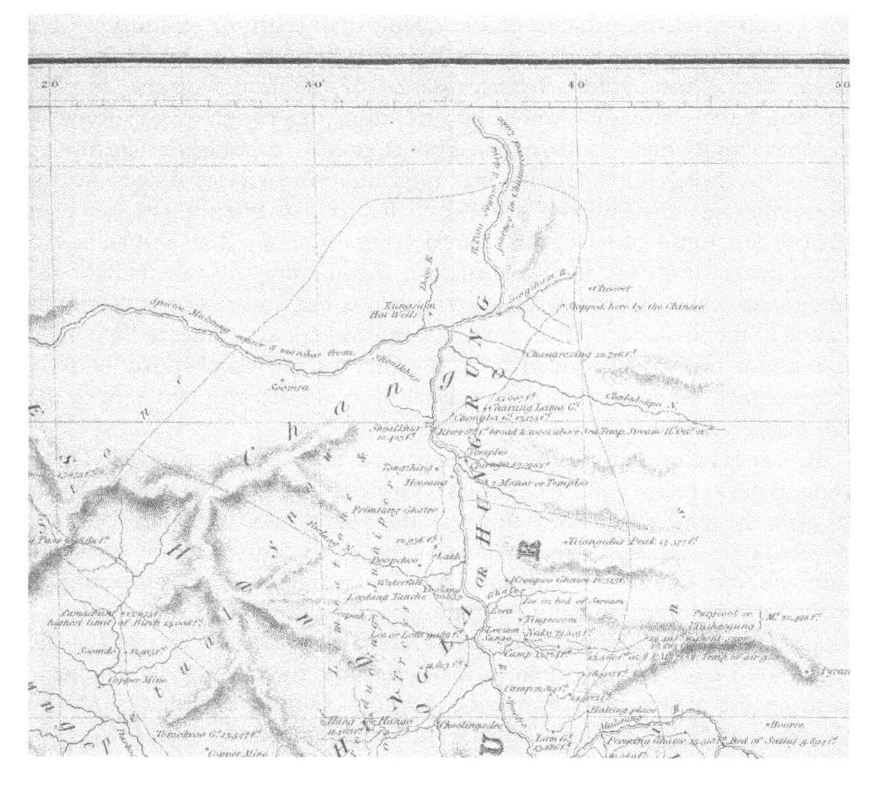

Figure 1.2 Detail from the upper-right corner of 'A Map of Koonawur', which was made following the EIC acquisition of Kumaon and Garhwal. Source: Alexander Gerard, *Account of Koonawur, in the Himalaya*, ed. George Lloyd (London: James Madden, 1841). With kind permission of the Syndics of Cambridge University Library.

surveyors were thus forced to check their instrumental ambitions at the Tibetan frontier.[74]

These political concerns continued into the 1840s, and even while the imaginative coherence of the Himalaya evolved as more travellers returned and more knowledge proliferated, the frontiers mostly remained assumed, ill-defined and resistant to instrumental fixing. This was still

[74] Gerard, *Account of Koonawur, in the Himalaya*, 105; William Moorcroft, 'A Journey to Lake Mánasaróvara in Ún-Dés, a Province of Little Tibet', *Asiatic Researches* 12 (1816): 375–534. See also Garry Alder, *Beyond Bukhara: The Life of William Moorcroft* (London: Century, 1985).

very much the case when Bengal Infantry officer Henry Strachey made another attempt to reach the sacred lakes of Rakas Tal and Manasarovar in 1846.[75] In his journal, Strachey records carefully prioritising his geographical and instrumental aims for the trip, noting that when operating in the borderlands, 'there was constant risk of an untimely end to our expedition, should we be detected, by the intervention of the Lhassan authorities'.[76] The shadow of Moorcroft's ill-received excursion in 1812 still hung over the frontier, even as Strachey employed similar tactics in attempting to disguise his appearance: 'I have of course adopted the Hindustani "*Dhab*" of costume, just enough to pass muster in the distance, and nothing more, as I have not attempted to disguise the Feringi [foreign] complexion of my face and hair'.[77] (These necessary antics prompted Joseph Hooker to write to his father at Kew that 'Strachey was gone (on a wild goose chase) to Lhassa disguised as a Tartar'.[78]) Strachey also took care to hide his scientific instruments, even if this meant compromising his cartographic ambitions: 'I was about to take bearings of this and other points when the alarm was given of a horseman ahead, which obliged me to pocket my compass ... depriving me as I afterwards found of a most valuable observation for my survey'.[79] As Figure 1.3 indicates, surveyors in the mountains often found themselves coming up against the very real, if ill-defined, limits of empire and of British imperial mastery.

'Hidden Histories' in the Mapping of the Mountains

In attempting to survey the lakes of Rakas Tal and Manasarovar, Henry Strachey – much as Webb, Hodgson, Herbert and Gerard, who had earlier measured the Himalaya as the world's highest mountains – relied heavily on his guides for both information about the route and navigating the complex politics of the frontier. Recovering and highlighting these roles and networks is one of the key aims of this book, and indeed these contributions are inseparable from the story of the Himalaya's ascendency over the vertical globe. It is nevertheless important to recognise at the outset that the networks of labour that made Himalayan exploration possible were highly heterogeneous. Expeditions were mobilised in

[75] Henry Strachey, 'Narrative of a Journey to Cho Lagan (Rakas Tal), Cho Mapan (Manasarowar), and the Valley of Pruang in Gnari, Hundes, in September and October 1846', *Journal of the Asiatic Society of Bengal* 17, part 2 (1848): 111, 129.
[76] Strachey, 'Narrative of a Journey to Cho Lagan', 152–53.
[77] Strachey, 'Narrative of a Journey to Cho Lagan', 132.
[78] Hooker to William Jackson Hooker, 20 January 1848, Kew Archives, JDH/1/10/39–42.
[79] Strachey, 'Narrative of a Journey to Cho Lagan', 146.

Figure 1.3 'Lake of Rakas-Tal or Tso-Lanak, 15,200 feet', which Richard Strachey noted was 'about 750 feet lower than the summit of Mont Blanc'. Here the surveyors were forced to adopt 'native' dress, indicative of the limits of imperial mastery on the high frontiers, even as it added to the 'othering' of the landscape and invoked adventure for audiences at home.
Source: Plate XIII from Richard Strachey, 'The Physical Geography of the Himalaya', 1854. British Library, Mss. Eur. F127/202 (The proofs of this work and the plates were completed, but it was apparently never published in its final form). With kind permission of The Society of Authors as agents of the Strachey Trust.

different ways at different times, occasionally through forced labour (sometimes known as the 'begar' system), or more often through patronage networks and recommendations.[80] The expedition party that accompanied Henry Strachey's mission to map Manasarovar and Rakas Tal provides an instructive illustration of this. In Strachey's account, his two chief brokers and guides, a Bhotia named 'Rechung or Rechu, Padhán of Kunti' and a Kumaoni named 'Bhauna Hatwál Khasiah', are mentioned

[80] Jayeeta Sharma, 'A Space That Has Been Laboured On: Mobile Lives and Transcultural Circulation Around Darjeeling and the Eastern Himalayas', *Transcultural Studies* 7, no. 1 (2016): 54–85; Bergmann, *The Himalayan Border Region*. On the complexities of the 'begar' system and colonial misinterpretations of it, see Jahoda, *Socio-Economic Organisation in a Border Area of Tibetan Culture*, 71–86.

by name, itself a relatively rare occurrence for indigenous assistants, and usually indicative of high social status amongst a hierarchical order of brokers and porters. In terms of recruitment, Strachey explained for Bhauna that 'having heard previously of his qualifications, I engaged him to accompany me'. Here Strachey indicates the way patronage recommendations were significant, and indeed later chapters feature brokers – like Pati Ram, the vizier of Sungnam – who made supplementary careers assisting multiple European surveyors and naturalists over many decades. Familial networks and connections were also important, and Strachey was assisted by Rechu's brother and son (named, respectively, Tanjan and Tashigal) and later by 'Anand, a young relation whom Bhauna has thought proper to bring with him, to assist in cooking dinner, etc: though as this is Anand's first visit to Húndés, or southern Bhote even, he is likely to be of small use in manual service'.[81] Strachey explained these sometimes complex familial and political negotiations in some detail: 'when first asked who were to accompany me, I said that I left Rechu to bring whom he chose from his own village, (as I thought the most simple and convenient plan), but the men of Kunti raised objections' so that only 'after much discussion, it was settled . . . that the service should be equally distributed (like the supply of baggage cattle, provisions, &c.) each village furnishing one man'.[82]

Local and regional politics were thus never far away, and Strachey continued that 'never having been to the lakes by the out-of-the-way route I am now taking, he [Bhauna] is nothing of a guide, but promises to be useful as informant generally, and negociator [sic] in case of any untoward collision with the Hunias; also as interpreter'.[83] Bhauna was a trader, widely travelled 'either on his own account or as agent for some of the Almora merchants', and 'in quest of Pearls and Coral and other merchandise for Húndés' he had 'been often to Jaipúr and sometimes as far as Calcutta and Bombay'.[84] Here scholars like Felix Driver are increasingly sensitive to the way that often 'local' knowledge 'could hardly be characterised as "local" or "indigenous" in any straightforward sense'.[85] Moreover, presenting South Asian expertise as 'local' or 'indigenous' might be seen as a particular kind of imperial work, placing it as hierarchically lesser than European science. This is an important consideration in the Himalaya, where guides accompanied and assisted

[81] Strachey, 'Narrative of a Journey to Cho Lagan', 131.
[82] Strachey, 'Narrative of a Journey to Cho Lagan', 131.
[83] Strachey, 'Narrative of a Journey to Cho Lagan', 129–30.
[84] Strachey, 'Narrative of a Journey to Cho Lagan', 129.
[85] Driver, 'Hidden Histories Made Visible?', 427. For a further reflection, see also Driver, 'Intermediaries and the Archive of Exploration', 15–16.

travellers for diverse reasons, and were notably varied: sometimes simply local youths hired to point out the way to the next village, at other times professional or semi-professional brokers travelling with an expedition from entirely different parts of the mountains (or in some cases even from the lowlands). Brokers and porters in the Himalaya could thus sometimes find themselves geographically, culturally and linguistically almost as far from home as the European travellers they were guiding. This was common enough to be deployed for rhetorical effect, and for example, when he was setting up camp near Hatu in 1822, Welsh EIC officer William Lloyd (1782–1857) noted that 'our servants from the plains, who had never seen snow before, looked at it with that indifference which is so peculiar a mark of the Hindoo character'.[86] In this, he was far from the only traveller in this period to use the first encounter of porters with snow to claim cultural superiority, even as moments like this complicate any simplistic ideas of a dichotomy between 'European' and 'indigenous' knowledge – not least because both categories were highly prone to fragmentation.

Whatever surveyors' insistence on superiority, their accounts of measuring the mountains are nevertheless rife with moments that can be read 'against the grain' and for 'countersigns' to recover the agency of South Asian guides and brokers, and to identify acts of resistance both large and small.[87] For example, at the lake of Rakas Tal (see Figure 1.4), Henry Strachey hoped to make a circuit to complete his map. However, fearing the chances of detection in the borderlands were too high, his guides engaged in creatively resisting this plan. As Strachey describes it, 'in the evening, Rechu, with a well-assumed air of distress, reported that both the ponies had strayed from our camp' though 'I have a strong persuasion that this was a contrivance of my worthy companions to put a spoke in the wheel of my *parkarma* [circuit]; for being rather sulky, I had not yet informed them of my consent to abandon that design'. Perhaps intending to save face for his European readers, Strachey insisted that 'their clumsy artifice would certainly not have stopped me, if I had resolved upon it', even as Rechu is here made to fit the trope of the lazy and duplicitous guide so often featured in European travel narratives. Meanwhile, desertion was not uncommon and while trying to recover the ponies, Strachey discovered two of his porters 'had walked off, as they pretended to enquire for mutton at Tokar, but in fact more probably straight back to Byáns, for they never showed themselves again to the end of our journey'.[88]

[86] Lloyd and Gerard, *Narrative of a Journey*, Vol 1, 162.

[87] See Stoler, *Along the Archival Grain*; Douglas, *Science, Voyages, and Encounters in Oceania, 1511–1850*.

[88] Strachey, 'Narrative of a Journey to Cho Lagan', 167.

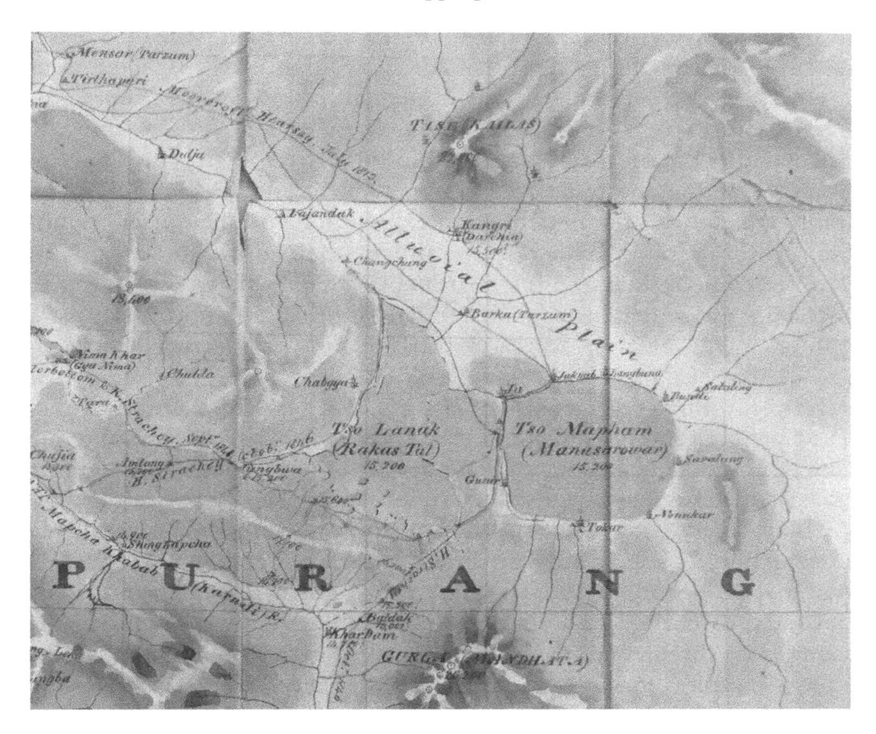

Figure 1.4 Section of Henry and Richard Strachey's 'Map of Kumaon and British Gurhwal' (1850), featuring the sacred lakes of Rakas Tal and Manasarovar. The peak of Kailash, far more important in South Asian cosmology than Everest or K2, can be seen in the top right. Like Alexander Gerard's earlier map, it peters out into 'blank space' at the edge of nominally British territory and geographical knowledge. Similarly, the three-dimensionality of the mountains is here only awkwardly flattened onto the two-dimensional page. Far from confident projections, maps of the Himalaya in this period often reveal the fragility of imperial claims and the imaginative incoherence brought about by the sheer scale and verticality of the mountains.
Source: British Library, Mss. Eur. F127/202. Courtesy of The Society of Authors as agents of the Strachey Trust.

Such moments also indicate how explorers often rendered brokers and guides more visible in accounts when things went wrong and they could be blamed. Indeed, the agency of Himalayan peoples could be played up or down, deliberately obscured or made visible, depending on the context and audience. Such practices were a central feature of exploration throughout the period, and indeed into the second half of the nineteenth century (as seen, for example, in the expedition of the Schlagintweit brothers, whose extensive acknowledgement of indigenous agency was often more explicit than their predecessors, even as it was carefully calculated to bolster their own scientific authority).[89]

Interactions along and across frontiers also offer other glimpses of the 'hidden histories' that underpinned exploration and measuring mountains in this period. Some of these anticipate the now famous story of Nain Singh, Kinthup and the so-called 'Pundits' who were sent out in the 1860s as 'human instruments' to measure parts of the Himalaya still politically inaccessible to Europeans. Fifty years earlier, some of the Bengal Infantry surveyors had attempted to develop similar practices.[90] Surveyor General of Bengal John Garstin, commenting in 1812 on an account from a *munshi* employed by William Webb, wrote that he could 'conceive they will all be of great use' and concluded that 'these native surveyors work hard, for small pay, they can penetrate into parts of the country inaccessible to Europeans and procure valuable information'.[91] John Hodgson similarly sent out Indian surveyors, in one instance towards Lake Manasarovar, which remained enticingly beyond the frontier.[92] Though it is not possible to recover the surveyor's name from the silences of the colonial archive, Hodgson notes that he was a Brahmin 'formerly in the service of Lieut. Webb, who he says taught him something of the use of a Compass, but I found him deficient & gave him some instructions'.[93] Here we again see the way these individuals might make careers, exploiting opportunities to work for multiple surveyors. The 'Brahmin' also apparently had the linguistic fluidity to operate in the borderlands, and a cover story: 'the man is intelligent and enterprising, writes Hindee, some Persian, & a little execrable English, & poses as a pilgrim & native doctor, with medicines for those who are so unfortunate as to become his patients'.[94] Reflecting on the surveyor's efforts, Hodgson went on to note that 'the Bramin says he had & used, a compass as far as Udsein ... & that it was there broken,

[89] See Brescius, *German Science in the Age of Empire*, 151–216.
[90] For the wider history of 'native explorers' in India, see Mathur, 'How Professionals Became Natives'.
[91] John Garstin to Charles Wright Gardiner, 7 March 1812, NAI/SOI/DDn. 128, f72.
[92] John Hodgson to Charles Crawford, 14 November 1813, NAI/SOI/DDn. 130, f99. For more on John Hodgson's schemes, see Mathur, 'How Professionals Became Natives', 27–30.
[93] John Hodgson to Charles Crawford, 14 November 1813, NAI/SOI/DDn. 130, f99–100. See also Reginald Henry Phillimore, *Historical Records of the Survey of India* (Dehradun: Survey of India, 1950), Vol 2, 353–54.
[94] John Hodgson to Charles Crawford, 14 November 1813, NAI/SOI/DDn. 130, f100.

I doubt his having had one & indeed ~~the whole~~ much of his story, but such as it is I send it for want of better'.[95] Hodgson's agency here is somewhat undermined, but he is still willing to admit that this information, however compromised, was better than leaving the map blank in such an important quarter. Such doubts were common, and Hodgson elaborated that 'knowing too much of the exaggeration & falsity of natives I dare not vouch for the correctness of this route' and though he had 'reason to believe that the man did go to Maunsir himself ... suspect that the remainder of the journey back he may have got by information'.[96] In instances like this, surveyors' inability to trust and manage their informants ultimately undercut confidence in the entire information order.

The employment of South Asian rather than European surveyors also did not entirely overcome the danger of causing diplomatic incidents in the borderlands, and Hodgson considered sending the Brahmin to Kashgar 'to get some idea of the distance & route to the nearest part of the Russian dominions' but reconsidered 'apprehensive that he might attempt to pass himself off as an authorised agent of Government & misbehave accordingly'.[97] Indeed, it was perhaps this last factor that led to the EIC largely suspending the development of programmes using Indian surveyors until the 1860s. This disbanding was ordered in no uncertain terms, and responding to Hodgson's dispatch from 'the Brahmin', Charles Crawford stated that the 'Government have notified to me that they wish to throw cold water on all natives being taught or employed in making geographical discoveries' and elsewhere noted that 'Govt. were anxious to prevent the Natives from obtaining, or being taught, any knowledge of the kind'.[98] However short-lived, these early experiments are nevertheless significant in light of the turning back to the 'Pundits' in the 1860s when mastery remained elusive, and for revealing the inherent hierarchies of labour and knowledge that underlay and sometimes undermined attempts to map the Himalaya onto the vertical globe.

[95] John Hodgson to Charles Crawford, 14 November 1813, NAI/SOI/DDn. 130, f101.

[96] John Hodgson to Charles Crawford, 14 November 1813, NAI/SOI/DDn. 130, f99–100.

[97] John Hodgson to Charles Crawford, 14 November 1813, NAI/SOI/DDn. 130, f101–2. This fear was perhaps warranted, and for a later incident that did spark a diplomatic furore, see Matthew Mosca, 'Kashmiri Merchants and Qing Intelligence Networks in the Himalayas: The Ahmed Ali Case of 1830', in *Asia Inside Out: Connected Places*, ed. Eric Tagliacozzo, Helen F. Siu and Peter C. Perdue (Cambridge, MA: Harvard University Press, 2015), 219–42.

[98] Charles Crawford to John Hodgson, 2 December 1813, NAI/SOI/DDn. 135, f38–39; Charles Crawford to George Fleming, 23 July 1813, NAI/SOI/DDn. 135, f15.

Conclusion

In 1856, Mount Everest was finally confirmed as the highest mountain in the world. This last part of the story, as the mountain designated on survey maps as Peak XV was announced by Surveyor General Andrew Scott Waugh, based on the calculations of Bengali mathematician Radhanath Sikdar (at the time 'Chief Computer' for the Great Trigonometrical Survey), and ultimately named for a man who never saw it, is one that has been told many times.[99] Rather than focus on this now well-known moment in Himalayan history, this chapter has instead considered the fraught process of measuring the mountains and mapping the frontiers in the preceding decades. Doing so has highlighted the significant imaginative, technological and political reorientations needed before Everest could be thrust into the spotlight as the definitive roof of the world. This is an honour it still retains today, however sullied by the chicanery around modern climbing and severe overcrowding on a small piece of terrain the relevance of which – now taken-for-granted – is merely that it happens to be higher than any other (and which from the perspective of the present is one of the main things that makes a mountain matter). As this chapter has detailed, establishing the supremacy of Everest always meant navigating global comparisons. In the first half of the nineteenth century, the Himalaya had to be grafted onto – even they forced the reconfiguration of – a vertical globe already populated in European thought and scientific theorising, especially by the Alps, so close to home, and the Andes, elevated by their role as the tableau for Humboldt's outsized influence. As the chapters that follow take up, these factors – from comparative confusions to Humboldt's gravitas – were recurring problems in early attempts to map the natural history of the mountains and draw them into a coherent scientific and imperial framework.

Equally importantly, this chapter has set the scene for what follows by demonstrating how measuring the mountains and their natural history was complicated by both frontier politics and dependency on the labour and expertise of Himalayan peoples. This has been introduced as, in part, a story of imperial insecurity, indigenous resistance and uncertain agency. At the same time, however much EIC surveyors' paranoia, dependence on local networks and self-policing of their instrumental practices often demonstrate the limits of imperial mastery in the period the Himalaya became the highest mountains in the world, these measurements had consequences that would resonate through the nineteenth century – and

[99] See for example Keay, *The Great Arc*, 158–72; Sen, *Astronomy in India, 1784–1876*, 91–97.

indeed, down to our postcolonial present. Wherever they are evoked, 'blank spaces' have an aspirational dimension for empires. The challenge presented by 'blank space' and the insecurity this generated was ultimately essential in clearing away indigenous presences and subsuming older local and regional topographies and borders, even if this could only be haphazardly achieved (indeed, access to the mountains in this period was highly uneven, as politics and topography funnelled European naturalists and surveyors along a sometimes embarrassingly limited number of well-trodden routes). This period nevertheless matters, as Kalyanakrishnan Sivaramakrishnan argues for Indian landscapes more broadly, because 'the observations and writings of official, scientific and commercial travellers in the early nineteenth century created the language and representational repertoire for the consolidation of empire'.[100] Ultimately, the imagining of the Himalaya as both the highest mountains in the world and an often insecure frontier had long-term consequences for the way the mountains were woven into the imperial tapestry as peripheral places. In delving into these questions further, Chapter 2 considers the use of scientific instruments in greater detail, and how they were made unstable in this period by the sensory derangement that was unavoidable when measuring on the roof of the world.

[100] Kalyanakrishnan Sivaramakrishnan, 'Science, Environment and Empire History: Comparative Perspectives from Forests in Colonial India', *Environment and History* 14, no. 1 (2008): 47.

2 Unstable Instruments

In October 1818, Scottish brothers Alexander and James Gilbert Gerard (1793–1835) found themselves trudging upwards after a sleepless night on the slopes of Reo Purgyil in Kinnaur. Dogged by fatigue, the brothers – both surveyors attached to the Bengal Infantry, Alexander as an officer and James as a surgeon – persevered in their ascent as they 'wished much to see the barometer below fifteen inches'.[1] As they climbed, they were assaulted by the disturbing symptoms of altitude sickness. These included near-constant headaches, difficulty in breathing, insomnia, loss of appetite and lethargy, and James Gerard suggested these were 'similar to the sedative effect of intoxication'.[2] These bodily debilitations haunted the Gerards' efforts at precise measurement, especially as they could not necessarily trust their senses. Indeed, in high mountains, scale and distance are difficult to judge. In his journal, Alexander continued, 'it was 4 p.m. when we gained the summit, so we had no time to make half the observations we wished'.[3] This was exacerbated because their 'hands were so numbed, that it was not until we had rubbed them for some time that we got the use of them'.[4] While Alexander fumbled with their theodolite, struggling to adjust it through bulky gloves, James had more success, getting readings from three homemade 'mountain' barometers, which all agreed at 14.675 inches. Even recording these precious measurements in the fieldbook was far from trivial: 'the ink froze, and I had only a broken pencil, with which I could write very slowly', a significant

[1] Gerard, *Account of Koonawur*, 290. Reo Purgyil is located in what is now Himachal Pradesh, and is variously also known as Purgeool, Parkyul, Riwo Phargyul, Leo Pargial, Tarhigang or Turheegung. Near-identical accounts of this ascent were published elsewhere as Alexander Gerard, 'Journal of an Excursion through the Himalayah Mountains, from Shipke to the Frontiers of Chinese Tartary', *Edinburgh Journal of Science* 1 (1824): 41–52, 215–25; Alexander Gerard, 'Narrative of a Journey from Soobathoo to Shipke, in Chinese Tartary', *Journal of the Asiatic Society of Bengal* 11, part 1 (1842): 363–91.

[2] James Gilbert Gerard, 'A Letter from the Late Mr J.G. Gerard', in *Narrative of a Journey from Caunpoor to the Boorendo Pass, in the Himalaya Mountains*, ed. George Lloyd, vol. 1 (London: J. Madden, 1840), 326.

[3] Gerard, *Account of Koonawur*, 291–92. [4] Gerard, *Account of Koonawur*, 292.

limitation given how crucial notetaking was in a world that placed hypoxic stresses on the mind and memory.[5] Whatever the difficulties attached to precise measurement in a world in which the senses were untrustworthy, the impetus to keep moving upwards was pressing. In building on Chapter 1, this chapter examines the use of scientific instruments in a context of growing insecurity around limited knowledge of the mountains. Here instrumental measurements taken in the mountains themselves became crucial not only to debates around the supremacy of the Himalaya over the Andes, but also to mapping and fixing newly acquired frontiers (particularly those with Nepal and Qing China acquired following the Anglo-Gurkha War of 1814–1816). In other words, the Gerards' ascent on Reo Purgyil occurred at a key moment not just for the science of measuring altitude, but also for the imaginative and political constitution of the Himalaya as the northern frontier of British India.

This chapter traces surveyors' and travellers' use of scientific instruments in the Himalaya across the first half of the nineteenth century. I focus especially on three types of instruments associated with measuring altitude – mountain barometers, boiling point thermometers and fieldbooks – and on the social performances required to render the knowledge they produced of the high mountains credible. In particular, I examine the moments in which limits were exceeded, and instruments were found to be inadequate, or broke down and were repaired. These are revealing of the importance instruments played in establishing scientific authority in a world in which the senses were unreliable.[6] The various roles of these instruments are evident in the way that, later and from the relative luxury of the lowlands, Alexander Gerard compiled an unpublished note titled 'Memoir of the Construction of a Map of Koonawur', in which he reflected at length on the technical aspects of the measurements he and his brother had taken on Reo Purgyil. Using the readings from the barometers (those from the theodolite were ultimately no good), he calculated that the high point he and James had reached was not less than 19,411 feet above the level of the sea.[7] As he then added, 'I have discussed the elevation of this station at some length, because the subject

[5] Gerard, *Account of Koonawur*, 292.

[6] For the significance of these moments, see Schaffer, 'Easily Cracked'; Charles W. J. Withers, 'Geography and "Thing Knowledge": Instrument Epistemology, Failure, and Narratives of 19th-Century Exploration', *Transactions of the Institute of British Geographers* 44, no. 4 (2019): 676–91. On other efforts to repair and modify instruments in India, see Jane Insley, 'Making Mountains out of Molehills? George Everest and Henry Barrow 1830-39', *Indian Journal of History of Science* 30, no. 1 (1995): 47–55; Chakrabarti, *Western Science in Modern India*, 39.

[7] Alexander Gerard, 'Memoir of the Construction of a Map of Koonawur', 1826, British Library, Mss. Eur. D137, f200.

is interesting, from the circumstance of no former travellers having attained such a height on the earth's surface'.[8]

In making this claim, Alexander Gerard was thinking specifically of Alexander von Humboldt, whose high point on Chimborazo during his famous ascent of 1802 made him the previous holder of this distinction. (Ironically, neither Humboldt's nor the Gerard brothers' high points were ever actually the altitude records they believed they were; indeed, we now know from the frozen bodies of the so-called 'ice maidens' and other evidence that Incas had reached the summit of Llullaillaco in the Andes several centuries earlier, which at 22,110 feet was significantly higher.[9]) Meanwhile, the Gerards were accompanied to their high point on Reo Purgyil by at least one Tartar, although in his account Alexander goes to considerable effort to downplay the co-presence of the brothers' South Asian companions. For example, he recorded that during the ascent, eventually the 'the man who carried the bundle of sticks [to make a fire in order to measure the boiling-point of water] sat down and said he must die, as he could not proceed a step further' so 'we accordingly left him, and after an ascent of 700 feet, attained the top . . . 19,411 feet above the level of the sea'.[10] The other porters who did share the Gerards' 'record' are thus elided, and such rhetorical sleight of hand is further evidence of the way the agency of Himalayan peoples could be played up or down in accounts depending on context and need. It is also notable that neither Humboldt's nor the Gerards' supposed 'record' high points were actually on the summits of Chimborazo or Reo Purgyil, and it was believed unlikely 'that the higher peaks can ever be ascended to be determined barometrically'.[11] By engaging with Humboldt, Gerard was thus seeking recognition for what was an essentially arbitrary record. As Humboldt himself chided, 'these mountain ascents, beyond the line of perpetual snow, however they may engage the curiosity of the public, are of very little scientific utility'.[12] Indeed, this is an important contrast with the more usual 'summit position' that was to become the goal and reward of Himalayan mountaineering later (and indeed already elsewhere, especially in the Alps).[13] Who could go the highest (and instrumentally verify

[8] Gerard, 'Memoir of the Construction of a Map', 1826, British Library, Mss. Eur. D137, f200. Early balloonists had, of course, reached greater heights *above* the earth's surface.

[9] Having lived in the Himalaya for millennia, it is not unlikely that *indigenes* had at some point been even higher, and Chinese travellers had probably been higher too. See Walt Unsworth, *Hold the Heights: The Foundations of Mountaineering* (London: Hodder & Stoughton, 1993), 191.

[10] Gerard, *Account of Koonawur*, 291.

[11] John Hodgson, 'Field Book of April 1816', NAI/SOI/Fdbk. 87, f70.

[12] Humboldt, *Views of Nature*, Vol 1, 236.

[13] See Hansen, *The Summits of Modern Man*.

it) was a question that was thus beginning to play out around the globe, but the lack of demonstrable scientific relevance meant this was a distinction that often continued to be regarded with a degree of (perhaps feigned) ambivalence.[14]

In considering new measuring practices, this chapter thus highlights one of the many ways that travellers in the Himalaya were compelled to compare and contrast their experiences with expectations arising from accounts of the Alps and the Andes. In so doing, it emphasises the laboriousness of the instrumental measurements necessary to impose – if incompletely – a form of universality that made comparisons possible.[15] New instruments and practices for measuring altitude accurately thus emerge as central to the story of a rising sense of global verticality in this period, and the Gerards' claiming of a new high point represents an illuminating episode. In Alexander Gerard's extended discussion of his instrumental practices, it is apparent that, in effect, the brothers had to ascend to their record-breaking height on Reo Purgyil in two different ways: first physically with their instruments, and later socially in their descriptions and defence of their methods. Questions around credibility were unavoidable in the high mountains, as devices originally conceived in Europe usually needed to be reconfigured – often in ad hoc ways – before they could be useful in spaces that placed immense stress on both bodies and instruments. Reconfigurations might be physical and reflect the terrain, enabling instruments to survive the rigors of mountain travel and labour conditions, or make them easier to operate in the face of bodily debility. However, these reconfigurations were just as often conceptual, and were linked to uncertainties around the true scale of the mountains and the need to measure what were, at the beginning of the nineteenth century, unprecedented heights. By focusing on the moments when the limits – of bodies, of technologies and of imperial mastery – were

[14] The Gerards' height 'record' would not be definitively broken again until 1855 by the Schlagintweit brothers, although Jean-Baptiste Boussingault very probably went higher on Chimborazo in 1831. For this and the continuing reflexivity towards Humboldt in the Himalaya, see Finkelstein, 'Conquerors of the Künlün? The Schlagintweit Mission to High Asia, 1854–57'. Joseph Hooker is also sometimes credited with going higher than Humboldt in 1848 or 1849, with the Gerards overlooked. See for example Browne, *The Secular Ark*, 44. Indicative of the difficulty of determining altitudinal priority, in 1827 James Gerard likely reached an even higher altitude of 20,400 feet (measured by two barometers). Unlike for the ascent of Reo Purgyil, details and sources are however obscure, perhaps explaining why this was never picked up by Humboldt or other later commentators. See James Gilbert Gerard, 'Observations on the Spiti Valley and the Circumjacent Country within the Himalaya', *Asiatic Researches* 18, no. 2 (1833): 254–55; Gerard, *Account of Koonawur*, 180.

[15] For the broader context, see Bourguet, Licoppe, and Sibum, *Instruments, Travel and Science*.

exceeded, often at the same time, this chapter thus illustrates the social practices and positioning upon which they relied.

In recent years, there has been a resurgence of scholarly interest in scientific instruments and in instrumental practice among both historians of science and historical geographers. Within this, maritime and astronomical instruments have received the most attention.[16] Studies of maritime instruments in particular have interesting parallels with this story of mountain instruments, with marine barometers prone to breaking and ongoing tensions around ship's officers owning their own instruments. However, in this chapter, I shift the focus away from ships and observatories and into the spaces of the high Himalaya to examine the spectrum of relationships between instruments, notetaking and bodies in a different sort of challenging environment. Doing so allows us to gain a fuller understanding of the rhetorical strategies required to establish credible knowledge of remote locations which strained the limits of both bodies and technologies. This follows from the way that, as Charles Withers has shown, 'concerns over method and conduct' in developing instrumental practices on a global scale were tied to tensions over the status of the field as a site of observation, such that 'inscription and regularity of performance was both a scientific and a moral necessity'.[17] Scholars have thus noted that 'part of the fundamental puzzle of the survey sciences was their apparent dependence on reliable action at a distance' in which globally oriented projects were forced to depend 'on extensive networks of highly variable and often undisciplined observers', and it was for this reason that 'the provision of standardized equipment might begin to address challenges of data reliability and accumulation'.[18]

Such provisions remained elusive in the early nineteenth-century Himalaya, however, even as the practices and strategies to insist on the reliability of instrumental knowledge multiplied accordingly, much as they did in remote and extreme environments across the globe. Here scholars have demonstrated how instruments served multiple rhetorical functions in the accounts of European explorers, as justifications for their

[16] See Fraser MacDonald and Charles W. J. Withers, eds., *Geography, Technology and Instruments of Exploration* (Farnham: Ashgate, 2015), 4–5. See also special issues of *Osiris* (Volume 9, 1994) and *Isis* (Volume 102, No. 4, 2011).

[17] Charles W. J. Withers, 'Science, Scientific Instruments and Questions of Method in Nineteenth-Century British Geography', *Transactions of the Institute of British Geographers* 38, no. 1 (2013): 168, 174. See also Michael Bravo, 'Precision and Curiosity in Scientific Travel: James Rennell and the Orientalist Geography of the New Imperial Age (1760–1830)', in *Voyages and Visions: Towards a Cultural History of Travel*, ed. Jaś Elsner and Joan-Pau Rubiés (London: Reaktion Books, 1999), 162–83.

[18] Simon Naylor and Simon Schaffer, 'Nineteenth-Century Survey Sciences: Enterprises, Expeditions and Exhibitions', *Notes and Records: The Royal Society Journal of the History of Science* 73, no. 2 (2019): 140.

travels and as guarantors of their scientific credibility. As Nigel Leask suggests, instruments 'represented talismans of authorial veracity', and separated scientific travellers from tourists.[19] This was nevertheless a complex relationship, and Dorinda Outram has pointed to the way that measuring with instruments did not automatically confer objectivity or universality in the first half of the nineteenth century, and 'even if he used instruments to extend and calibrate sense impressions, what the explorer himself saw was crucial to establishing the truth status of his observations' and thus 'instruments were thought of as enhancing human sense impressions rather than replacing them'.[20] As this chapter demonstrates, however, embodied instrumental practice in the Himalaya needs to be looked at through the lens of the limits of the senses as much as the disciplining or extension of them. Ultimately, the human and non-human elements need to be analysed together. Even beyond the social considerations, instruments could be metal and glass, or flesh and blood, and field-notes might carry the traces of both bodily debility and congealed ink.[21]

While the physiological and psychological trials surveyors faced in the high mountains were real, this chapter nevertheless demonstrates how they often needed to emphasise bodily hardship and sensory deprivation, and even exaggerate the idiosyncrasy of their surroundings, in order to elevate their authority (even as this invocation of peculiarity in turn complicated claims to be producing universal knowledge). Indeed, accounts of instrumental practice in the Himalaya across the first half of the nineteenth century are intertwined with descriptions of, and an insistence on, challenges that could only be solved by those with direct experience of high mountain spaces. Surveyors worked to establish their credibility by showing instances when scientific and instrumental practices could not be transferred unmodified from those developed for the lowlands and hope to be effective. The use and reconfiguration of instruments to comprehend the vertical globe thus required an insistence on distance from metropolitan centres, reflecting a disconnect between the realities of scientific practice as understood not only between Europe and India, but also between Calcutta and the high spaces of the Himalaya. Indeed, this chapter furthers one of the central themes of this book, which is the way scientific practice in the Himalaya – with its combination of challenges, including assaults on the senses, logistics of porterage, border

[19] Leask, *Curiosity and the Aesthetics of Travel Writing*, 73.

[20] Outram, 'On Being Perseus', 283–89. See also John Tresch, 'Even the Tools Will Be Free: Humboldt's Romantic Technologies', in *The Heavens on Earth: Observatories and Astronomy in Nineteenth-Century Science and Culture*, ed. David Aubin, Charlotte Bigg, and Otto Sibum (Durham: Duke University Press, 2010), 270.

[21] Bourguet, 'Landscape with Numbers', 7.

politics and ongoing inadequacies in instrument design – was often construed as displaced not only from London, but also from India's 'centres in the periphery' like Calcutta and Bombay.

Even if the conditions experienced in the Himalaya might have been less idiosyncratic than surveyors often insisted that they were, extreme mountain environments nevertheless did test the relationships between instruments, notetaking, bodies and authority.[22] Underlying this was the problem of uninstrumentally mediated senses in high places (see Figure 2.1). As John Hodgson, one of the most prominent of the Bengal Infantry surveyors (and later Surveyor General of India), mused, 'whether it be from the changes in the atmosphere on high mountains, or the inconvenience of being exposed to severe cold & high winds, I find my observations never agree a fourth part so well as on the plains'.[23] The ability to see greater distances and a lack of referents like trees or buildings meant that scale and proportion were extremely difficult to judge and 'everyone knows the extreme vagueness and liability to error in judging of the heights and distances of mountains merely by the eye'.[24] Nor were European senses the only ones on trial, as Alexander Gerard reveals in his description of the Shatul Pass: 'it is reckoned by the people of the country far more lofty than Boorendo; but the difference of elevation is only 450 feet' though 'it is not surprising that a few hundred feet should create a belief of a much greater altitude, since their ideas are formed upon local circumstances, such as the distance of the ascent, absence of trees, and quantity of snow, added to the difference of level from which they set out'.[25] Impaired judgement applied not only to sight, but also to hearing as 'the diminution of the intensity of sound in a rarefied atmosphere is a familiar phenomenon to those who are accustomed to ascend very high mountains'.[26]

Despite the sensory derangement that came with high altitude, the human body itself might on occasion still be fashioned directly, an instrument, if a potentially highly unreliable one. This was most famously so in the case of Alexander von Humboldt, who records enthusiastically deploying his body as a barometer during his ascent of Chimborazo in 1802.[27] Such impressions were of dubious trustworthiness, however, as Scottish

[22] For the problems of estimating scale in mountainous spaces, see Bourguet, 'Landscape with Numbers'. See also Cosgrove and Della Dora, *High Places*, 9–12.

[23] John Hodgson, 'Field Book of April 1816', NAI/SOI/Fdbk. 87, f73.

[24] James Baillie Fraser, *Journal of a Tour through Part of the Snowy Range of the Himala Mountains, and to the Sources of the Rivers Jumna and Ganges* (London: Rodwell and Martin, 1820), 329.

[25] Lloyd and Gerard, *Narrative of a Journey*, Vol 2, 286.

[26] 'Miscellaneous Notices', *Gleanings in Science* 1 (1829): 374.

[27] For more on instrumentalised bodies, see Livingstone, *Putting Science in Its Place: Geographies of Scientific Knowledge*, 75.

Figure 2.1 'Jumnoo 24,000 feet from Choonjerma Pass 16,000 feet. East Nepal'. Plate from Joseph Hooker's *Himalayan Journals* (1854).[28] This image – featuring tiny surveyors flanked by immense spires – reflects the sublime aesthetic associated with mountains in this period. However, it also hints at the sometimes overwhelming scale of the Himalaya, and the challenges of measurement in a sense-scrambling world.

surgeon and naturalist Hugh Falconer demonstrated when he 'found the elevation to be 15,822 feet', which was 'considerably less than I imagined, as many of our party were attacked with the symptoms of distress about the head which extreme altitude brings on'.[29] Instruments and fieldbooks had the potential to serve as anchors in an environment known for playing tricks on the eyes and placing hypoxic stresses on the mind, but this potential was limited when they too were often strained to the point of failure. In the Himalaya, the senses could thus not necessarily be relied upon to provide a trustworthy interface with the environment, and the social performances required to establish credible knowledge were complicated accordingly. Instruments could extend the senses, mediated by the notetaking practice necessary to record them, but this was an unstable complex. Instruments, written inscriptions and bodies, when functioning together, had the potential to bring about coherence in the mountains; but because they were all

[28] Hooker, *Himalayan Journals*, Vol 1, Plate V.
[29] Hugh Falconer to Thomas Currie, 18 April 1839, British Library, IOR/F/4/1828/75444, para 17.

prone to failure, the accumulation of this coherence was haphazard and contested.

This chapter analyses the tensions around instrument use in the Himalaya in four sections. The first examines responses to damaged and destroyed instruments, focusing especially on moments in which barometers were found to be not functioning or inadequate, and how modifications and repairs might be made or not in the mountains. This is followed by a discussion of surveyors' fieldbooks and inscriptive practices, which brings together the dual problems of the fragility of senses and the fragility of instruments in high mountain spaces. I then consider the problems – both conceptual and physical – with instruments designed in Europe but intended for the unprecedented heights of the newly crowned highest mountains on the globe. Finally, I analyse the recurrence of these issues in a new type of instrument designed especially for measuring altitude – the Wollaston 'thermometrical barometer' or hypsometer – and the way it was compromised by its designer's failure to understand both the scale of the Himalaya and labour conditions in Asia. I thus argue that the staggered recognition of the true scale of the Himalaya ultimately reveals multiple levels of displacement in the understanding of scientific practices, that is, between actors in the mountains, in Calcutta, and in London. Indeed, if surveyors and critics alike had hoped that operating in situ in the mountains with precision instruments would quickly dispel doubts raised about the scale of the Himalaya and its newfound status as the 'roof of the world', then they would be sorely disappointed.

Fragile Instruments, Fragile Methods

Returning to Alexander Gerard on the windswept side of Reo Purgyil, it is apparent that when recording that he had reached a new high point on the surface of the earth, he was well aware that this claim was not without the need for some pre-emptive defence. In a justification in which competitiveness and scientific utility blur, Gerard wrote that 'as everything depends upon the accuracy of the instruments employed, I shall observe, that the barometers used by my brother and myself in 1818 ... were manufactured by a native of India, and every precaution was taken to ensure precision'.[30] These instruments, made from tubes blown by a local craftsman and fitted up by the brothers with scales made from fir rods, had become necessary replacements when a set of mountain barometers they had ordered from London had been smashed in the process of being shipped from Europe. They were nevertheless excellent instruments, and

[30] Gerard, *Account of Koonawur*, 161.

when later compared, the Gerards' makeshift barometers were found in one case to differ 'only two feet from that deduced in the following year by Dollond's mountain barometer; and the discrepancies are rarely thirty feet'.[31] This remark points, however, to a key problem with the brothers' barometers, specifically that the recognizable name of Dollond – one of the most respected of the London instrument makers – conveyed an authority that the Gerards' India-made devices never could. Indeed, Alexander Gerard felt compelled to return to the hypoxic and frightening heights in 1821 to confirm some of his earlier readings, this time with a pair of Dollonds acquired especially for the purpose, even while showing that they were not necessarily more accurate or precise.[32] There was thus a tension between instruments from the workshops of reputable metropolitan artisans that provided credibility but were ill-suited to the scale and challenging topography of the Himalaya, and India-made, modified and repaired instruments that functioned well in mountainous spaces, but lacked rhetorical claims to authority.

Similarly, scholars have shown that the credibility of instrumental claims usually depended on the status of the observers as much as the instruments.[33] This chapter focuses especially on the practices of a group of little-studied actors, represented alongside the Gerard brothers by many of the East India Company (EIC) surveyors introduced in Chapter 1, including William Webb, John Hodgson, James Herbert and Henry and Richard Strachey. These individuals were all members of the Bengal Infantry – employed as army officers and surgeons – but seconded to surveys in the Himalaya, where they combined many of the characteristics of professional technicians with amateur scientific interests, albeit with variable success. While these surveyors sometimes travelled alongside the army, they usually operated autonomously within the tracts they had been assigned to survey. Moreover, for the most part limited to traverse rather than trigonometrical surveys, their activities were not part of the Great Trigonometrical Survey proper, and these early Himalayan surveys were not subsumed into this larger project until closer to the middle of the century. Barometrical measurements were nevertheless becoming crucial, not only for confirming altitudes in their own right, but also for establishing baselines for geometrical surveys (which continued to inspire the greatest degree of confidence). Alexander Gerard was aware of this, noting that the Reo Purgyil station 'being a principal point,

[31] Gerard, *Account of Koonawur*, 164.

[32] See Gerard, 'Memoir of the Construction of a Map', 1826, British Library, Mss. Eur. D137, f200–1.

[33] See among others, Steven Shapin and Simon Schaffer, *Leviathan and the Air-Pump: Hobbes, Boyle, and the Experimental Life* (Princeton: Princeton University Press, 1985).

and I believe the greatest height ever attained on the earth's surface, either in India or any other country, I was at some pains to determine it ... by trigonometry'.[34] However, applying trigonometrical methods in the Himalaya had its own rhetorical issues around refraction (as detailed in Chapter 1), was arduous and resource intensive and in many cases was simply unfeasible in the first decades of the nineteenth century. In this context, Matthew Edney has demonstrated the 'epistemological confusion which characterized the relationship of the Great Trigonometrical Survey with the Company's other mapmaking activities', even if they shared the same political and economic imperatives for constituting frontiers and minutely mapping imperial domains.[35]

While conveying less authority than trigonometry, taking altitudes by barometer was thus becoming the preferred method by the 1820s, and the best option when logistics or terrain prevented geometrical methods, having the advantages of relative simplicity and reasonable accuracy. Using mercury-filled tubes to measure air pressure was a well-established practice, dating back to the seventeenth century, and barometers explicitly intended for measuring height – 'mountain' barometers – had been in use since the eighteenth century, with Horace-Bénédict de Saussure carrying one to the summit of Mont Blanc in 1787, William Kirkpatrick using them in the Himalayan foothills of Nepal in 1793 and Humboldt deploying them on Chimborazo in 1802.[36] The biggest limitation of barometers, however, was that they were inherently fragile and easily damaged (or destroyed) by the rigours of travel. Rates of attrition were high, and as Alexander Gerard recorded: 'two barometers were left at Soobathoo [as controls], and out of the fourteen which we took with us, only two returned in safety'.[37] Beyond their fates in the mountains, barometers were often broken during shipping from Europe, and arrived in India already unserviceable. John Hodgson lamented, for instance, that his new barometers had all arrived smashed to pieces and pleaded that, 'whenever barometers are sent, there should be to each at least 6 spare tubes filled in England by the maker & these should be carefully packed in separate cases of copper or wood lined with flannel'.[38]

[34] Gerard, *Account of Koonawur*, 176.

[35] Matthew Edney, *Mapping an Empire: The Geographical Construction of British India, 1765–1843* (Chicago: University of Chicago Press, 1997), 116–18. On revenue and traverse surveys, see Burnett, *Masters of All They Surveyed*; Arnold, *Science, Technology and Medicine*, 22–29.

[36] For earlier developments in measuring altitude using barometers, see Theodore S. Feldman, 'Applied Mathematics and the Quantification of Experimental Physics: The Example of Barometric Hypsometry', *Historical Studies in the Physical Sciences* 15, no. 2 (1985): 127–95; Bourguet, 'Landscape with Numbers'; Jouty, 'Naissance de l'altitude'.

[37] Gerard, *Account of Koonawur*, 162.

[38] John Hodgson, 'Field Book of May 1817', NAI/SOI/Fdbk. 91, f170. Makers of marine barometers faced similar problems, as the sloshing of the mercury in rough seas also tended to break tubes.

Even those that made it to India intact could still be destroyed on their way from Calcutta to Himalayan staging grounds like Saharanpur or Subathu. James Herbert, for example, recorded that a long-awaited barometer he had 'looked forward to' was damaged on the *dawk* journey from Calcutta, though by reboiling it he was able 'to restore its value'.[39] No sooner had it been repaired, however, than it was smashed again, this time in a definitive fashion – 'broken by the carelessness of the servant who had charge of it and rendered utterly useless' – a loss 'much to be regretted independent of its pecuniary value' as it was irreplaceable so far from the workshops of London.[40] While at Saharanpur, Herbert was forced to attempt to manufacture an alternative, using 'tubes which are constructed here by Native Glass men'.[41] In these instances, Indian instrument makers were usually reduced to rote labourers rather than artisans, with surveyors appropriating credit for providing the instruction, which was also essential to bolstering the credibility of the final product.[42]

The fragility of barometers was partially addressed by developing methods for repairing and replacing tubes in the field. However, as Hodgson recorded in his fieldbook of May 1817 – while being rocked by earthquakes as he struggled to become the first European to reach the source of the Ganges – boiling tubes was a vexing endeavour and 'none but a professed artist can expect to succeed in this difficult business, once in ten times'.[43] The difficulty of boiling tubes in the mountains was frustrating given this had the advantage of overcoming the lag time of potentially more than a year when ordering instruments over from Europe (not to mention the ongoing problem of pre-filled tubes arriving already broken). In these moments though, the embodied observers might prove as fallible as the instruments, and Hodgson admitted to being 'too much tired to attempt to boil the mercury in the tubes today', while the 'frequency of the earthquakes made us very anxious to get out of our dangerous situation in the bed of the river'.[44] Hodgson made use of his imperfect

[39] Herbert to Charles Lushington, 7 February 1827, British Library, IOR/F/4/1068/29191, f31-2.
[40] Herbert to Charles Lushington, 7 February 1827, British Library, IOR/F/4/1068/29191, f32.
[41] Herbert to Charles Lushington, 7 February 1827, British Library, IOR/F/4/1068/29191, f33.
[42] 'Literary and Philosophical Intelligence', *Asiatic Journal* 11 (1821): 377. Occasionally Indians might be considered artisans, for example Mir Mohsin, who worked as instrument maker in the Surveyor General's office for the better part of two decades. See Sen, *Astronomy in India, 1784–1876*, 85–89.
[43] John Hodgson, 'Field Book of May 1817', NAI/SOI/Fdbk. 91, f169. A version of this account was also published as John Hodgson, 'Journal of a Survey to the Heads of the Rivers, Ganges and Jumna', *Asiatic Researches* 14 (1822): 60–152.
[44] John Hodgson, 'Field Book of May 1817', NAI/SOI/Fdbk. 91, f167, f171.

barometer anyway, insisting that it was still of value and 'with the unboiled mercury there must be an error but I should not think it can affect the height more than 200 feet & generally not 100 feet & as under the present circumstances we cannot do more'.[45] In examples like these, as Charles Withers, Innes Keighren and Bill Bell argue, the fact that observers 'had *tried*, if not succeeded, was nevertheless central to their self-positioning as credible, scientifically minded observers'.[46]

Even when boiled successfully, questions over whether a barometer was of the same value once it had been repaired or modified were unavoidable. French traveller and naturalist Victor Jacquemont, after vilifying an assistant for breaking one of his barometers, consoled himself by noting that anyway 'it was no longer comparable with the standard of the Paris Observatory, since I had changed the tube', and he had subsequently never trusted the instrument fully again.[47] These moments are revealing, leading to what Simon Schaffer has called the importance of 'managing states of disrepair' in instrumental practice. This was especially critical in far-flung, displaced locations like the high Himalaya, where replacements and expert repairs were not readily available.[48] In turn, managing disrepair required emphasising the dangers and difficulties faced in order to justify the use of less -than- perfect instruments, and to establish the credibility of any claims made using them. The fieldbooks of the Bengal Infantry surveyors are rife with descriptions of improvised and ad hoc instruments, complete with almost desperate insistence by their observers that the readings they were producing were of value.

Given the difficulties with barometers, one possibility was to turn to a different class of instruments entirely. While barometers were becoming popular, determining altitude by calculating the falling temperature of the boiling point of water had long been known – 'an experiment exhibited in every class where natural philosophy is taught' – as a reasonably accurate method of determining elevation above sea level.[49] The smaller boiling point thermometers were somewhat less fragile than the larger barometer tubes, cheaper and relatively straightforward to use. On the other hand, the boiling point method could never deliver as high a degree of accuracy, reliability or precision. According to Alexander Gerard, John Hodgson

[45] Hodgson, 'Field Book of May 1817', NAI/SOI/Fdbk. 91, f169.
[46] Keighren, Withers, and Bell, *Travels into Print*, 97.
[47] Victor Jacquemont, *Voyage dans l'Inde pendant les années 1828 à 1832* (Paris: Firmin Didot frères, 1841), Vol 2, 156. ['Il n'était déjà plus comparable avec l'étalon de l'observatoire de Paris, puisque j'en avais changé le tube'.] By the 1840s, such procedures were more commonplace, see for example Richard Strachey, 'Narrative of a Journey to the Lakes Rakas-Tal and Manasarowar, in Western Tibet, Undertaken in September, 1848', *The Geographical Journal* 15, no. 4 (1900): 165–67.
[48] Schaffer, 'Easily Cracked', 709. [49] Gerard, *Account of Koonawur*, 177.

was 'the first person in India who thought this method sufficiently accurate for determining heights', and his opinion carried significant weight once, his constitution shattered from the brutal work of the survey, he moved from the mountains to Calcutta to take up the job of Surveyor General in 1821 and had the opportunity to apply his direct experience of the high places.[50] Indeed, the boiling point method was enthusiastically advocated by Hodgson, who acknowledged that even if it was 'only approximative', this was to a 'very desirable degree in many cases'.[51] (It also had, as English orientalist scholar James Prinsep noted, the benefit of allowing one to simultaneously practise science and brew oneself a reinvigorating cup of tea.[52]) The boiling point was thus often considered a secondary method and a useful backup, such as when a parties' barometers had all been broken or were with straggling porters and not to hand. Indeed, thermometers were only approximative instruments and even if they 'very seldom indeed gave the altitude 300 feet different from the barometer', the rhetorical work required to establish their credibility exceeded that of barometers.[53]

Thermometers nevertheless came to be widely used, especially by those who came later with more modest instrumental ambitions, and who often relied more on previous travellers than their own practices. As Thomas Thomson wrote in 1852: 'the heights of places given in the work have been derived from very various sources. Those in the earlier part are chiefly from the extremely accurate observations of the Gerards; for others I have to thank my fellow travellers'.[54] Indeed, as he went on to confess of his own numbers, most were derived 'from my own observations of the boiling-point of water, and do not therefore pretend to great accuracy. Still the thermometer which I used (by Dollond) was a very good one' and he pronounced himself satisfied that his elevations 'may be depended upon to within three or four hundred feet as an extreme error'.[55] Thermometers and barometers were also often used in parallel, and Joseph Hooker, even with access to improved mountain barometers in the 1840s, thought that 'the use of the boiling-point thermometer for the determination of elevations in mountainous countries appear[ed] to me to be much underrated', even if often overcome by practical difficulties at very high

[50] Gerard, *Account of Koonawur*, 177. Hodgson was Surveyor General from 1821 to 1823, and again from 1826 to 1829.

[51] James Herbert and John Hodgson, 'Description of Passes in the Himalaya', *Asiatic Journal* 9 (1820): 589.

[52] James Prinsep, 'Table for Ascertaining the Heights of Mountains from the Boiling Point of Water', *Journal of the Asiatic Society of Bengal* 2 (1833): 200.

[53] Gerard, *Account of Koonawur*, 307. [54] Thomson, *Western Himalaya and Tibet*, iv.

[55] Thomson, *Western Himalaya and Tibet*, iv–v.

altitudes.[56] A key issue was the necessity of having firewood to hand (and the ability to make a fire in an extreme environment) to boil the water. Indeed, the journal of Henry Strachey, in which he took elevations in the manner 'common with ill-equipped private travellers' is littered with remarks like 'fuel being scarce and Bhotias dilatory, I was unable to boil the thermometer here' or there 'was so little fuel forthcoming that I could not boil my Thermometer'.[57] His brother Richard commented on this issue, and the way it might be partially addressed by ad hoc modifications: 'the greatest difficulty is constantly met with in managing the fire in the rarefied air and violent winds of high passes' and noted that 'my brother's apparatus – which he tells me was the only one that he found thoroughly efficient and satisfactory in all exposures – was made by himself from a small tin lantern, which was fitted with a spirit lamp'.[58]

In spaces where terrain and logistics frequently exposed the inherent fragility of the available apparatuses of science, self-sufficiency and the ability to repair or modify instruments with rudimentary supplies, or find Indian craftsmen who could, was critical to the required skill set of the Bengal Infantrymen. As a pseudonymous contributor to the *Asiatic Journal* of 1818 wrote, if a young surveyor: 'received some instruction in this country that might enable him to replace a screw, or any similar defect, in an instrument, to replace the glass tube to a barometer, in filling a spare tube with quicksilver, it may become' something 'of the greatest importance in a distant survey, for it would be in vain, then, to think of aid from the mathematical instrument-makers residing in Europe, or even of any that might, or might not happen to dwell in Calcutta'.[59] For those operating in locations displaced not only from London but also far removed from Calcutta, innovation and self-reliance were considered essential. As John Hodgson argued, admittedly from a position of pride as he stepped down from his second brief tenure as Surveyor General, 'the best geographers have not been issued from the learned universities and academies' and instead 'are, and will be, the officers of the native army, Captains and subalterns, men accustomed to march from one extreme of this vast country of Hindoostan to the other', and learning 'in the school of necessity & experience how to adapt means

[56] Hooker, *Himalayan Journals*, Vol 2, 453.

[57] Henry Strachey, 'Explanation of the Elevations of Places between Almorah and Gangri given in Lieut. Strachey's Map and Journal', *Journal of the Asiatic Society of Bengal* 17, part 2 (1848): 527; Strachey, 'Narrative of a Journey to Cho Lagan', 147, 178.

[58] Richard Strachey, 'The Physical Geography of the Himalaya', c.1851 (unpublished), British Library, Mss. Eur. F127/200, f403-4.

[59] [Amicus], 'Remarks on the Himalaya Mountains', *Asiatic Journal* 5 (1818): 323–24.

to ends'.[60] In the 1810s, members of the government in Calcutta proposed employing professionals rather than seconded army officers for the Himalayan surveys, but Hodgson was dismissive because those specialist engineers 'who possess sufficient local knowledge to make them useful as surveyors have more pleasant and profitable duties open to them', and suggested that 'nothing but necessity or a strong bias towards the science ever made any man serve in the dangerous and arduous work of surveying'.[61]

Beyond the risks and physical hardships were significant expenses, especially those associated with the large parties that surveyors had to travel with to transport their instruments. Costs meant limitations, and Alexander Gerard, as a working surveyor rather than a gentlemanly traveller, complained that he had an allowance of only 250 rupees a month, 'whilst the wages of the porters alone, for my baggage and instruments, (exclusive of my own Servants & expenses) amounted to upwards of 400 Rps per month, I had not the means of extending my journey longer'.[62] The instruments themselves also represented a financial burden, often purchased and owned by surveyors personally, with significant potential to be lost, stolen or broken in the mountains. William Webb wrote in 1815 that he had 'no instrument, of any description, belonging to Government, and have, including barometers, expended nearly £1200 on instruments, a great proportion of which arrived (or have since been) broken'.[63] Though official instruments existed and 'there were occasions when Government instruments were issued on loan or payment', the tools supplied by the government tended to be of inferior quality 'sent by contractors trying to maximise their profits' and in some cases 'so bad & rough' as to be hardly worth the cost 'of carriage for them'.[64] James Prinsep remarked, for example, that 'on some standard thermometers in the Surveyor General's office ... we found the boiling point erroneous *two degrees*', a margin of error that meant heights could be off by as much as 1,000 feet.[65] The expenses for essential but inherently fragile instruments were partially reimbursed through the Company's unwieldy surveying allowances

[60] Hodgson to Duncan Montgomerie, 30 November 1826, NAI/SOI/DDn. 220, f219; Hodgson to William Casement, 24 January 1829, NAI/SOI/DDn. 231, f262-3.
[61] Hodgson to Duncan Montgomerie, 30 November 1826, NAI/SOI/DDn. 220, f260-1.
[62] Gerard, 'Memoir of the Construction of a Map', 1826, British Library, Mss. Eur. D137, f206.
[63] Webb to Colin Mackenzie, 8 October 1818, NAI/SOI/DDn. 150, f69.
[64] Phillimore, *Historical Records of the Survey of India*, 1954, Vol 3, 211–22; Hodgson to Colin Mackenzie, 1 July 1816, NAI/SOI/Memoir 60, f273. For more on issues around inferior instruments being sent to India, see Simon Schaffer, 'The Bombay Case: Astronomers, Instrument Makers and the East India Company', *Journal for the History of Astronomy* 43, no. 2 (2012): 151–80.
[65] Prinsep, 'Table for Ascertaining the Heights of Mountains', 198.

system, but this was rarely sufficient.[66] The protracted struggles of the Bengal Infantrymen to make administrators in Calcutta and London understand the instrumental challenges they faced in the high mountains highlights the ongoing imaginative disconnect between those in the Himalaya and those lower down on the vertical globe. These concerns also further illustrate the tensions around the professional status of Himalayan surveyors as employees of the EIC, and their amateur grafting of scientific interests onto official duties, pointing to the sometimes uneasy relationship between the EIC and the unofficial and idiosyncratic networks of scientific patronage that functioned within it. These tensions were exacerbated by insecurity around the lack of information about the high frontiers, but for knowledge to make its way out of the mountains it first had to be written down. This was far from straightforwardly achieved in a world in which bodies, senses and instruments were all pushed to their limits and beyond, as the next section explains.

Notetaking in the Mountains

In August 1822, James Gerard once again found himself high in the Himalaya, this time atop the Shatul Pass, which he calculated by barometer to be 15,500 feet above the sea. He had returned to the Shatul because he was hoping to recover some fieldbooks and instruments (a telescope and a thermometer) which had been lost during a prior expedition. In September 1820, they had vanished into the Himalayan snows after the boy carrying them and another porter had frozen to death at midday, the result of exposure and wind chill during an unexpected blizzard.[67] Gerard described his search for the fieldbooks in a letter that was later published, writing that after first discovering the 'bones and clothes of the Brahmin who carried a bundle of sticks' he continued down until:

> We came upon the body of the little boy who carried the field-book and all the papers of the route. He was half buried under the snow. He lies at 13,500 feet. We searched in vain for traces of the books, so that they are for ever lost. This being a chief object of my tour, and one I had much at heart, it made me look forward to the rest of it with less interest, but I had determined to ascertain the correct elevation of the cave, and continued descending.[68]

[66] See for example 'Captain Alexander Gerard Recommences His Survey of Malwa and Rajputana', 1822–27, British Library, IOR/F/4/1017/27954.
[67] Gerard, *Account of Koonawur*, 93.
[68] Gerard, 'A Letter from the Late Mr J.G. Gerard', 313–14. Lloyd and Gerard, *Narrative of a Journey*, Vol 2, 121. They had better luck with the telescope, which was put back into service and was in fact the very same one Alexander Gerard used to comfort himself the

The freezing conditions had left the body preserved – 'with all his clothing on, and his corpse untouched' – a reminder, not that Gerard needed one, of the fragility of bodies in an extreme environment.[69] Gerard's lament also points to the way that fieldbooks, and the instrumental data they contained, could have especial, almost talismanic, significance to the surveyors, and the loss of information comes across as more poignant than the loss of life. Instruments, notetaking practices and bodies, which needed to work together to establish the credibility of claims about the high mountains, were all fragile and prone to failure in high places. The social performances required to overcome these were somewhat different for fieldbooks than they were for barometers, as this section demonstrates, focusing on the moments when both writing materials and inscriptive practices broke down. Measurements using precision instruments were of little value if the data they produced could not be successfully transmitted out of the mountains, and the fieldbooks – fragile and idiosyncratic – were the sometimes shaky cornerstone of the relationship between bodies, senses and instruments in the high Himalaya.

In 1818, and only shortly before becoming Surveyor General himself, John Hodgson was scolded by the Surveyor General's Office in Calcutta for failing to lodge his fieldbooks in a timely manner. In response, he noted that all his fieldbooks were 'digested and compiled from the rough notes taken as I proceed and written briefly in the open air and often in the rain and snow so as frequently to be hardly intelligible to any but the writer' and moreover 'they are interspersed in various books with miscellaneous matter of tentative calculations, extracts and tables copied from scientific works' such that they are 'almost illegible owing to the circumstances under which they were written'.[70] Fieldbooks, being made of ephemeral materials like paper, card and parchment, were by their nature highly vulnerable. As well as being effaced by snow and rain or in river crossings, they might be lost entirely, as in the case of James Gerard's fieldbooks on the Shatul, or Herbert, who ruefully had to report that 'the total distance to the [Gunas] pass I have no means now of ascertaining, for the last few leaves of the route-survey ... were afterwards blown from the book on the stormy summit of the pass'.[71] As much as a broken barometer might render a surveying expedition a waste of time, the inability to write or failure of writing materials might have the same effect.

shughars he spotted were not border guards (see Chapter 1). Lloyd and Gerard, Vol 2, 120–21.

[69] Gerard, 'A Letter from the Late Mr J.G. Gerard', 314. See also Colley, *Victorians in the Mountains*, 220.

[70] Hodgson to Charles Lushington, 12 December 1819, NAI/SOI/DDn. 145, f69-70.

[71] Phillimore, *Historical Records of the Survey of India*, 1954, Vol 3, 42.

Indeed, like barometers and thermometers, fieldbooks would be developed as a technology of exploration, so that by 1854 the authors of a general 'Hints to Travellers' could recommend that 'writing and drawing materials, stationery, scales, tapes, and register-books, should be carried in convenient cases – water-tight, if possible'.[72]

The inclusion of fieldbooks alongside other scientific instruments in this chapter follows recent interest in the practices of writing observations in the field, and the relationships between memory, writing and editing practices. Here Marie-Noëlle Bourguet has shown that 'the rationale for travel note-taking derived from the twin dangers of an unruly observation in the field and an unreliable memory'.[73] Bodily movement and inscription were intertwined, and good notetaking practice involved disciplined, daily writing coupled with an attention to intelligibility.[74] However, this was more often the ideal than the reality in the face of the extreme topography, and Herbert confessed that often while travelling in the mountains, 'the fatigue is so great that it is impossible after arriving at the ground to set down immediately, to copy field books, or protract'.[75] Bodily limitations meant that rough field jottings usually needed to be transcribed and edited before they were useful, and most calculations were not performed on the spot but worked up later. Notetaking discipline was also important in a more morbid sense, as fieldbooks could, assuming they were intelligible, serve to transmit data in a manner that might transcend the deaths of their makers. This is evident from a moment in which it went wrong. In 1842, having recovered several of James Herbert's fieldbooks following his untimely death, the assistant commissioner of Kumaon, John Hallett Batten (1811–1886), lamented that several were 'badly written, and parts of them are very obscure. One of the vols. is written topsy-turvily, *i.e.* one set of observations are recorded on one side of a page, and another set on the other, and large *lacunae* intervene'.[76] As he continued, 'luckily this volume relates to Kumaon, and British Gurhwal, tracts with which I am intimately

[72] Henry Raper and Robert FitzRoy, 'Hints to Travellers', *Journal of the Royal Geographical Society of London* 24 (1854): 329. See also Keighren, Withers, and Bell, *Travels into Print*, 43.

[73] Marie-Noëlle Bourguet, 'A Portable World: The Notebooks of European Travellers (Eighteenth to Nineteenth Centuries)', *Intellectual History Review* 20, no. 3 (2010): 381.

[74] See Charles W. J. Withers and Innes M. Keighren, 'Travels into Print: Authoring, Editing and Narratives of Travel and Exploration, c.1815–c.1857', *Transactions of the Institute of British Geographers* 36, no. 4 (2011): 565.

[75] James Herbert, 'Field Book of August 1818', NAI/SOI/Fdbk. 97, f37. For a similar case, see Philipp Felsch, 'Mountains of Sublimity, Mountains of Fatigue: Towards a History of Speechlessness in the Alps', *Science in Context* 22, no. 3 (2009): 341–64.

[76] John Hallett Batten, 'Letter to Henry Torrens, 8 February 1842', *Journal of the Asiatic Society of Bengal* 11, part 1 (1842): 583.

acquainted, and my local knowledge enables me to decypher the names of places, and connect the threads of the narrative'. As he concluded: 'I assure you that nobody at Calcutta can possibly interpret the volume in question'.[77] In these volumes, lack of discipline and good notetaking practice diminish the value of Herbert's books, and increase the difficulty of extracting useful knowledge. Batten's social performance nevertheless rests on explicitly claiming distance from those in Calcutta, and asserting that only someone with direct experience of the displaced spaces of the Himalaya might be able interpret the imperfectly inscribed notes (and if those in Calcutta would struggle, what hope for interpreters in London?). Here, as in other social performances by the Bengal Infantrymen, it is a deliberate insistence of the idiosyncrasy of Himalayan spaces that elevates Batten's contribution to the increasingly pressing questions around the true scale of the Himalaya.

Reliable and immediate inscription and disciplined notetaking practice were especially critical in an environment that induced extreme fatigue and the stresses of altitude sickness. As Joseph Hooker recorded while descending the Yangma Valley in 1848: 'lassitude, giddiness, and head-ache came on as our exertions increased, and took away the pleasure I should otherwise have felt in contemplating by moonlight the varied phenomena, which seemed to crowd upon the restless imagination'.[78] However, as he continued, 'happily I had noted everything on my way up, and left nothing intentionally to be done on returning'. Indeed, he further explained that 'it is impossible to begin observing too soon, or to observe too much: if the excursion is long, little is ever done on the way home' when 'the bodily powers being mechanically exerted, the mind seeks repose, and being fevered through over-exertion, it can endure no train of thought'. In what was a similar rhetorical strategy to reassure the reader of the reliability of his observations, he also claimed that he 'always carried my note-book and pencil tied to my jacket pocket, and generally walked with them in my hand'.[79] Against the sensory assaults of the mountains, fieldbooks could be examined and held up as talismans of authority, and as bulwarks against later failings of memory, editorial meddling or hindsight rejigging. Indeed, fieldbooks might have noteworthy afterlives, such that transparency in method, recognition of potential sources of error and the provision of the raw uncorrected measurements – 'for those persons who may wish to re-calculate them' – were important considerations.[80]

[77] Batten, 'Letter to Henry Torrens', 583.
[78] Hooker, *Himalayan Journals*, Vol 1, 247.
[79] Hooker, *Himalayan Journals*, Vol 1, 247.
[80] Gerard, *Account of Koonawur*, 165.

This was even more so in case of the Bengal Infantrymen, as working surveyors rather than well-connected naturalists like Hooker, and who sometimes awkwardly combined the characteristics of amateur enthusiasts and professional technicians.[81] The tension around the surveyors' status was very apparent, as Chapter 1 demonstrated, when it came to the controversy surrounding the true height of the Himalaya, and their supremacy over the Andes. As James Herbert put it, 'no determination of heights will ever satisfy the curious in Europe that is not accompanied with ample details as to the original observations as well as a full exposition of the methods of calculation'.[82] The fieldbooks of raw data, alongside careful justifications of the instrumental practices developed and the conditions they were used under, thus became essential. Hodgson and Herbert went on to sum up their findings by placing an emphasis on their authority by distance and their ability to operate in a physiologically extreme and sense-scrambling world: 'while we deprecate the theorists pronouncing too decidedly on the value of results, which may appear to him, much too discordant' actually, 'we feel confident that in the eyes of the practised observer, who will consider the nature of our instruments, and the difficulties with which we had to contend, these very discrepancies will prove our strongest claim to his confidence'.[83] The making available of fieldbooks was a rhetorical strategy for claiming credibility, even if this also implies that interpreting the data might rely on first-hand experience of particular places. Herbert and Hodgson are, like Batten, claiming distance from not just London but also Calcutta, and thereby establishing their privileged ability to produce knowledge of displaced spaces from within them. Their challenge to the 'theorists' is nevertheless muted because, in carefully laying out the particulars of their practices and instruments, it is ultimately the acceptance and approbation of these gentlemen that they are seeking.

Beyond the precious data they carried – the *raison d'être* of heading into such extreme environments at all – fieldbooks were also crucial to Himalayan surveyors (specifically those answering to the Bengal Presidency) in a more practical sense, in that they needed to be furnished to be reimbursed for surveying allowances. Regulations stated that fieldbooks had to be sent back monthly, though this was often an impossible ask in the mountains. Sometimes, Hodgson resorted to sending a summary: 'I call it a field book to entitle me to my allowance as I am in debt for the

[81] See Endersby, *Imperial Nature*. For a fuller discussion of these ambiguities, see also Chapters 4 and 5.

[82] Herbert to Charles Lushington, 7 February 1827, British Library, IOR/F/4/1068/29191, f130-1.

[83] Hodgson and Herbert, 'An Account of Trigonometrical and Astronomical Operations', 211.

expenses of my extra hill carriage'.[84] Surveyors frequently complained at the impracticality of having copies made in the mountains, and the limited availability of trained scribes.[85] The problem was that the regulations were designed for the plains, and failed when they came to the mountains where labour and porterage were more expensive, copyists scarcer and repairing or obtaining replacement instruments a much slower process. In an extended critique of the fieldbook regulations, Hodgson cited as his most glaring example of this disconnect that 'by the regulations all surveys are to be discontinued during the rainy season, and the full allowance of a surveyor cease' however, he continued 'it is in that season alone, when the snow is to a certain degree melted, that we can best explore these deep recesses of the Himalaya, where the rivers originate'.[86] Fieldbook regulations were unpopular and were ongoing sources of friction between those in the mountains and those in Calcutta, further highlighting the disconnect in the understanding of the conditions and bodily hardships associated with scientific practice in the high mountains. By the time Hodgson descended from the mountains to take up the post of Surveyor General in 1821 and had the opportunity to apply his direct experience to redressing the regulations, the problem was urgent. As William Webb argued, 'no person would undertake the fatigues, risks, and exposure of those alpine journies, with the chance of being fined in the amount of his establishment'.[87] There were signs of reform, however sluggish, by the 1840s, as seen for example by the way Richard Strachey was 'authorised to obtain on the public service any instruments that can be spared by the officers who have them in charge'.[88] He was also allowed to 'purchase or obtain from England' any instruments otherwise unborrowable but essential to successfully prosecuting the task and 'the Lieutenant Governor will authorise the payment for them on the public account, if the expense be moderate', but even such qualified largesse remained far from the norm.[89]

The Problem of Scale

Beyond developing practices for day-to-day repairs, intelligible notetaking and the repurposing of inadequate devices, it was quickly apparent to the Bengal Infantry surveyors gaining access to the high Himalaya in the

[84] John Hodgson, 'Field Book for September & October 1816', NAI/SOI/Fdbk. 91, f7.
[85] Hodgson to William Casement, 6 November 1827, NAI/SOI/DDn. 231, f77.
[86] Hodgson to Lieut. Colonel Young, 12 July 1818, NAI/SOI/DDn. 152, f66-7.
[87] William Webb to Colin Mackenzie, 8 October 1818, NAI/SOI/DDn. 150, f71.
[88] J. Thornton to Richard Strachey, 14 February 1848, British Library, IOR F/4/2356/124635, f18.
[89] J. Thornton to Richard Strachey, 14 February 1848, British Library, IOR F/4/2356/124635, f18-9.

1810s that the instruments offered by London artisans had notable limitations, both conceptual and physical. Central to this was the need for instruments that would read low enough to give altitudes for the highest elevations such instruments had ever been deployed at.[90] Indeed, Alexander Gerard wrote of standard barometers – even those marketed as 'mountain' barometers – with exasperation as 'all I have seen are not adapted for the measurement of very high places'.[91] He continued that while one 'went as low as 20 inches', it would nevertheless 'be of no use amongst the Himalaya, where I have travelled many hundred miles without having the barometer above 19½ inches, and once so low as 14.675 at the height of 19,450 feet'.[92] Gerard thus leverages his authority from experience, elevating himself from a mere data gatherer to a scientific pioneer. As a contributor to the Calcutta-based journal *Gleanings in Science* reveals, scales were an ongoing problem even in 1829: 'when I ordered my brass scales from Dollond, I had them divided down to 17 inches, which I imagined, then, would be the limit of their travels upward'.[93] However, 'the mercury sunk 1½ inch below this point' and the observer was forced to try and compensate for this by carving marks below the scale with a penknife in the hope of extending it later.[94] In this instance, the instrument's unsuitability for the task was literally inscribed into it. There was an ongoing disconnect between instruments designed in Europe (perhaps for the Alps) and those needed for the highest mountains on the globe. That instruments capable of measuring very high places remained hard to come by even in the 1820s is indicative of the sustained imaginative failings of those outside India (and in Gerard's opinion those in Calcutta without direct experience of the mountains) when it came to the true scale of the Himalaya. This was a problem that could only be rectified with instruments especially conceived for the highest places and, after his experiences on Reo Purgyil, Alexander Gerard ordered barometers to his own specification, 'the scale which by means of a vernier shews 1/1000 of an inch [and] extends so low that altitudes of 24,000 feet may be measured by it'.[95] Although surveyors and explorers in the first half of the nineteenth century never reached

[90] Contemporary balloonists faced similar problems, needing instruments that would read even lower than mountain barometers to serve as altimeters. See Gerard L'Estrange Turner, *Nineteenth-Century Scientific Instruments* (London: Sotheby Publications, 1983), 234.

[91] Alexander Gerard, 'Remarks upon Barometrical Heights', NAI/SOI/Fdbk. 113, f132.

[92] Alexander Gerard, 'Remarks upon Barometrical Heights', NAI/SOI/Fdbk. 113, f136.

[93] 'On the Most Eligible Form for the Construction of a Portable Barometer', *Gleanings in Science* 1 (1829): 316.

[94] 'On the Most Eligible Form for the Construction of a Portable Barometer', 316–17.

[95] Gerard, 'Memoir of the Construction of a Map', 1826, British Library, Mss. Eur. D137, f202.

altitudes anywhere near that high, they were becoming aware of the possibility that human beings and the apparatus of science had much higher yet to go.

Theoretical limitations with barometers were also reflected in the way that the barometric formulas needed to be adjusted for the particular climates and latitudes of the Himalaya, and to produce accurate readings for very high (and at the time the formulas had been devised, unprecedented) altitudes, a process that was ongoing in the first half of the nineteenth century. Raw measurements from barometers always needed to be corrected, especially for temperature, and for other causes of error relating to time of day, latitude, season and extremes of local weather. One of the most common of ways of doing this was 'Dr Maskelyne's method' which, as Gerard argued, 'always gives the altitudes of very elevated places too little, because the equation for the latitude is not taken into account'.[96] Most critically, barometrical readings needed to be checked against equivalents taken at the same time of day from instruments kept at lower stations, 'as without corresponding observations, the results of Barometrical measurement are likely to be erroneous'.[97] Calcutta and Saharanpur provided base measurements against which other variations in the higher mountains could be compared, such that refinements to these standards might mean recalculating earlier heights from the raw readings preserved in fieldbooks.[98] Figures for height were not static, and the surveyors' original fieldbooks might prove more valuable than their worked-up and published calculations. As James Prinsep wrote in 1833, referring to the curious behaviour of diurnal oscillation at extreme altitudes: 'the determination of the zero or no oscillation altitude, may probably be obtainable from the journals of Captain Gerard or his brother, Dr. J. G. Gerard'.[99] Operating in the Himalaya did have some natural advantages, as 'in these climates ... the Barometer is so much more regular in its indications than in Europe', although this perhaps made establishing credibility among metropolitan instrument makers who were used to less stable barometrical outcomes even more difficult.[100]

Woefully inappropriate scales, unsatisfactory formulas, fragile tubes and adjustment screws that were difficult to operate with frozen fingers

[96] Alexander Gerard, 'Remarks upon Barometrical Heights', NAI/SOI/Fdbk. 113, f128.

[97] Herbert, 'An Account of a Tour', 414.

[98] Hodgson and Herbert, 'An Account of Trigonometrical and Astronomical Operations', 319–20.

[99] Patrick Gerard, 'Abstract of a Meteorological Journal, Kept at Kotgarh', ed. [James Prinsep], *Journal of the Asiatic Society of Bengal* 2 (1833): 619.

[100] Herbert, 'An Account of a Tour', 414.

and bulky gloves were merely part of a smorgasbord of problems with instruments designed in London or Paris by those with no experience of Himalayan conditions. As Alexander Gerard wrote, 'it appears to me that in regard to barometers and indeed most instruments, accuracy has generally been sacrificed to portability, and instruments of various kinds which are only suited for some parts of Europe are daily sent to India where they are often useless'.[101] Similarly, in the first volume of the journal *Gleanings in Science*, edited by James Herbert and to which the Gerards and other Bengal Infantry surveyors were contributors, it was noted that 'it is extraordinary that in making instruments for the Indian market, the artists of London will not advert to the difference in the habits of the two countries'.[102] He was alluding, as well as, the topographical factors, to the differing labour conditions in India. Himalayan surveyors had the ability to relatively cheaply co-opt local bodies to carry their instruments, and as Hodgson argued, in 'instruments intended for India solidity should be considered; we want those which will do their work effectually, & and are not anxious that they should be small & easily portable, as we can always here find means of carrying them'.[103] This emphasis on durability also implies the potential for rough treatment at the hands of 'careless' porters. For professional, working surveyors – rather than gentlemen savants – ornamental or aesthetic qualities were of little value, and instruments only needed to be functional, without 'the bungling and expensive contrivances which are applied to instruments' of science, as fabricated by London artists, in the elaborate and useless finish given to them'.[104]

While lauding the local knowledge of displaced locations possessed by the Bengal Infantrymen, Hodgson was explicit that this included an understanding 'not only of the language but of the customs, prejudices and peculiar feelings of the natives'.[105] The necessity of large parties of hill porters to carry instruments and supplies, and the omnipresence of South Asian guides, brought a further dimension to the social performances required to establish the credibility of instrumental practices. Interactions between Himalayan peoples and instruments were usually recorded by the surveyors through standard tropes of curiosity, awe and superstition. These tropes allowed explorers to represent themselves as 'agents of a technologically inspired modernity'.[106] Alexander Gerard

[101] Alexander Gerard, 'Remarks upon Barometrical Heights', NAI/SOI/Fdbk. 113, f135.
[102] 'On the Most Eligible Form for the Construction of a Portable Barometer', 314.
[103] John Hodgson, 'Field Book of May 1817', NAI/SOI/Fdbk. 91, f170.
[104] 'On the Most Eligible Form for the Construction of a Portable Barometer', 318.
[105] Hodgson to William Casement, 24 January 1829, NAI/SOI/DDn. 231, f261.
[106] Kennedy, *The Last Blank Spaces*, 157.

described the Tartars and their opinion of his collection of instruments in this manner: 'they are very inquisitive and curious, and were constantly asking questions about the reflecting circle, sextants, barometers, and the astronomical telescope; the latter pleased them most, and I had frequently to shew the same objects to thirty or forty different people'.[107] Meanwhile, Joseph Hooker employed similar rhetoric when he wrote of reascending a pass to 'to verify my observations', suggesting that 'the Tibetan Sepoys did not at all understand our ascending Bhomtso a second time' and the 'instruments perplexed them extremely, and in crowding round me, they broke my azimuth compass. They left us to ourselves when the fire I made to boil the thermometers went out, the wind being intensely cold'.[108] Of course, it is difficult to know what the Tibetans truly made of the surveyors' instruments. The way these interactions are recorded inevitably tell us far more about the European writer than about his South Asian assistants, even while they are a tacit reminder of the imperial consequences for those being surveyed.

New Instruments, Old Issues

In bringing together several themes of this chapter, it is worth briefly considering a new sort of instrument – one specifically designed for measuring altitude – that was arriving in India just as the Gerards were struggling to make their way up Reo Purgyil, namely: Francis Hyde Wollaston's 'thermometrical barometer' (hypsometer). Its fate in the Himalaya provides a summation of the challenges – both real and rhetorical – facing the Bengal Infantrymen in spaces in which instruments, notetaking practices and bodies were all prone to failures. In this period, taking rough altitude readings using a regular thermometer and tin shaving cup had sometimes been worthwhile in the absence of anything more suitable (as was practised by the Gerards, Hodgson, Herbert and Webb, and later in the 1840s by Hooker and Thomson). However, the growing necessity of precisely measuring altitude led to the emergence of specific boiling point apparatuses that purported to increase accuracy and durability, of which the Wollaston was the most prominent example. The device took its temperature reading from the steam, which was more regular than the boiling water itself, and came fitted with a vernier (see Figure 2.2).

In choosing the name 'thermometrical barometer', Wollaston was perhaps deliberately making the claim that his device could compete

[107] Gerard, *Account of Koonawur*, 108–9.
[108] Hooker, *Himalayan Journals*, Vol 2, 174–75.

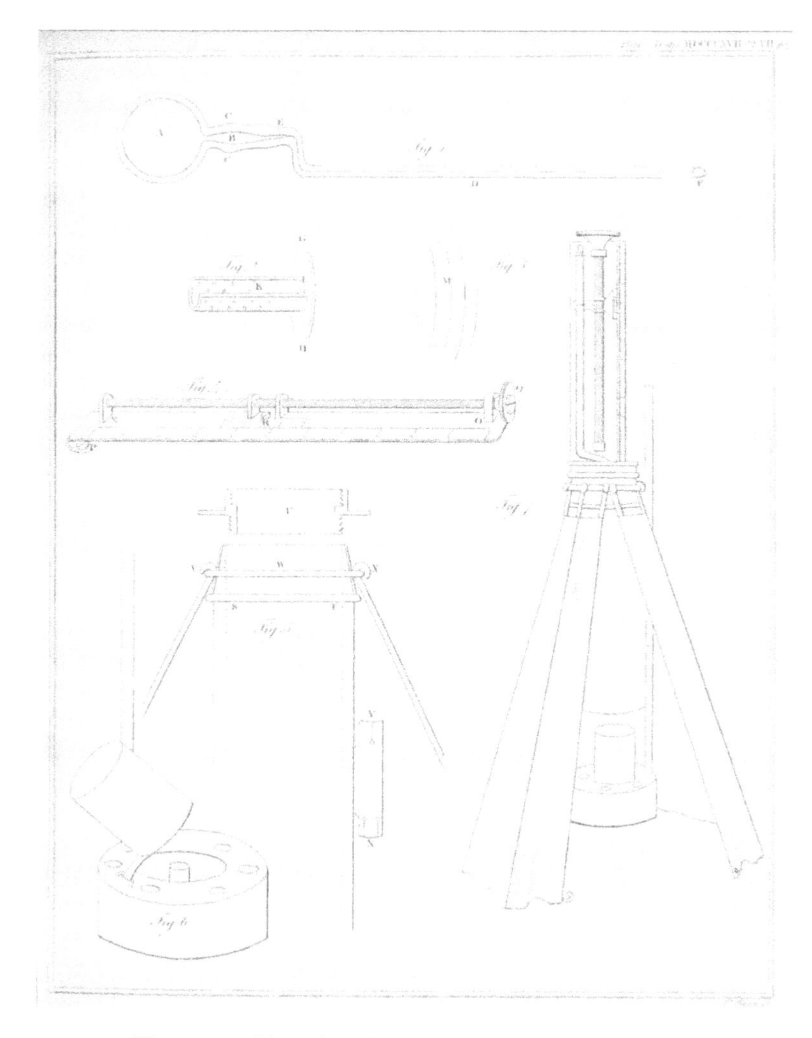

Figure 2.2 Plate detailing Francis Hyde Wollaston's 'Thermometrical Barometer', as published in the *Philosophical Transactions* (1817).[109] Such intricate devices proved prone to breakdown in the mountains, and difficult to operate with frozen fingers and bulky gloves.

[109] Francis John Hyde Wollaston, 'Description of a Thermometrical Barometer for Measuring Altitudes', *Philosophical Transactions of the Royal Society of London* 107 (1817): 183–96.

directly with barometers in terms of accuracy, portability and cost-effectiveness. Indeed, this device represents an attempt to evolve boiling point thermometers from an instrument of approximation into an instrument that was both accurate and precise. One of the Wollaston's main selling points was also that it was supposedly less prone to breaking than barometer tubes. An 1823 advertisement for this new thermometer in *The Calcutta Journal of Politics and General Literature* quoted extensively from Hodgson's misfortunes with broken barometers in promoting its superiority.[110] Hodgson himself was initially optimistic when he heard about the thermometrical barometers, noting 'I think that Dr. Woolaston's improved thermometer will supersede the mountain barometers altogether. It has every advantage'.[111] However, once he had actually had the chance to examine one, he became markedly less convinced, stating rather bluntly that 'Woolaston's thermometrical barometers are of no use, you cannot take them out of their case without breaking, they will not do; besides there are none in Calcutta'.[112] This criticism was echoed by James Prinsep, who thought that 'the error into which Wollaston fell was an attempt at too great sensibility. His instrument is beautiful in a laboratory, where it will serve to shew minute variations in the index error . . . but for rough work out of doors, accuracy must in some measure be sacrificed to strength and portability'.[113] Even in designing an instrument specifically intended to measure mountains in situ, the metropolitan artisan, based in London and at multiple removes from the spaces of the high mountains, could not imagine the world faced by Himalayan surveyors on a daily basis.

This was doubly apparent given that, as with barometers, Wollaston's boiling point devices suffered lingering issues with scales, legacies of the way the instrument was initially tested on Mount Snowdon in Wales. As James Prinsep continued, Wollaston and his assistants 'drew up a table of the value of the degrees between 214° and 202° in feet . . . but, as this range only extends to an altitude of 5405 feet, it is evidently quite insufficient for the traveller in India, who may ascend to 18,000 feet and still see *Snowdons* towering above his head'.[114] Wollaston seemed to be aware of this issue, insisting that the 'instrument, though adjusted now to my own particular use by the quantity of mercury in the thread, is capable of measuring any greater altitude, even Mt. Blanc or Chimboraço'.[115] Writing in 1825,

[110] 'Mountain Barometers', *The Calcutta Journal of Politics and General Literature* 1, no. 26 (1823): 415–16. The extracts cited were from Hodgson, 'Journal of a Survey'.
[111] Hodgson and Herbert, 'An Account of Trigonometrical and Astronomical Operations', 209–10.
[112] Hodgson to Richard Wilcox, 28 August 1827, NAI/SOI/DDn. 220, f292.
[113] Prinsep, 'Table for Ascertaining the Heights of Mountains', 197.
[114] Prinsep, 'Table for Ascertaining the Heights of Mountains', 197.
[115] Wollaston, 'Description of a Thermometrical Barometer', 192.

however, Herbert suggested that the thermometrical barometer was still 'not made of sufficient extent as yet to comprehend within its range the boiling points' encountered in the Himalaya.[116] He also thought that the method of calibration described by Wollaston was impractical in the high mountains and seemed more suited to 'the easier journies through civilized Europe'.[117] From his Himalayan vantage point, Herbert was frustrated by the way that London savants and instrument makers were still struggling to address the need for instruments that had scales appropriate for mapping the upper reaches of the vertical globe, and that were rugged enough for Himalayan travel. With his direct experience of the Himalaya and position of distance, he had no qualms in pointing out these failings. Of course, this was itself a social performance, and in critiquing Wollaston, Herbert was simultaneously insisting on his own ability to produce reliable knowledge of the Himalayan frontier.

The problems with fragility and scales were at least partially addressed in later models of the thermometrical barometer, and its eventual eligibility for the Himalaya is borne out by the way that Wollaston's devices were used by English surgeon and botanist William Griffith (1810–1845) and Victor Jacquemont in their Himalayan travels in the late 1820s and 1830s.[118] Griffith also noted in 1842 that several were 'furnished to the Surveying Officers of the Army of the Indus', but as he also explained, 'they contain the readings off of the Thermometer, Barometer invented by Dr. Woollaston ... with the substitution of an ordinary Thermometer for his delicate one'.[119] Such moments indicate that fragility and ad hoc reconfigurations were an ongoing concern. At the same time, even as instruments continued to break and be repaired, by this time some of the imaginative disconnect around scales had been alleviated. Although many of the challenges of surveying in the mountains remained unsolved, as later nineteenth-century stories only too clearly show, by the mid-century it was no longer impossible to imagine that the Himalaya held within its fastness the loftiest places on the globe.

Conclusion

As surveying in the Himalaya became increasingly important, both to science and to the constitution of imperial frontiers, the instrumental

[116] Herbert, 'An Account of a Tour', 412. [117] Herbert, 'An Account of a Tour', 412.

[118] William Griffith, *Journals of Travels in Assam, Burma, Bootan, Afghanistan and the Neighbouring Countries*, ed. John McClelland (Calcutta: Bishop's College Press, 1847), 339, 371; Jacquemont, *Voyage dans l'Inde*, Vol 1, 342–43.

[119] William Griffith, 'Tables of Barometrical and Thermometrical Observations, Made in Affghanistan, Upper Scinde, and Kutch Gundava, during the Years 1839-40', *Journal of the Asiatic Society of Bengal* 11, part 1 (1842): 49.

practices used to measure the highest mountains on the globe came under significant scrutiny. Here the surveyors' management of 'states of disrepair' was achieved with only varying degrees of success, and sometimes compromises between accuracy and practicality, and between name-brand and India-made instruments, were necessary in an environment hostile to the human body and mind, as well as to the apparatuses of science. Surveyors nevertheless insisted on the idiosyncrasy of the high mountains and the challenges to instrumental practice they faced within these spaces, in order to leverage their ability to overcome them for authority. There is thus undoubtedly much to be gained by examining 'the integration of instrumentation with human performance and the establishment through such performances of trustworthy accounts of remote phenomena'.[120] However, we need to keep in mind the extent to which these social performances had multiple audiences, and played out against the ongoing challenges provided by a disconnect between the realities of instrumental practice as understood between not just metropole and colony, but between those in the mountains, those in Calcutta and those in London. Barometers, thermometers and fieldbooks played key rhetorical roles and functions in establishing the reliability of knowledge produced in a hypoxic and sense-scrambling world, and a world which the EIC still little understood, but as beginning to fear did not provide the secure frontier its jagged aspect seemed to convey. Instruments, bodies and notetaking practices, and the moments their limits were exceeded, thus provide a valuable window into the scientific, political and imaginative remaking of the Himalaya. Through this window, we see surveyors struggling to assert their ability to produce reliable knowledge in a world in which one might stand at the height of the summit of Mont Blanc and still see Snowdons towering above.

Returning one last time to the Gerards on the side of Reo Purgyil, it seems that their claim of reaching a new high point on the vertical globe (and ultimately the credibility of their instruments and observations) was acknowledged, if perhaps a little grudgingly, by Alexander von Humboldt, whose altitude record on Chimborazo they had apparently beaten (albeit only slightly).[121] The Gerards' attitude to their achievement (shared, however poorly acknowledged, with their Tartar guides)

[120] Bourguet, Licoppe, and Sibum, *Instruments, Travel and Science*, 7.

[121] Humboldt could never quite bring himself to unequivocally accept this, noting the Gerards had gotten 'as high, if not 117 feet higher, than I ascended the Chimborazo'. Humboldt, *Views of Nature*, Vol 1, 236. This equivocation comes through in translation, see Alexander von Humboldt, *Ansichten Der Natur*, 3rd ed. (Stuttgart: J. G. Cotta, 1849), Vol 1, 56, 123; Vol 2, 50. Humboldt's source was the overtly patriotic James Bell and John Bell, *Critical Researches in Philology and Geography* (Glasgow: James Brash, 1824), 144.

nevertheless reveals something of an ambiguous attitude towards altitudinal supremacy at this time. Whether the Gerards high point was (as they thought) or was not (as we now know) a new altitude record does not perhaps matter all that much anyway, except in the context of the later nineteenth-century preoccupation with summiting higher and higher peaks, which has persisted until the present, remaining most explicitly embodied by Mount Everest. The Gerards were keen to see their barometers read lower than 15 inches, and undoubtedly later put considerable effort into instrumentally verifying their high point and proving this, indicating their willingness to engage in an essentially arbitrary imperial contest. That this record was nevertheless not as important to them as it might have been later in the century is apparent from the understated way Alexander Gerard records this 'accomplishment' in a technical memoir on mapmaking.[122]

Commentators in Europe were not always as reticent, and in 1824 James and John Bell summoned mock outrage in a xenophobic screed against foreign participation in Himalayan scientific discoveries: 'What! Are those men who pierced the rugged defiles, climbed the steep ascents, and scaled the lofty ridges of the Heemalleh, and stood on higher ground than was ever trod by a Saussure, a Condamine, or a Humboldt,* to be thus deprived of the legitimate reward of their toils' so that 'their laurels may adorn the temples of Frenchmen and Germans?' (Here the footnote proudly proclaimed that the Gerards had ascended 'to the height of 19,411 feet, or 118 feet higher than what Humboldt attained in his ascent of Chimborazo'.[123]) This was in many ways, however, the exception that proved the rule. Perhaps most tellingly, nobody appears to have been overly anxious to rush out and beat their 'record'. Indeed, these exchanges reveal quite a different dynamic to the often-flamboyant self-promotion attached to mountaineering later in the century (and the limited fanfare presents a similar contrast to the bombastic press attention that would become a feature of Victorian exploration more widely). This reflects the equivocal social status of many of those – like the India-bound Gerards – who were accessing the Himalaya at this time, as well as a still only nascent valorising of altitude, compounded by the uncertainties involved in calculating it accurately. Reaching new heights was one thing, but reliably instrumentally verifying them in a challenging and sense-scrambling environment was another, and arguably where for Alexander Gerard the true contest lay.

[122] The claim for reaching 19,411 feet was repeated in a journal published posthumously in 1841, though again in an undramatic fashion. Gerard, *Account of Koonawur*, 176.

[123] Bell and Bell, *Critical Researches in Philology and Geography*, 144.

Beyond these personal interests, arguments were also more urgently being made for the 'great practical utility, which may be derived to geography from a knowledge of the true position and elevation, of several snowy peaks in the *Himālāya* chain'.[124] It was thus for both science and empire that the Bengal Infantry surveyors lugged (or rather, employed South Asian porters to lug) a panoply of fragile instruments into the Himalaya to accurately record the elevations, shapes and locations of what were only just coming to be acknowledged were by far the highest mountains on the globe. Even if the scale was sometimes different, Himalayan surveyors were not, of course, alone in grappling with the problems of accurately measuring altitude, and similar issues with instruments were playing out in other mountains on other continents. Throughout the first decades of the nineteenth century, the necessity of mastering the unstable complex of instruments, notetaking practices and bodies to produce credible knowledge was thus attached to an increasing recognition – and laborious imposition of – global commensurability onto mountain environments. However, instrumental practice was far from the only thing made unstable by altitude. Indeed, Chapter 3 looks in more detail at scientific and environmental uncertainties around altitude sickness, and how this complicated both expedition sociability and the development of medical topographies of high mountains like the Himalaya.

[124] William Webb, 'Memoir Relative to a Survey of Kumaon', *Asiatic Researches* 13 (1820): 294.

3 Suffering Bodies

Altitude affects the body and the mind in unusual ways. The higher up one climbs, the more apparent these affects become, even as the cause remains invisible. In 1822, East India Company (EIC) surgeon James Gerard was ascending the Shatul Pass, some 15,500 feet above the level of the sea. He recorded that 'the smallest attempt to make an effort threw us back. The extreme labour we had in getting up the last 500 feet cannot be described', and 'anxiety and slight sickness deprived us of using our arms when inclined to break off a chip of rock by the blow of a hammer; respiration was free, but *insufficient*: our limbs could scarcely support us'.[1] Difficulty breathing, headaches, somnolency, loss of appetite, lethargy and hypoxia, all awaited those venturing into the high mountains. Movement itself was difficult, let alone swinging a hammer in the pursuit of scientific knowledge, or raising a pen to record reliable observations about the still largely unmapped northern frontiers of the EIC's burgeoning Indian empire. As Gerard went on to conclude, he had 'never experienced so decided a proof of the existence of an agent inimical to the principles of animal life'.[2] Later, he tried to make some sense of the symptoms he suffered, and his account is interwoven with an attempt to scientifically explain the – at the time – very little understood effects of high places on the body. 'The cause here is not quite obvious, nor are those extraordinary symptoms of prostration of strength, anxiety, and mental imbecility satisfactorily explained' he noted, continuing to suggest that 'while we cannot hesitate to refer the primary and immediate agent to the thinness of the air, or more properly, the diminished pressure ... the effects are so capricious and irregular as to be at variance with the idea of a constant cause'. This uncertainty, he concluded, 'leads many to disbelieve the existence of even any one symptom, and those who have by accident resisted the impression in crossing the mountains, remain unalterable in their conviction'.[3] In his account, Gerard thus describes a medical

[1] Gerard, 'A Letter from the Late Mr J.G. Gerard', 308.
[2] Gerard, 'A Letter from the Late Mr J.G. Gerard', 326.
[3] Gerard, 'A Letter from the Late Mr J.G. Gerard', 320–21.

topography of high mountains that is lacking in fundamental coherence. He correctly identifies the rarefaction of the atmosphere as the cause of his suffering, but ultimately struggles to reconcile this with the extreme inconsistency with which symptoms were experienced.

In the context of early Himalayan exploration these inconsistencies are central to the story. Tensions around the 'capricious and irregular' distribution of the symptoms of altitude sickness were exacerbated by the way that the European travellers who entered the high Himalaya in the first half of the nineteenth century rarely – that is, if ever – did so alone. As James Gerard continued, 'all my people have also been affected in different ways, some with sickness, others with headache', but he tellingly clarified that 'everyone is not equally affected'.[4] European explorers, surveyors and travellers were thus forced to compare and contrast their bodily performances against those of their Bhotiya, Tartar and Lepcha guides and porters. This dependence on Himalayan labour, combined with the invisible and insidious effects of altitude sickness, served to intensify expeditionary relationships: between explorers and guides, employers and employees, Europeans and South Asians. In this chapter, I examine the resulting tensions by considering the politics of bodily comparison that developed around altitude sickness at several different scales: in the way bodies, European and South Asian, experienced altitude sickness; in the way comparisons between bodies affected cross-cultural interactions within expedition parties; in the way these experiences and comparisons were represented in written accounts; in the way these comparisons were made globally, especially with the Alps and the Andes; and in the way these practices ultimately helped constitute medical topographies of high mountains as peripheral spaces and aberrant environments in relation to lowland norms.[5]

This chapter nevertheless demonstrates that comparativity operated differently for different actors, and that there was an inherent politics of access to comparisons, between European explorers and their South Asian guides, as well as between colonial actors in the high mountains, those in the lowlands of Calcutta and those in Europe. In thinking about the politics of bodily comparison in the face of an invisible force, seasickness provides a helpful analogy, and indeed contemporary travellers made frequent allusions. In both cases the cause was invisible, even if clearly related to a particular environment. Symptoms were also experienced with high degrees of variability, and with seemingly little regard for social

[4] Gerard, 'A Letter from the Late Mr J.G. Gerard', 326.

[5] For other 'mountain' diseases like goitre, and an overview of medical topographies in South Asia more generally, see Arnold, *Science, Technology and Medicine*, 75–81.

hierarchies around fitness, age, gender or race.[6] The crucial difference, however, was that after millennia of seafaring, seasickness was a known and accepted phenomenon, even if medical understandings of it remained limited. On the other hand, travelling to very high altitudes was something largely new (at least for European travellers) at the end of the eighteenth century.[7] In this chapter, I thus consider social relationships in an environment that was unfamiliar and hostile to many of the actors. In keeping with the aims of the book more broadly, I do so not so much from questions of epistemology and hybrid knowledge, but from an approach grounded in everyday practice, and the cross-cultural relationships that these expeditions relied on to function.

This approach reveals that altitude sickness exaggerated existing tensions within expedition parties because it risked inverting expected hierarchies around race and gender in relation to bodily performance. This chapter thus traces the way European travellers responded – consciously or unconsciously – by developing tropes that they could use in journals and travel narratives in (sometimes almost plaintive) attempts to reassert their assumed superiority. Throughout, I examine responses to altitude sickness particularly through the concept of self-fashioning, that is, the various ways that travellers not only experienced and sought to scientifically explain, but also represented their bodily experiences, both individually and collectively. I consider explorers' self-fashioning on two different levels: in terms of everyday performances directed at different members of expedition parties, and in terms of accounts written for audiences beyond the mountains, who might have little or no experience of altitude. In the latter case, I show how explorers exploited the uncertainty around altitude sickness to describe their bodily performances in a variety of self-serving ways, and attempted, if not always successfully or convincingly, to fashion their imperial and scientific selves through tropes like masculine heroism and duty to empire.

The travel narratives and journals that form the basis of this chapter were written by an eclectic grouping of (with a handful of exceptions) European, and indeed largely British, surveyors and travellers. The most extensive set of accounts come from EIC employees, especially Bengal Infantryman seconded to surveys in the mountains. Others, particularly towards the mid-century, are provided by a wider range of travellers who

[6] See Tamson Pietsch, 'Bodies at Sea: Travelling to Australia in the Age of Sail', *Journal of Global History* 11, no. 2 (2016): 209–28. On bodily comparison in South Asia more broadly, see David Arnold, 'Race, Place and Bodily Difference in Early Nineteenth-Century India', *Historical Research* 77, no. 196 (2004): 254–73.

[7] With some exceptions in South America, see John B. West, *High Life: A History of High-Altitude Physiology and Medicine* (Oxford: Oxford University Press, 1998), 10–19.

visited the mountains to pursue a combination of scientific, imperial, economic and leisure interests (though these interests did not yet include mountaineering in any recognisably modern sense). This source base adds a further dimension to the politics of comparison by mediating what sorts of comparisons were available, and to whom. Those who wrote were able to constitute high mountain environments and their experiences of them through their narratives and textual productions, an opportunity not usually afforded to the people who carried their loads, shared their tents and campfires and struggled to breathe alongside them. In tracing the strategies that European travellers used to try to assert their bodily superiority over their South Asian companions, it is important to recognise (as discussed in Chapter 1) that the networks of labour that made Himalayan exploration possible were highly diverse. Indeed, expedition parties often included South Asian members who were not in any meaningful sense 'local'.[8] Porters – and even guides – could sometimes be recruited from the lowlands, complicating any simplistic comparisons between European and South Asian bodies (even if this dichotomy often appears in contemporary travel accounts). This makes the moments when travellers *did* identify the differences between the bodies of uplanders and lowlanders all the more important, and in what follows I trace how these distinctions would eventually become essential to both scientific understandings of altitude and perceptions of imperial possibilities in the Himalaya. In this chapter, I nevertheless want to suggest that the inherent uncertainties around altitude were also open to exploitation by guides and porters, even if performed in different ways and for different audiences. Given the limitations of the colonial archive, such suggestions can of course only be tentative, even if reading certain moments against the grain does suggest that guides and porters were sometimes able to navigate the uneven distribution of symptoms to exert agency, in particular by resisting often unpleasant and dangerous labour conditions.

While I use 'altitude sickness' for convenience throughout this chapter, the term itself had yet to enter usage in the Himalaya. 'Mountain sickness' and '*mal des montagnes*' were sometimes used, reflecting climatic and environmental associations, but also a lack of codification as a coherent illness. Early efforts to address the knowledge gap around altitude were usually anecdotal, though as the century progressed travellers increasingly turned to quantitative methods and self-monitoring (if often in an ad hoc and opportunistic fashion) by taking pulses and counting inspirations (or rate of breathing). In this vein, historians of science have been

[8] For this issue in histories of exploration more broadly, see Driver, 'Hidden Histories', 427.

productively examining the human (and non-human) body as both a site of a scientific practice and as a scientific instrument, and have argued for greater recognition of the embodied nature of scientific knowledge.[9] Scholars point, for instance, to the way that 'a human body whose walking pace and perceptual skills have been trained and disciplined is also functioning as an instrument' with a key example being the so-called 'Pundits' such as Nain Singh.[10] Studies of altitude or respiratory physiology represent a much narrower subset within scholarship, and otherwise excellent work has focused almost exclusively on the second half of the nineteenth and the twentieth centuries, that is, the period of systematic and often institutionally sponsored scientific studies of altitude.[11] Scholars like Sarah Tracy and Alex McKay have emphasised that these often had an imperial dimension, concerned with long-term acclimatisation in colonial labour contexts.[12] Respiratory physiology has also been productive in complicating the distinction between 'laboratory' and 'field' practices, most notably in the work of Vanessa Heggie.[13]

This chapter thus complements the existing scholarship, even as it diverges from it, by historicising and contextualising an earlier phase in the development of scientific understandings of altitude physiology. The politics of bodily comparison also looked different before 1850 because the debilitations experienced in high places were often still mysterious and terrifying, and the effect of the rarefaction of the air on bodies was not yet a given. Significantly for self-fashioning, the long-term adaptation of mountain peoples – most explicitly embodied today by mountaineering Sherpa – was also yet to be widely recognised. Beyond the narrower history of altitude physiology, the bodily comparisons this chapter traces

[9] See Christopher Lawrence and Steven Shapin, *Science Incarnate: Historical Embodiments of Natural Knowledge* (Chicago: University of Chicago Press, 1998).

[10] Bourguet, Licoppe and Sibum, *Instruments, Travel and Science*, 7.

[11] Interesting exceptions are provided by Jorge Lossio, 'British Medicine in the Peruvian Andes: The Travels of Archibald Smith M.D. (1820–1870)', *História, Ciências, Saúde – Manguinhos* 13, no. 4 (2006): 833–50; Felsch, 'Mountains of Sublimity, Mountains of Fatigue'. The non-anglophone sphere has had more to say, see for example Oswald Oelz and Elisabeth Simons, *Kopfwehberge: Eine Geschichte Der Höhenmedizin* (Zurich: AS Verlag, 2001); Cayetano Mas Galvañ, 'Los primeros contactos de los europeos con las grandes altitudes', *Revista de Historia Moderna. Anales de la Universidad de Alicante* 29 (2011): 139–68.

[12] Alex McKay, 'Fit for the Frontier: European Understandings of the Tibetan Environment in the Colonial Era', *New Zealand Journal of Asian Studies* 9, no. 1 (2007): 118–32; Sarah W. Tracy, 'The Physiology of Extremes: Ancel Keys and the International High Altitude Expedition of 1935', *Bulletin of the History of Medicine* 86, no. 4 (2012): 627–60.

[13] Vanessa Heggie, 'Experimental Physiology, Everest and Oxygen: From the Ghastly Kitchens to the Gasping Lung', *British Journal for the History of Science* 46, no. 1 (2013): 123–47; Heggie, *Higher and Colder*.

are also significant because they played out at the beginning of the colonial encounter for many Himalayan peoples with an increasingly expansionist EIC. At a time when surveyors and administrators were attempting to appropriate the Himalaya into both a regional imperial framework and a global scientific order, these expeditionary interactions laid the groundwork for long-term political, economic and labour relationships to empire.

More broadly, the early decades of the nineteenth century represent a key moment in the formation of imperial medical topographies of high places. Indeed, tensions around bodily comparisons in high mountains were far from unique to the Himalaya. Grappling with the problem of altitude sickness in the early nineteenth century also required drawing on imperial networks to make comparisons with different mountain ranges. While travelling in the Himalaya, explorers were simultaneously imagining other parts of the vertical globe, occasionally drawing from personal experience, but more usually from published descriptions of other late eighteenth and early nineteenth-century ascents. This chapter thus further points to the development of ideas of verticality, and increasing recognition that the high Himalaya, the high Andes and the high Alps might represent medically commensurate environments. There is sometimes a juxtaposition though, and the recognition of the commensurability of mountain environments in South America and Europe could serve to increase, as much as to alleviate, the uncertainty around altitude sickness. Comparisons made it clear that high spaces disagreed with bodies everywhere, but curious inconsistences in the heights in which symptoms appeared in different places sometimes led initially to more confusion rather than coherence.[14] Confusion stemmed from several factors, both real and imagined. These included the differences between tropical and temperate mountains and the idiosyncrasies of travel in different ranges (which could lead to vastly different amounts of acclimatisation time), as well as issues stemming from different accepted modes of self-fashioning, and the unevenness with which information allowing comparisons travelled over imperial and scientific networks. Recognising the intensification of comparisons in high places – on a variety of scales by a variety of actors, both locally and globally – thus allows this chapter to dissect the early phases of the formation of what was an inherently global science, but it also demonstrates why this globality must be traced through disconnection as much as connection.

[14] See the reports compiled in Paul Bert, *La pression barométrique: recherches de physiologie expérimentale* (Paris: G. Masson, 1878), 3–367.

This chapter examines the early history of altitude physiology in four sections. The first considers the extent of the uncertainty around the effects of altitude on human and non-human bodies in the early nineteenth century, and contextualises the politics of bodily comparison in relation to both upland frontiers and lowland colonial concerns round health, acclimatisation and air. The second section then turns to a discussion of long-standing indigenous understandings of altitude sickness (as arising from the poisonous miasmas of plants or a *Bis*), and the ways these were developed as a trope in European accounts. Next, I analyse the different ways the comparative performances of European and South Asian bodies were recorded in travel narratives, and use these as a lens into the complexities of self-fashioning. Finally, I examine the various experimental approaches around quantification and the instrumentalisation of bodies (by measuring pulses and breathing) that were employed to try and parse the seemingly incessant contradictions in the way symptoms were experienced, and the way these ultimately point to the gradual emergence of high mountains as medically coherent but aberrant spaces.

Encountering High Places

Though almost never recorded in writing, upland populations had long developed understandings of altitude sickness. French physiologist Paul Bert, in the historical overview that opens his monumental study, *La pression barométrique: recherches de physiologie expérimentale* (1878), records an abundance of local terms for mountain sickness, collected from all corners of the globe, including, 'the *veta*, the *puna*, the *mareo*, the *soroche* of the South Americans, the *bis*, the *tunk*, the *dum*, the *mundara*, the *seran*, the *aïs* of the mountaineers of Central Asia, the *ikak* of the natives of Borneo'.[15] The earliest written descriptions of altitude sickness are usually accredited to Chinese sources: specifically, tales of the Great and Little Headache mountains recorded in 37–32 BCE, and the monk Faxian's account of his travels in Kashmir and Afghanistan from the fifth century CE.[16] Here we also see variations of the *Bis* – that is, a plant-based explanation – tracing back to antiquity, with Chinese pilgrims

[15] Bert, *La pression barométrique,* 341. ['la veta, la puna, le mareo, le soroche des Sud-Américains, le bis, le tunk, le dum, le mundara, le seran, l'aïs des montagnards de l'Asie Centrale, l'ikak des naturels de Bornéo'.]

[16] See Daniel Gilbert, 'The First Documented Report of Mountain Sickness: The China or Headache Mountain Story', *Respiration Physiology* 52, no. 3 (1983): 315–26; Michael Ward, 'Mountain Medicine and Physiology: A Short History', *The Alpine Journal* 95 (1990): 191.

reporting that strong-scented leeks caused the headaches they experienced while traversing the '*Tsung-Ling*' (or 'Onion Mountains') of the Karakoram.[17] Similar ideas also arose in other parts of the vertical globe than the Himalaya, including in the Andes, where the poisoning of the air was sometimes attributed to mineral- rather than plant-based emanations, such as from buried antimony, suggesting the widespread utility of miasmatic explanations.[18]

Of European accounts, one of the earliest and most interesting comes from Spanish Jesuit José de Acosta's *Historia natural y moral de las Indias* published in 1590, which included a description of the infamous high-altitude silver mine at Potosi. Acosta's account was quite prescient in identifying symptoms and speculating that the thinness of the air was the cause, as well as making explicit analogies with seasickness.[19] Corresponding experiments with air pumps and atmospheric pressures saw recognition of the rarity of the air (if not understanding of its implications) in the seventeenth century, and Europeans became increasingly fascinated with mountains across the eighteenth century. It was not, however, until the last decades of the eighteenth century that these two interests began to properly intersect, perhaps no more famously – once again – than in the body of Alexander von Humboldt, who recognised the lack of oxygen (which had only been 'discovered' as relatively recently as the 1770s) as the cause of the headaches and nausea he experienced on Chimborazo in 1802. During this ascent, he famously fashioned his body as an instrument, recording the increasing debilitation he suffered, even if he was not immune to conflating symptoms like bloodshot gums and eyes.[20] In parallel with the strange debilitations of the body experienced in high mountains were those encountered during early experiments with hot- air balloons. In the case of ballooning, however, the rapidity of ascents – and hence lack of acclimatisation time – meant the particulars differed significantly. While commentators in Europe sometimes made explicit connections in the first decades of the nineteenth century, those in the Himalaya did not yet do so, indicative of the unevenness with which comparisons travelled and were applied.[21]

[17] Alexander Cunningham, *Ladak, Physical, Statistical, and Historical; with Notices of the Surrounding Countries* (London: W.H. Allen, 1854), 2.

[18] Bert, *La pression barométrique*, 35–42. The idea of a poisoned wind also occurred in other contexts, such as the jungle lowlands of the Nepal Terai.

[19] See West, *High Life*, 10–19.

[20] See Douglas Botting, *Humboldt and the Cosmos* (London: Sphere Books, 1973), 151–55; Michael Dettelbach, 'The Stimulations of Travel: Humboldt's Physiological Construction of the Tropics', in *Tropical Visions in an Age of Empire*, ed. Felix Driver and Luciana Martins (Chicago: University of Chicago Press, 2005), 43–58.

[21] For more on altitude physiology and ballooning, see West, *High Life*, 49–58. For explicit comparisons, see among others, M. Rey, 'Influence Sur le corps humain, des ascensions

Turning to the Himalaya, explicit descriptions of the symptoms of altitude sickness, let alone attempts to explain them, are surprisingly rare before 1800. Europeans who travelled over Himalayan high passes in the eighteenth century – including several Jesuits and the EIC-sponsored trade missions of George Bogle and Samuel Turner to Tibet – tended not to describe recognisable symptoms of altitude sickness in their travel accounts at all, even though they must have experienced them.[22] While descriptions of suffering do occur, these are usually conflated with the deleterious effects of wind, cold and exhaustion, and altitude was not clearly separated out as a discrete phenomenon. Even as the nineteenth century dawned and Himalayan travellers began more consistently recognising the unique effects of altitude, the high variability with which symptoms were experienced meant that considerable ambiguity remained. French naturalist Victor Jacquemont exemplifies this, even in the 1830s repeatedly claiming not to have experienced the same ill effects high in the Himalaya that he himself had experienced at much lower elevations in the Alps (and he was among the only travellers at this point to have visited both).[23] As he recorded, 'the effect, if it depends solely on the rarefaction of the atmosphere, should be the same at the same height in all regions of the globe', and yet in his opinion this appeared not to be the case.[24] Jacquemont did, however, frequently dismiss the symptoms of headache and nausea he felt at high altitudes as the result of extreme exertion, and cold and hunger affecting his digestive system, rather than primarily the rarefaction of the atmosphere (loss of appetite is, itself, a common symptom of altitude sickness).[25] At a time when the vertical globe was still being mapped out, the idea of an equivalency of altitude symptoms – and that the same effects might be experienced at the same elevations in the Himalaya, Andes and Alps – remained debated. In general, symptoms seemed to be felt lower in the Alps than in the Himalaya, and in temperate rather than tropical mountain ranges.[26] Travellers later pointed out that the necessity of making long approaches through the foothills meant that Himalayan explorers

sur les hautes montagnes', *Nouvelle annales des voyages et des sciences géographiques* 19, no. 3 (1838): 228, 242.

[22] Bert, *La pression barométrique*, 141–42; West, *High Life*, 2, 49.

[23] Jacquemont, *Voyage dans l'Inde*, Vol 2, 260.

[24] Jacquemont, *Voyage dans l'Inde*, Vol 2, 101. ['L'effet, s'il dépend uniquement de la raréfaction atmosphérique, devrait être le même à la même hauteur dans toutes les régions du globe'.]

[25] Jacquemont, *Voyage dans l'Inde*, Vol 2, 288. This is not to say that Jacquemont was unaware of the rarity of the air (indeed, elsewhere he discusses the effect of short-term acclimatisation), only that he was unconvinced it was the principal cause of his debility.

[26] See Bert, *La pression barométrique*, 7.

tended to be better acclimatised, which might explain the discrepancies.[27] Collectively, examples such as these remind us that experiences of altitude in the Himalaya could not be addressed without also evoking other mountains, even as Jacquemont demonstrates that the exact correlations might be confusing and contested. In the first decades of the nineteenth century, however, global comparison of differences (perceived and actual) thus often only added to the uncertainty around altitude.

This was further intensified by the way that European encounters with altitude sickness were bound up with the machinations of empire. As discussed in previous chapters, increasing scientific interest in the high passes and peaks was intertwined with the constitution of the Himalaya as a high mountain frontier. Physiological challenges and possibilities for movement in the high mountains thus had implicit military implications. As William Webb recorded while operating near Tibet: 'I considered that to pass churlishly along the frontier prying into its passes, and reconnoitring would be more likely to excite and to confirm than to allay their jealousy already kindled' though he anyway had to call a halt because 'the extreme labor and great difficulty of respiration experienced in the last undertaking has occasioned a general sickness in my Camp'.[28] This concern with upland frontiers also echoes recent scholarly interest in the concept of 'Zomia', which is helpful for thinking through the overlap of social and environmental factors in Himalayan exploration.[29] As proposed in its original form by Willem van Schendel, this metaregion consisted of uplands across Asia, including much of the Himalaya.[30] Both the idea of Zomia and its scope have subsequently been debated and revised, with perhaps its most famous adoption being that by James Scott (albeit with a more limited focus on Southeast Asia).[31] The creation and study of this new region was explicitly intended to highlight areas and peoples 'otherwise neglected as merely peripheral, exotic, or backward' in the traditional area studies divisions of South, Central, East and Southeast Asia.[32] For Scott and others, it has subsequently been about giving agency back to marginalised peoples and describing the non-state spaces

[27] Wood, *Narrative of a Journey*, 362.
[28] Webb to C.J. Doyle, 7 June 1816, British Library, IOR/F/4/552/13384, f4, f7.
[29] See Michaud, 'Editorial – Zomia and Beyond', 213.
[30] Willem Van Schendel, 'Geographies of Knowing, Geographies of Ignorance: Jumping Scale in Southeast Asia', *Environment and Planning D – Society & Space* 20, no. 6 (2002): 647–68.
[31] James C. Scott, *The Art of Not Being Governed: An Anarchist History of Upland Southeast Asia* (New Haven: Yale University Press, 2009).
[32] Michaud, 'Editorial – Zomia and Beyond', 199; Scott, *The Art of Not Being Governed*, 1–39.

given little attention in traditional histories that follow the contours of empires and nations.

For all this, Zomia is not without its limitations and has not been uncontroversial. It is, after all, entirely artificial, and just as there has never been a Zomia, there are no 'Zomians'. As Jean Michaud has pointed out, concepts like Zomia 'have never been needed by the subjects themselves' because 'that scale of things simply does not make sense, either practical or symbolic, for highlanders'.[33] The application of Zomia to the Himalaya has its own more specific caveats. As Sara Shniederman argues, for reasons both 'empirical and political', the Zomia concept may not map onto the Himalaya as readily as we might like.[34] This is partly because, rather than being state-free, the Trans-Himalaya has seen powerful indigenous states over the centuries, especially in Tibet, but also in Nepal, Bhutan and Afghanistan (and which, as I have shown, matter to the story of measuring the Himalaya because they dictated a highly asymmetrical access to the mountains). Shniederman nevertheless goes on to argue for the value of 'Zomia-thinking', even if the Himalaya are better conceived 'as a "multiple-state space," comprised of the territory of all of the nations and states in question, yet transcending the individual sovereignty of any single state'.[35] Christoph Bergmann meanwhile builds on Shneiderman's formulation, considering 'High Asia as a continuous zone and an agentive site of political action' in order to 'argue that confluent territories and overlapping sovereignties are key to understanding imperial frontiers in the Himalayan region'.[36]

As an analytical tool for moving beyond traditional political or national histories, Zomia thus has considerable appeal, sufficient to at least partially overcome objections over its artificiality and lack of intelligibility to the actors living within its imagined borders. Here, 'Zomia-thinking' helps to frame the perspective of the Himalaya when viewed from the subcontinent and vice versa (even while highlighting the homogenizing tendencies within these perspectives, and the sometimes drastically different topographies, ethnographies, biogeographies, climates, political configurations and histories, which had to be conflated into a coherent 'Himalaya' that could be mapped onto the vertical globe). Ultimately, it points to the mutual formulation of the Himalaya and the subcontinent,

[33] Michaud, 'Editorial – Zomia and Beyond', 212–13.
[34] Sara Shneiderman, 'Are the Central Himalayas in Zomia? Some Scholarly and Political Considerations across Time and Space', *Journal of Global History* 5, no. 2 (2010): 290.
[35] Sara Shneiderman, 'Himalayan Border Citizens: Sovereignty and Mobility in the Nepal–Tibetan Autonomous Region (TAR) of China Border Zone', *Political Geography* 35 (2013): 28.
[36] Bergmann, *The Himalayan Border Region*, 17.

and the way that the mountains could, in a sense, only exist when constituted simultaneously and in contrast with the lowlands. In this context, altitude sickness might be considered, to use James Scott's term, a 'friction of terrain'. Here Scott's assertion that 'the degree of friction represented by a landscape cannot simply be read off the topography. It is, to a considerable degree, socially engineered and manipulated to amplify or minimize that friction' is useful in thinking about the way cross-cultural labour relations shaped the way altitude sickness was recorded in travel accounts, and the way these, in turn, determined imperial ambitions with regard to the mountains.[37] In this chapter, drawing on 'Zomia-thinking' to re-centre the story on upland spaces and destabilise the horizontal as the default point of reference is thus helpful in two opposing respects: firstly, in recovering Himalayan people as agentive participants in expeditions, and secondly, in explaining how experiences of altitude sickness contributed to the peripheralisation of the Himalaya – socially, politically and environmentally – in relation to lowland imperial ambitions.

These concerns also surface in relation to the way that early understandings of altitude sickness had to be worked out with reference to European climatic theories of race and disease. While eluding any simplistic generalisations, the nineteenth century saw a growing emphasis on racial difference (and fixity), coupled with a more explicit European belief in the superiority of their 'climate, culture and constitutions'.[38] As Alan Bewell has argued, the confrontation with new diseases in this period, and often differing levels of susceptibility between coloniser and colonised, 'played a key role in producing difference' and led to the creation of new racial and cultural myths.[39] Historians of medicine have, in turn, shown how this brought about insecurity at a time when the subcontinent was coming to be seen as inherently pathological, spurring a growing pessimism about the possibilities for acclimatising European bodies to the 'tropics'.[40] Some early travellers in the Himalaya – including James Gerard – were trained and employed as surgeons, and investigations into the medical topographies of the mountains were tangibly linked to these lowland concerns. Comparisons were expressed especially in

[37] Scott, *The Art of Not Being Governed*, 166.

[38] For an overview, see Pratik Chakrabarti, *Medicine and Empire: 1600–1960* (Basingstoke: Palgrave Macmillan, 2014), 57–62.

[39] Alan Bewell, *Romanticism and Colonial Disease* (Baltimore: Johns Hopkins University Press, 1999), 17.

[40] See especially Mark Harrison, *Climates and Constitutions: Health, Race, Environment and British Imperialism in India, 1600–1850* (Delhi: Oxford University Press, 1999). See also Wendy Jepson, 'Of Soil, Situation, and Salubrity: Medical Topography and Medical Officers in Early Nineteenth-Century British India', *Historical Geography* 32 (2004): 137–55.

relation to miasmas, effluvia and poisonous air (and sometimes water), which were all significant contemporary imperial concerns.[41] Discussions around the debilitating effects of high altitude thus provide a contrast to developing ideas of the 'hygienic' properties of mountain air, something that would eventually be extended to the cultural characteristics and morals of mountain peoples.[42] The notion of the benefits of 'pure' mountain air for health was also a primary impetus for the construction of hill stations such as Shimla, Mussoorie and Darjeeling, especially from the 1820s onwards.[43] Such romantic, sublime and picturesque associations were not lost on the European travellers that feature in this chapter, but where this story departs from narratives of the hills as refuges from miasmas and tropical disease is in the difference between the foothills and the high mountains. The lowlands, increasingly seen as diseased and debilitating, might be categorised as environmentally distinct from the hill sanatoria. However, as one continued higher, invigoration turned again to debilitation (albeit for different reasons), creating a 'friction of terrain' with implications for colonisation and frontier security. In relation to developing medical topographies of mountains, this indicates the recognition of a difference between moderate altitudes and the upper reaches of the vertical globe. Not all high places were equal, and as much as plants increasingly needed to be placed into the multiple zones in which they thrived or withered, so too did bodies.

These concerns also went beyond the human, and Himalayan travellers paid frequent attention to the effects of altitude on non-human bodies, adding a further dimension to the politics of comparison. These observations were especially applied to the horses and yaks that plied the mountain paths (see for example Figure 3.1, which reminds us that these were spaces in which human and non-human bodies had long coexisted). Attention to the effects of altitude on non-human bodies was nevertheless anecdotal rather than systematic in this period, even as it was clear that animals also suffered, that symptoms were not qualitatively different to those afflicting human bodies, and that they originated from the same cause. For example, while examining the permeability of the frontier in Ladakh, William Moorcroft recorded that difficulty breathing 'extended to the animals, particularly the horses; but the yaks were not wholly exempt, and we were obliged to halt repeatedly

[41] See Alison Bashford and Sarah W. Tracy, 'Introduction: Modern Airs, Waters, and Places', *Bulletin of the History of Medicine* 86, no. 4 (2012): 495–514.

[42] See especially Bishop, *The Myth of Shangri-La*, 46–48. See also Nicolson, *Mountain Gloom and Mountain Glory*.

[43] Dane Kennedy, *The Magic Mountains: Hill Stations and the British Raj* (Berkeley: University of California Press, 1996); Sharma, 'A Space That Has Been Laboured On'.

Figure 3.1 'Kylass Mt. Road to Mansarowar Lake. Mr Moorcroft and Capt. H. and Chinese Horsemen' (1812). Watercolour by Hyder Young Hearsey of himself and William Moorcroft in local disguise while trying to sneak past Qing border guards into Tibet and visit Lake Manasarovar in 1812. This incident was later cited as a reason for the necessity of exercising restraint when it came to surveying at or beyond the Tibetan frontier (see also Chapter 1). This image both draws on and reinforces the 'exotic' and sublime tropes that typify visual representations of the Himalaya in this period. © British Library Board WD350.

to give the cattle relief'.[44] Other travellers occasionally considered the relationship between fauna and altitude more broadly, such as Scottish surgeon and botanist Thomas Thomson, who in the 1840s wondered at the many species of birds wheeling above: 'large ravens were circling about overhead, apparently quite unaffected by the rarity of the atmosphere, as they seemed to fly with just as much ease as at the level of the sea'.[45] He also thought it 'very remarkable' to find fish in mountain streams as high as 15,500 feet 'inasmuch as it would certainly not have been very surprising that air at that elevation should, from its rarity, be

[44] William Moorcroft and George Trebeck, *Travels in the Himalayan Provinces of Hindustan and the Panjab; in Ladakh and Kashmir; in Peshawar, Kabul, Kunduz, and Bokhara*, ed. Horace Hayman Wilson (London: John Murray, 1841), Vol 1, 398–99.

[45] Thomson, *Western Himalaya and Tibet*, 435.

insufficient for the support of life in animals breathing by gills'.[46] Attempts to delineate altitude sickness as a scientific phenomenon thus required accounting for the bodies of both humans and non-humans. That both suffered peculiar debilitations at altitude ultimately strengthened the case for considering uplands as medically distinctive environments, even as it added to the 'friction of terrain' circumscribing movement and imperial control in high places.

Indigenous Explanations

If the Himalaya represented an uncertain environment for European explorers at the turn of the nineteenth century, they had, of course, long been coherent spaces to the people who lived there. As Alexander Gerard noted:

> It is worthy of remark, that the Koonawurees estimate the height of mountains by the difficulty of breathing they experience in ascending them, which, as before noticed, they ascribe to a poisonous plant; but, from all our enquiries, and we made them almost at every village, we could find nobody that had ever seen the plant, and from our own experience we are inclined to attribute the effect to the rarefaction of the atmosphere, since we felt the like sensation at heights where there were no vegetable productions.[47]

As well as implying that Humboldt's use of his body as a barometer on Chimborazo was far from unprecedented, Gerard refers to the most widespread explanation given by Himalayan people to European travellers for the debilitating effects of high places, namely: they resulted from the noxious emanations of plants, which produced a poisoned wind. Across the period of this study, the engagement of European and South Asian lowlanders with altitude coexisted with long-standing indigenous explanations that attributed the headaches, nausea and hypoxia encountered in the high passes to the poisonous miasmas of plants. This idea was reported by travellers across the full span of high Asia, from Afghanistan to Bhutan, albeit with significant local variations and nuances, and was most commonly referred to as the 'Bis-ki-huwa', or simply the 'Bis'. English traveller 'Mrs' Hervey, who made extensive trips to Kashmir, Tibet and Tartary at the mid-century, wrote, '"Bischk," or Bikh, is "poison," and "Hâwa" signifies "wind," so the expression is literally translated by, "*Wind of Poison.*"'[48]

[46] Thomson, *Western Himalaya and Tibet*, 165. [47] Gerard, *Account of Koonawur*, 296.
[48] [Mrs] Hervey, *The Adventures of a Lady in Tartary, Thibet, China and Kashmir* (London: Hope, 1853), Vol 1, 133. Despite her expansive three-volume account, little is known about Hervey. Often accompanied by a long-suffering personal servant named Ghaussie, and without any European companions, she made extensive trips to Kashmir, Tibet and

As Gerard's account suggests, European observers were, on the whole, dismissive of the idea of a poisoned wind. In one sense, the *Bis* might be seen as a colonial story, an imperfect facsimile of an idea travellers received from their guides, and moulded into a trope that could be used to dismiss superstitious indigenous knowledge, keep the 'others' in their place and serve as a foil for their own physiological inquiries. This section, which considers the *Bis* only through extant colonial sources, is not intended to reconstruct an indigenous medical topography of the Himalaya.[49] Rather, I am interested in the way the *Bis* was recorded in travel accounts, and contributed to the imaginative constitution of the mountains as distinctive spaces of exotic, sublime experience. In thinking about the story of the *Bis* as a European shorthand, I follow Michael Bravo and Sverker Sörlin, who argue for paying attention to 'narrative as a technology of travel' and that 'narratives, rather than notions of local knowledge, can reveal the complex linkages between cultures of science, travel, and observation, and the conditions that sustain their circumscribed domains of power'.[50] However, if tracing the idea of the *Bis* through European accounts is on the one hand the genealogy of a colonial story, it can also be used to highlight the extent to which Himalayan exploration was dependent on pre-existing networks, and the way that altitude sickness operated as a 'friction of terrain'. Indeed, because the uneven distribution of symptoms was open to exploitation, explorers had to account for the *Bis* if they were to successfully marshal the labour necessary for high mountain travel.

Though sceptical, European travellers sometimes investigated the plants to which the *Bis* was attributed, perhaps compelled by the way these stories coincided with the growing necessity of understanding the distribution of plants in three dimensions. Indeed, both plants and bodies increasingly needed to be located by attending to a vertical scale, and, for a time, it seemed possible that the relationships between bodies and air, and air and plants, might be linked. However, as Gerard alluded, the most significant evidence arranged against the *Bis* was that the correlation between plant distribution and altitude sickness did not add up. Scottish artist and traveller James Baillie Fraser (1783–1856) recorded

Tartary between 1850 and 1852 (some to areas never previously visited by Europeans). Hervey, Vol 1, 46; Vol 2, 298. For more, see Rosemary Raza, *In Their Own Words: British Women Writers and India, 1740–1857* (Oxford: Oxford University Press, 2006), 263; Brigid Keenan, *Travels in Kashmir: A Popular History of Its People, Places and Crafts* (Gurgaon: Hachette India, 2013), 144–56.

[49] For a brief precis of Tibetan sources on altitude sickness (and their limitations), see McKay, 'Fit for the Frontier', 119–22.

[50] Michael Bravo and Sverker Sörlin, *Narrating the Arctic: A Cultural History of Nordic Scientific Practices* (Canton: Science History Publications, 2002), 4, 24–25.

while near Gangotri that 'what proved the fact that all this was the effect of our great elevation, was, that as we lowered our situation, and reached the region of vegetation and wood, all these violent symptoms and pains gradually lessened and vanished'.[51] While Fraser had been directed towards flowers, Hervey was informed by her guides that a species of moss was to blame: 'they believe the wind becomes poisonous, by blowing over a certain plant of a moss species, which grows abundantly on all high mountains in Tartary, and is found when all other vegetation ceases'.[52] As Hervey alludes, mosses had been discovered at heights far exceeding the limits of other forms of flora, and were sometimes thought to still be able to release their noxious emanations while buried by snow, thereby potentially offering an accounting for symptoms of altitude sickness at elevations where there was no apparent vegetation.

Observations of plant distribution, examination of specimens and enquires made among local informants continued to be sought across the first half of the nineteenth century. Of the various species of plants and mosses examined as potential sources of the *Bis*, one widely discussed suspect was *Aconitum Ferox*. An extract from the root was presumed to be the source of the notorious Himalayan substance known as Bikh (which essentially translates as 'the poison').[53] Widely available in the bazaars despite attempts to ban it, surgeon and naturalist John Forbes Royle also reported the probably apocryphal idea that it was used in attempts to poison water sources during the Anglo-Gurkha War.[54]

Meanwhile, Scottish surgeon George Govan (1787–1865), who was also the first superintendent of the Saharanpur Botanic Garden from 1817 to 1823, extensively considered the connection between altitude sickness and plants, writing that 'the lofty *Aconite*, the well-known poisonous effects of which, when taken internally, seem to have given rise to a belief among the natives, that it poisons the air in its vicinity' however, this was 'an opinion for which I never could discover any foundation, unless it may be found in the lofty elevation of the belt, inhabited by this showy plant, where *occasionally* (certainly not *always* or uniformly) the disagreeable effects usually ascribed to the rarity of the air are experienced by travellers'.[55] From his vantage in the Himalaya foothills, Govan went on to suggest that his informants attributed 'this belief to the

[51] Fraser, *Journal of a Tour*, 449. [52] Hervey, *The Adventures of a Lady*, Vol 1, 134.

[53] See David Arnold, *Toxic Histories: Poison and Pollution in Modern India* (Cambridge: Cambridge University Press, 2016), 57–63. Though reflecting extensively on Bikh, this otherwise excellent account makes no mention of the connection to altitude sickness.

[54] Royle, *Illustrations*, Vol 1, 48.

[55] George Govan, 'On the Natural History and Physical Geography of the Districts of the Himalayah Mountains between the River-Beds of the Jumna and Sutluj', *Edinburgh Journal of Science* 2, no. 3–4 (1825): 282.

circumstance of the plants always occurring at very high elevations', having 'never met with it much below where the barometer stood at 19 inches'.[56] The idea that Bikh was the source of bodily debility was, as Govan acknowledges, not an unreasonable one, even as his investigations established its improbability. The connection between *Aconitum* and altitude was perhaps strengthened by the way Bikh poisoning, which induced asphyxia, was superficially similar to the symptoms of mountain sickness (though this was investigated not so much experimentally as through the acquisition and comparison of specimens).[57]

Often implied as being simplistic, credulous and unscientific, representations and understandings of the phenomena of mountain sickness and the *Bis* among Himalayan peoples were nevertheless more complex and nuanced than many European commentators acknowledged. Indeed, while reading these sources 'against the grain' has its limits, there are occasional moments that suggest that the *Bis*, as characterised in European travellers' accounts, failed to adequately encompass the indigenous explanation it came to represent. John Batten, at the time Assistant Commissioner of Kumaon, recorded for example that 'the natives do not attribute the effects indiscriminately to "nirbisi," or aconite – and indeed the worst oppression is felt above the reach of all vegetation', and ultimately '"Bish ke howa" (The poisoned air) is the general expression for the cause of the oppression, though it is true that certain plants are often quoted as the root of the evil'.[58] Batten thus suggests that the *Bis* was a term that might encompass the phenomena of altitude sickness more broadly, and implies that it did not always rely on plants as an explanation. James Fraser also hints at the complexity inherent in understandings of mountain sickness provided by his guides, both of whom were natives of the mountains, namely: Goving Bhisht, 'a man of high caste and considerable consequence', and Kishen Sing, 'a favourite servant of the late rajah [of Garhwal] '.[59] As Fraser wrote: 'they cannot account for this phenomenon, but believe it to proceed from the powerful perfume of myriads of flowers in the small valleys and on the hill sides; but they do not seem quite satisfied with this solution of the difficulty themselves'.[60]

Fraser's record of his relationship with Bhisht and Sing, however one-sided, also suggests that guides and porters might have exploited

[56] Quoted in Nathaniel Wallich, *Plantae Asiaticae Rariores* (London: Treuttel and Würtz, 1830), Vol 1, 37.
[57] See Wallich, *Plantae Asiaticae Rariores*, Vol 1, 36.
[58] James Manson, 'Capt. Manson's Journal of a Visit to Melum and the Oonta Dhoora Pass in Juwahir', ed. John Hallett Batten, *Journal of the Asiatic Society of Bengal* 11, part 2 (1842): 1163.
[59] Fraser, *Journal of a Tour*, 398, 405–6. [60] Fraser, *Journal of a Tour*, 435.

uncertainties around the *Bis* to resist unpleasant and potentially perilous labour conditions. On one occasion, Fraser suspected his guides were exaggerating in an attempt to dissuade him from proceeding via a difficult route by talking 'wildly of a serār or wind from the mountains, pregnant with this mysterious poison'.[61] So as not to lose fourteen days travel time, Fraser wanted to take the risk, having 'observed how prone these people, particularly Goving Bhisht and Kishen Sing, were to exaggerate difficulties and the length of the road, and to throw obstacles in the way'.[62] Altitude sickness and the *Bis* might thus have sometimes allowed guides to exert agency, in this instance in manoeuvring to resist a dangerous journey. Indeed, sometimes Fraser suspected members of his expedition parties might be faking or exaggerating symptoms, on one occasion as a scapegoat for other self-inflicted ills:

> We experienced much trouble to-day from our coolies, who were, probably, many of them the same that we saw so much intoxicated, and busily engaged in dancing for the two previous days: they were with difficulty urged on. . . . They told us that they were affected by the Serān, or poison in the air, from the flowers above noticed; and though I believe that their situation may in some degree be referred to drunkenness and excess, and something may be allowed for laziness, still their general behaviour and appearance indicated a good deal further that could not be accounted for.[63]

This is inflected with Fraser's increasingly antagonistic relationship with his expedition party, but he is not entirely unforgiving, and even while castigating his porters for supposed moral failings, he is willing to concede that altitude is at least partially responsible for their debility. On another occasion though, a porter seemed to collapse 'to all appearance senseless, and totally heedless of the arguments, both verbal and manual'; however, 'there was no doubt that he thus feigned illness, for his pulse and breathing were perfectly regular and good; and the people of Comharsein who were with us were perfectly aware of the trick'.[64] This is a fascinating moment, and one that might easily be read in terms of resistance. Even as European travellers were working out how to navigate the politics of bodily comparison and fashion their own experiences of altitude sickness, moments like this thus suggest that they also had to contend with the performances of others.

Travel Narratives and the Problem of Comparison

If the high Himalaya were spaces in which the performance of European and South Asian bodies were inevitably placed in situations of direct

[61] Fraser, *Journal of a Tour*, 435. [62] Fraser, *Journal of a Tour*, 435.
[63] Fraser, *Journal of a Tour*, 440. [64] Fraser, *Journal of a Tour*, 195.

comparison, the question for European travellers became how to record these comparisons in travel accounts and journals. Recent scholarship has demonstrated that mountains can be productive spaces for examining social relationships, and the way that, for example, mountaineering could blur class but codify gender relationships in the Alps.[65] Similarly, as Marie-Noëlle Bourguet, Christian Licoppe and H. Otto Sibum have argued, while on Mont Blanc, Swiss naturalist Horace-Bénédict de Saussure developed a hierarchy within his expedition party based on altitude, pulse and nausea, a hierarchy in which 'fortunately (and somewhat suspiciously)' de Saussure outperformed his hired guide. They suggest that while 'traditional hierarchies were reasserted on this occasion' it was nevertheless the case that 'the social organisation corresponding to barometric measurements could sometimes transgress and reshape cultural and anthropological boundaries'.[66] Concern about the potential of high mountain spaces to destabilise expected hierarchies around bodily performance haunts the accounts of Himalayan travellers, where the additional factor of race – itself a highly unstable category in this period – presented a complication not experienced in the same way in the Alps.[67] Indeed, embodied knowledge production looks different when it has to account for the bodies of 'others' in a context of uncertainty.

In written accounts, however, unlike in everyday negotiations over loads and routes, explorers had much greater control over the politics of comparison. It is difficult to generalise about tropes of self-fashioning in Himalayan accounts across the first half of the nineteenth century, and there are more exceptions than rules. However, more often than not, travellers seem to explicitly or implicitly suggest that Himalayan peoples did not necessarily do better at altitude, and in fact usually did worse than the European explorers they were hired to guide.[68] Though Europeans and South Asians are shown to suffer together, considerable rhetorical effort was made in travel narratives to ensure hierarchies around bodily performance were not inverted. Performing superiority was nevertheless complicated by the way that, as historians have demonstrated, the human body could be deployed to bestow authority on scientific accounts, and to

[65] See Reidy, 'Mountaineering, Masculinity, and the Male Body'. See also Hansen, 'Partners: Guides and Sherpas in the Alps and Himalayas'.

[66] Bourguet, Licoppe, and Sibum, *Instruments, Travel and Science*, 12–13. Within this volume, see also Marie-Noëlle Bourguet, 'Landscape with numbers'.

[67] See Arnold, 'Race, Place and Bodily Difference'; Sujit Sivasundaram, 'Race, Empire and Biology before Darwin', in *Biology and Ideology: From Descartes to Dawkins*, ed. Denis R. Alexander and Ronald L. Numbers (Chicago: University of Chicago Press, 2010), 114–38.

[68] Interestingly, this trend was often reversed in accounts from South America. See Bert, *La pression barométrique*, 333–34.

arrogate 'field' observations over those from the 'armchair' or the 'laboratory'.[69] In a similar vein, Dorinda Outram has examined the 'authentication of the explorer's travels by the trials of his body', and here the body's vulnerability is key, without which 'the explorer could not manifest in his own person the moral economy which made his reporting acceptable as authentic knowledge'.[70] This section thus argues that there was an ongoing tension between representing the 'trials of the body' in travel accounts to establish a privileged position for producing credible knowledge of remote locations, while also avoiding upsetting social hierarchies around race and bodily performance.

This tension is evident in the writing of Alexander Gerard, who in describing his ascent to the high point on Reo Purgyil (as discussed in Chapter 2) spoke of how he and his brother James 'overtook our people not a mile from our halting place. We had infinite trouble in getting them to go on, and were obliged to keep calling out to them the whole way, at one time threatening, and at another coaxing them'. However, these assertions were followed by a rather frank admission that 'to tell the truth, however, we could not have walked much faster ourselves, for we felt a fullness in the head, and experienced a general debility'.[71] In these kinds of literary sleights-of-hand, Gerard manages to imply that European masculinity prevails, even while demonstrating that the brothers themselves suffered and interacted heroically with a challenging environment. Questions around self-fashioning were exacerbated by the way explorers' authority over their guides and porters was often fragile. While reflecting on the possibilities for movement in the borderlands near the frontier with Tibet, Gerard recorded: 'we were so completely exhausted at first, that we rested every hundred yards; & had we not been ashamed before so many people, some of whom we got to accompany us after much entreaty, we should certainly have turned back'.[72] This line appears in an unpublished report to the EIC, and in a published version of the same incident, an additional sentence was inserted, to the effect: 'we observed the thermometer every minute almost, in order to show the people we were doing something'.[73] Here the brothers attempted to mask bodily weakness with an instrumental performance that amounted to feigned scientific practice, and in moments like this, we are reminded that the European surveyors' supposed superiority was far from assured in the extreme environment of the high Himalaya. Shame

[69] Hevly, 'The Heroic Science of Glacier Motion', 67–68, 86.
[70] Outram, 'On Being Perseus', 290, 292. [71] Gerard, *Account of Koonawur*, 291.
[72] Alexander Gerard, 'Remarks regarding the Geological Specimens collected in 1821', British Library, Mss. Eur. D137, f20-1.
[73] Lloyd and Gerard, *Narrative of a Journey*, Vol 2, 32.

and an inability to risk loss of face motivated them to force their struggling bodies to keep moving upwards. The Gerards needed to convince the guides – whose enthusiasm for the task was already tenuous – that their suffering had a purpose, even if this purpose was not one that would necessarily have made sense to the guides. This also reminds us that travellers' letters, journals and reports were by no means free of self-fashioning, and were anyway often written with publication in mind; indeed, this example is representative of the way that there is usually little to distinguish between how bodily comparisons were recorded in unpublished and published materials in this context.

Though less frequent, accounts do sometimes clearly portray guides outperforming their European employers, even if this tended to be excused by gestures towards long-term acclimatisation. Bengal Infantryman James Manson (1791–1862), while attached to a mineralogical survey of the Himalaya in the 1820s, described that above 17,000 feet, 'without the assistance of two men (Bhoteahs) accustomed to travel at such elevations, and a *jabbu* (an animal bred between the Tartar *yak* and common cow), to whose tail I tied myself . . . I should never have reached the summit of the pass' and moreover, 'even with their combined aid I did not accomplish it without very severe fatigue. This sensation is experienced by the natives, though in a less degree'.[74] As well as providing an arresting image of his dependency and the suffering of non-humans, Manson depicts the 'accustomed' bodies of his Bhotiya companions as performing better than his own. Recognition of the adaptation of mountain people to their high-altitude homes nevertheless appears unevenly in accounts, and European observers were only intermittently engaged with questions of long-term acclimatisation (despite the parallels with contemporary insecurities around the adaptation of the European body to 'the tropics').[75] As James Gerard wrote, those 'who either breathe a highly-rarefied air, or are accustomed to ascend their steep sides, suffer much less than those who inhabit a lower zone and denser atmosphere'.[76] In this instance he was comparing upland and lowland South Asian bodies, but on another occasion when European bodies were part of the equation, he implied that living in high mountains did not necessarily provide the advantages (genetic and otherwise) that we now recognise in high-altitude populations, writing that '[we]

[74] [James Manson], 'On the Distress and Exhaustion Consequent to Exertion at Great Elevations', *Gleanings in Science* 1 (1829): 330.

[75] See Harrison, *Climate and Constitutions*; Arnold, *The Tropics and the Traveling Gaze*. For studies of long-term acclimatisation in the twentieth century, see Heggie, *Higher and Colder*, 125–59.

[76] Gerard, 'A Letter from the Late Mr J.G. Gerard', 321.

nevertheless outdid the villagers, who accompanied us, and reside at the height of 12,000 feet'.[77] Whatever the contradiction here, he ultimately thought this question did not apply in the highest reaches of the mountains, writing of the people of Kanawar: 'I have not learnt whether they are subject to occasional indisposition, such as that I experienced' but 'it is indisputable that, beyond a certain height, the effects of the rarefied air upon the functions of animal life are permanent, and neither custom nor constitution can bear up against them'.[78] Though recognising the possibility of placing bodies and plants into different vertical zones, the role of long-term acclimatisation is here discounted. The implications of these questions were, however, never benign, and laid the groundwork for the imperially motivated systematic studies of high-altitude populations of the later nineteenth century.[79]

It is only rarely that we get journals or travel accounts dealing with altitude sickness in this period written by South Asian travellers. Even these tend to be mediated by their production for European audiences, such as that of the Kashmiri (though Delhi born) Brahmin traveller and later diplomat Mohan Lal (1812–1877). Indeed, his account of his travels with Scottish officer and diplomat Alexander Burnes (1805–1841) and James Gerard to Afghanistan in 1832 was composed in English, and conforms to many European travel writing tropes.[80] In this, Lal recounted a tale from 'Babar's Memoirs', which claimed that 'the famous pass of Hindu Kush is so high, and the wind so strong, that the birds, being unable to fly, are obliged to creep over the top' and as a result 'they are often caught by the people, who kill and roast them for dinner. This is said by Dr. Gerard to be probably owing to the thinness of the air at that great elevation'.[81] Lal thus notes the existence of older stories around altitude but, through his association with Gerard, presents himself to a European audience as scientifically informed and dismissive of such myths. The writings of the *munshi* Mir Izzet Ullah, although based on reconnaissance conducted in the employ of William Moorcroft, provide an interesting contrast. Ullah kept a journal (in Persian, later translated by Horace Wilson) in which he attributes problems at altitude to impure water, suggesting that in the Karakoram, 'the water was also so

[77] Gerard, 'A Letter from the Late Mr J.G. Gerard', 322–23.
[78] Gerard, 'A Letter from the Late Mr J.G. Gerard', 322.
[79] See McKay, 'Fit for the Frontier'.
[80] For more on Lal see Bayly, *Empire & Information*, 133–39, 230–34; Mathur, 'How Professionals Became Natives', 27–59.
[81] Mohan Lal, *Travels in the Panjab, Afghanistan, Turkistan, to Balk, Bokhara, and Herat; and a Visit to Great Britain and Germany* (London: W. H. Allen, 1846), 75–76.

unwholesome, producing short breathing'.[82] While describing the route to Yarkand, he also describes a variant of the *Bis*: 'here begins the *Esh* – this is a Turkish word, signifying Smell; but, as here used, it implies something the odour of which induces indisposition; for from hence the breathing of horse and man, and especially of the former, becomes affected'.[83] These are, however, far from indigenous perspectives, and both Lal and Ullah were, like their European employers, lowlanders for whom altitude sickness was an unpleasant novelty. While these examples demonstrate that South Asian travellers might have been able to draw on different cultural myths relating to high places, they are thus of only limited use in recovering uplanders' understandings of altitude.

In examining tensions around self-fashioning, it is also important to consider that the accounts of altitude sickness discussed in this chapter were overwhelmingly written by men, and that in these accounts the bodies they describe (both their own and those of their South Asian companions) were never explicitly not male. This is not to say that these were spaces in which men and women did not suffer together (as is evident, for example in Figure 3.2). Though perhaps more rather than less often homosocial affairs, that women were sometimes present in expedition parties is also confirmed by the written accounts. While negotiating a particularly treacherous and vertigo-inducing path while travelling to Lake Manasarovar in 1812, William Moorcroft noted that several of his bearers lost their nerve, but 'one woman carried four burthens at different times for her less courageous companions'.[84] When it comes to Himalayan travel accounts written by women up to the mid-century, Hervey's *The Adventures of a Lady in Tartary, Thibet, China and Kashmir* (1853) is perhaps unique in explicitly describing symptoms of altitude sickness. Her three-volume journal of travels reflects a consolidated understanding of the bodily debility one was supposed to experience in high mountains, and while crossing the Hannoo Pass, for example, Hervey gives a visceral description, including a 'terrible nausea, like to nothing else in its overpowering nature but sea-sickness.[85] In terms of self-fashioning, there is little to suggest she was less anxious about showing weakness than her male counterparts,

[82] Mir Izzet Ullah, 'Travels beyond the Himalaya', trans. Horace Hayman Wilson, *Journal of the Royal Asiatic Society* 7, no. 14 (1843): 298. Bayly is thus right to suggest that Ullah's writings 'represent a halfway point between the Islamic travelogue and British topography'. Bayly, *Empire & Information*, 76.
[83] Ullah, 'Travels beyond the Himalaya', 296.
[84] Moorcroft, 'A Journey to Lake Mánasaróvara', 385.
[85] Hervey, *The Adventures of a Lady*, Vol 2, 368.

Figure 3.2 'Women coolies of Kanawar from above Kanum. Foot of Ranung Pass, May 28, 1853'. The artist's decision to feature these women is in contrast with the more limited visibility usually given to women's labour in the written accounts from this period. Watercolour by Conway Shipley.
Source: From a bound volume titled 'India, Tibet and Kashmir'. With kind permission of the Central Asia Library of The Henry S. Hall, Jr. American Alpine Club Library.

and while social relationships around gender have proven fruitful in the context of mountaineering in the Alps, these questions can be less satisfactorily examined in the early period of European exploration in the Himalaya.[86] Thinking about contemporary bodily experiences and performances of seasickness in relation to gender suggests some avenues of comparison, but this chapter can ultimately only acknowledge that these were spaces in which both European and South Asian women also suffered, and that the politics of bodily comparison must have had to account for this.

[86] See Clare Roche, 'Women Climbers 1850–1900: A Challenge to Male Hegemony?', *Sport in History* 33, no. 3 (2013): 236–59.

When it came to assessing the credibility of bodily performances in this period, the role of self-fashioning was not necessarily overlooked, and travellers' assertions about their personal experiences of the effects of altitude were not always accepted uncritically. On occasion, travellers were suspected of deliberately downplaying the effects. For example, when Joseph Alexander Weller, at the time Junior Assistant Commissioner of Kumaon, offhandedly referred to 'a bad night's rest' while on a shooting trip to the Unta Dhura Pass in 1842, his editor John Batten remarked in a footnote that 'probably the rarity of the air may have had a greater effect on our traveller than ... he seems inclined to admit'.[87] The unevenness with which symptoms were felt left significant scope for members of expedition parties to try to disguise their symptoms. As James Fraser remarked: 'after reaching that place [the Bamsooroo Pass] no one was proof against this influence' and it 'was ludicrous to see those who had laughed at others yielding, some to lassitude, and others to sickness, yet endeavouring to conceal it from the rest', even as he went on to claim: 'I believe I held out longer than any one; yet after passing this gorge every few paces of ascent seemed an insuperable labour, and even in passing along the most level places my knees trembled under me'.[88] This fits neatly into the trope of implying the superiority of European bodies while not discounting one's own suffering, but it also reveals the complexities of overlapping and contested performances within the expedition party, and the way the inconsistency of symptoms made performances an ever-present concern. The incessancy of self-fashioning thus added to the confusion around developing medical topographies, and is a reminder of why we need to pay particular attention to disconnection and imperfect comparison in the delineation of global sciences in this period.

Himalayan explorers never really resolved the problem of how to record the symptoms of altitude in their own bodies and those of their South Asian companions in the first half of the nineteenth century. Generally, travel writing tropes allowed for maintaining hierarchies by asserting that European bodies performed as well or better, with exceptions sometimes excused by acclimatisation. Even if this reflected reality and not self-fashioning, it is possible these discrepancies partially stemmed from a difference in motivation, with wages likely less of an impulse to drag oneself upwards when compared to scientific and exploratory fervour or

[87] Joseph Alexander Weller, 'Extract from the Journal of Lieut. J. A. Weller ... on a Trip to the Bulcha and Oonta Dhoora Passes', *Journal of the Asiatic Society of Bengal* 12, part 1 (1843): 101.

[88] Fraser, *Journal of a Tour*, 449.

a duty to empire. As was occasionally acknowledged, performance was also strongly correlated with exertion, and there was a significant difference in susceptibility depending on whether an individual was mounted or not, and the weight of the load they carried.[89] More significantly, these tropes also fail to account for the way that porters and guides might have exploited the 'capricious and irregular' distribution of symptoms to resist sometimes brutal labour conditions. Indeed, as the various performances of Fraser's porters suggest, it is unlikely that European travellers were the only ones exploiting the politics of comparison around altitude for their own ends.

Quantification and the Instrumentalisation of Bodies

As evident in the way an exasperated James Fraser resorted to taking the pulse of a porter he suspected of feigning sickness, Himalayan travellers increasingly turned to methods for quantifying bodily performance. In deploying their bodies as instruments to read changes in atmospheric pressure, they followed the example of early travellers in the Alps and the Andes, including most famously de Saussure and Humboldt.[90] Measuring pulses and rates of breathing added a new dimension to the politics of comparison, seeming to offer an opportunity to make sense of the wildly differing symptoms, and of the real and perceived (not to mention performed) differences between members of expedition parties, even if this was only implemented in an ad hoc rather than systematic, statistical fashion in this period. James Manson offers one of the most extensive early attempts to quantitatively account for altitude sickness in the Himalaya, measuring his pulse and counting the frequency of his breathing using a 'watch with a second hand', recording, for example: 'ascended the whole without being obliged to stop to take breath. Pulse never exceeding 140 in a minute, nor the number of inspirations 32'.[91] Manson also measured and compared the bodies of his companions: 'I found on standing still after a little bit of steep ascent, that my pulse beat at the rate of 160 in a minute. A seapoy's, (a hill man,) who was with me, beat at the rate of 172'.[92] His identification of the soldier as a 'hill man' rather than a lowlander suggests Manson was paying attention to acclimatisation, and he is also clearly grappling with the correlation between altitude sickness and exertion. This final section

[89] See, for example, Hooker, *Himalayan Journals*, Vol 2, 167.
[90] Bert, *La pression barométrique*, 44, 92.
[91] Manson, 'Capt. Manson's Journal', 1163, 1177.
[92] Manson, 'Capt. Manson's Journal', 1163.

of the chapter examines attempts at quantification, and the ways these fed into the development of a Himalayan medical topography that could be compared globally. Taken as a whole, these point to an increasing, though haphazardly acquired, degree of scientific coherence to the high Himalaya across the first half of the nineteenth century.

Scottish naval officer John Wood, while in the Pamirs employed on a survey of the Indus River, examined the pulses of all his party, writing that 'to my surprise found that the pulses of my companions beat yet faster than my own'.[93] He also went one step further, including a table of comparisons (see Figure 3.3). Both Wood's 'surprise', and perhaps his assertion that his own body was better adapted than his companions reflect self-fashioning, but he did acknowledge that the difference in loads carried meant that these comparisons were not entirely fair, or free of other variables such as fatigue. It is noteworthy that aside from race and occupation, the only other variable he includes is general fitness; age and gender, though implicit, are never addressed directly. Wood was nevertheless adamant about the body's potential for instrumentalisation, continuing that from this point onwards 'I felt the pulses of the party whenever I registered the boiling point of water' and insisting that 'the motion of the blood is in fact a sort of living barometer by which a man acquainted with his own habit of body can, in great altitudes, roughly calculate his height above the sea'.[94] Just as the Kinnauras had trained themselves to observe their breathing to determine their rough height above sea level, Wood suggests that the rise and fall of the pulse offered the potential to quantify experiences of altitude, even if he acknowledges that these readings could only ever be approximate.

	Throbs.	Country.	Habit of body.
My own . . .	110	Scotland	spare
Gholam Hussein, Munshi .	124	Jasulmeree	fat
Omerallah, mule-driver .	112	Afghan	spare
Gaffer, groom . .	114	Peshawuree	spare
Dowd, do. . .	124	Kabuli	stout

Figure 3.3 'Upon Pamir the pulsations in one minute'.
Source: From John Wood's *Narrative of a Journey to the Source of the River Oxus* (1841).[95]

[93] Wood, *Narrative of a Journey*, 363. [94] Wood, *Narrative of a Journey*, 363.
[95] Wood, *Narrative of a Journey*, 363.

Assertions of the ability of the body to act as a sort of 'living barometer' are not uncommon in travellers' accounts. Before crossing the Niti Pass, William Moorcroft had questioned his guide, Amer Singh – a local and 'the son of the *Seyana* [headman]' of the frontier village of Niti – and was told that the mountains were 'not so high as many in *Garwal*'.[96] Moorcroft was sceptical of this information, which was of elevated importance for the way it was entangled with the delineation of the frontier, remarking that 'from the view which I have had of them, it appears to me that they are higher' and once across he noted that 'the general difficulty of breathing experienced by us in passing them comes in confirmation of this opinion'.[97] In moments where readings diverged, those from the boiling point thermometers and barometers were nevertheless preferred, and bodily sensations became subservient to precision devices of metal and glass. Bengal Infantry surveyor William Webb meanwhile considered other variables that might mean bodily measurements were unreliable. As he wrote in 1819: 'but even considering *my own* sensations as affording no competent evidence, on account of the weak state of my health, I cannot for a moment doubt the existence of this effect'.[98] Webb discounts the readings provided by his own body, even while he is adamant of the value of self-monitoring and of the body's ability to produce scientific knowledge. Other difficulties in self-monitoring were more mundane, and as Victor Jacquemont noted while above 18,000 feet: 'I wanted to count the beats of my pulse, but both my hands were completely numb and insensitive'.[99] Measuring was not always easily or reliably achieved in an extreme environment, even if attempts at quantification seemed to offer the potential to understand the wildly different symptoms or, on a more basic level, remove doubts around self-fashioning and the downplaying or exaggeration of symptoms in the context of everyday expeditionary practice.

Far from passively entering mountainous spaces, Himalayan travellers actively sought out remedies and strategies to cope with the effects of high altitude. James Gerard, for example, recorded that he had been told by 'an intelligent servant', who had accompanied William Moorcroft, 'of

[96] Moorcroft, 'A Journey to Lake Mánasaróvara', 404, 414.
[97] Moorcroft, 'A Journey to Lake Mánasaróvara', 414.
[98] Webb, 'Extract of a Letter from Captain William Spencer Webb, 29th March, 1819', 65.
[99] Jacquemont, *Voyage dans l'Inde*, Vol 2, 288. ['J'aurais voulu compter les pulsations de mon pouls, mais j'avais les deux mains entièrement engourdies et insensibles'.] Elsewhere, after making a successful reading, Jacquemont used his pulse to again suggest the rarity of the air was not a main cause of his suffering in the Himalaya. Jacquemont, Vol 2, 297–98.

fatal consequences from the want of due precaution. He says that the passage of the lofty range should be made while fasting, and recommends frequent doses of emetic tartar during the journey'.[100] For his part, Moorcroft recorded various remedies collected from his Bhotiya guides, though he does not indicate whether he actually tried them: 'the natives recommend a small quantity of coarse sugar to be eaten whilst we are mounting, and speak highly of the power of the kind of spar found near the snow reduced to powder and mixed with water, in diminishing the distressingly quickened action of breathing'.[101] While James Gerard was accompanying Alexander Burnes on a reconnaissance expedition to Bokhara, he was told by a mullah named 'Nujeeb' that he and Burnes 'should *eat onions* in all the countries we visited; as it is a popular belief that a foreigner becomes sooner acclimated from the use of that vegetable'.[102] This echoes the idea of the 'onion mountains' and the way the *Bis* was sometimes ascribed to wild leeks, but it also points to the existence of well-developed local understandings of short-term acclimatization. In this sense, it is notable that the onions were prescribed specifically as a remedy for foreigners, implying that the mullah perceived a difference between the bodies of locals and those of the European travellers.

As well as seeking out remedies, and monitoring physiological effects by taking pulses and counting inspirations, some Himalayan travellers also sought to theorise the causes behind their debility. James Gerard, for example, discussed the way the change in pressure was key to the underlying problem:

> As respiration cannot be performed in a vacuum, we should consider that, at the height of 18,480 feet, the exhaustion is already half made, and . . . the progressive action becomes here an arithmetical series, reducible to an experiment in natural philosophy, where each succeeding stroke of the piston of an air-pump appears to draw the hand placed on the aperture closer and closer, till the pressure above so much overbalances that below, as to be insupportable to the person without risk of detriment. At 18,480 feet, the barometer, in the mean state of the air, stands at 15 inches, so that here we breathe an atmosphere half the density of that at the level of the sea; how then can we be surprised at the effects?[103]

[100] Gerard, 'A Letter from the Late Mr J.G. Gerard', 323.

[101] Moorcroft, 'A Journey to Lake Mánasaróvara', 399. Moorcroft also liberally employed other drugs, including 'ten grains of Calomel, three of James' Fever Powder and two of Dr Robinson's Brown Pills', more in hope than in reasonable expectation of their efficacy. See also Charles Allen, *A Mountain in Tibet: The Search for Mount Kailas and the Sources of the Great Rivers of India* (London: Little, Brown, 1982), 89.

[102] Alexander Burnes, *Travels into Bokhara* (London: John Murray, 1834), Vol 1, 105.

[103] Gerard, 'A Letter from the Late Mr J.G. Gerard', 323–24.

Gerard compares high altitude spaces with laboratory experiments using a vacuum, evoking well-known demonstrations using air pumps. These reflect both his own attempt to parse his experiences and a method of recording and describing which might be explicable to an audience that had no direct first-hand contact with altitude sickness. He also sets up a parallel between experiences in the field and imagined laboratory experiments that could potentially explain what happened to the body in high places. Even if he did not have the resources to carry out such experiments himself, Gerard nevertheless identifies the trajectory that altitude physiology would take later in the century when Paul Bert and his ilk began experimenting with pressure chambers large enough to encompass a human body.[104] Further advances in the scientific understanding of altitude would await a new era of both laboratory experiments and systematic studies in the field, and this chapter thus ends where the scholarship on altitude physiology to date usually begins.

Conclusion

In the early decades of the nineteenth century, the physiological effects of altitude became both a subject – if anecdotally and haphazardly – of scientific enquiry and a (sometimes literal) headache for Himalayan exploration. In a context of attempts to delineate and map the EIC's newly acquired high frontiers with Tibet, and growing (if largely illusory) concerns with Russian activities in Central Asia, altitude sickness also presented a perplexing natural phenomenon that needed unravelling. Here altitude sickness amounted to both a physiological and social 'friction of terrain', circumscribing the mobilisation of labour and imperial possibilities in the high mountains. (Like other 'frictions of terrain' it would eventually be reduced at the behest of lowland states, when 'distance-demolishing technologies' such as roads decreased reliance on labour in parts of the mountains, but not until the twentieth century.[105]) Ultimately, understanding the unequal and asymmetrical politics of comparison around altitude in this period matters, because imperial administrators, naturalists and returning travellers used the accounts this chapter has traced to formulate nascent medical topographies of the high Himalaya. These, in turn, worked to constitute the upper reaches of the mountains in relation to lowland norms, as aberrant environments and economically and politically peripheral spaces. However, 'Zomia-thinking' also indicates that these

[104] Heggie, 'Experimental Physiology'. Humboldt had also gone to the bottom of the Thames in a diving bell in 1827, where 'he was forcibly reminded of his ascent of Chimborazo'. See Botting, *Humboldt and the Cosmos*, 223.

[105] Scott, *The Art of Not Being Governed*, 166.

divisions could be exploited by uplanders, and indeed imperial control in the Himalaya remained contested and incomplete, especially when compared with an increasingly expansionist and unforgiving British Empire in the lowlands.

This chapter has focused on some of the ways that the constant and necessary co-presence of South Asian bodies shaped the way explorers understood altitude sickness, and developed both tropes of self-fashioning and medical topographies of the high Himalaya. The inevitability of comparison in these spaces, exaggerated by the 'capricious and irregular' way symptoms were distributed without due regard for conventional hierarchies around bodily performance, intensified cross-cultural relationships. High mountains are thus especially productive for examining the inherent social and racial tensions in the labour regimes that made European exploration possible. This comparativity played out on multiple levels, and in different ways for different actors, at both the scale of the local and of the global. Indeed, engagement with the high Himalaya occurred simultaneously in this period with an increasing recognition of the commensurability of mountain environments, and a growing sense of a global verticality. This is nevertheless far from a story of the seamless accumulation of knowledge. While it was increasingly clear that high places disagreed with bodies everywhere, as in the case of Jacquemont, curious inconsistences in the heights at which symptoms appeared could lead initially to more uncertainty.

By the middle of the nineteenth century, some of this uncertainty around altitude sickness had lessened, even while complexities around partial pressures and the inherent variability of symptoms remained unresolved. As more and more accounts were compiled, travellers nevertheless came to expect that they would likely suffer in particular ways when venturing into the high Himalaya. English naturalist, and later director of Kew Gardens, Joseph Hooker dealt with altitude sickness in some detail, and in typifying mid-century Himalayan travellers' understandings provides an appropriate place to leave this story. In his lavish tome *Rhododendrons of the Sikkim Himalaya*, he also repeats reports of the *Bis*, revealing the continuing utility of this trope. Discussing *Rhododendron setosum*, for example, Hooker recorded that this was 'the "*Tsallu*" of the Sikkim-Bhoteas and Thibetians, who attribute the oppression and headaches attending the crossing of the loftiest passes of Eastern Himalaya'.[106] Meanwhile, in his *Himalayan Journals* (1854), arguably the most famous of the nineteenth-century Himalayan scientific travel narratives, Hooker

[106] Joseph Dalton Hooker, *Rhododendrons of the Sikkim Himalaya* (London: Reeve, Benham and Reeve, 1849), part II, no. 21.

included an appendix titled 'On the weight of the atmosphere in Sikkim; and its effects on the human frame'.[107] In writing about recording pulses and taking into account exertion and time since eating, Hooker provides a snapshot of the monitoring and recording practices around altitude sickness that had developed by the middle of the century. He also dealt with both short- and long-term acclimatisation (recognising the former but not the later), and in his account there is a confidence, and a sense of familiarity. Altitude sickness was thus no longer mysterious and terrifying, and Himalayan exploration and the delineation of the frontier could continue apace. However, familiarity was by no means mastery, and the unevenness of the distribution of symptoms continued to provide scope for self-fashioning. Medical topographies remained incomplete and in flux, even as imperial scientific networks increasingly saw high mountains being recognised as commensurate environments with particular implications for human and non-human bodies.

Into the second half of the nineteenth century, tropes around performance and self-fashioning in relation to altitude sickness, far from vanishing, merely took different forms. In 1883, William Woodman Graham (c.1859–c.1932) made one of the first climbing trips to the Himalaya 'more for sport and adventures than for the advancement of scientific knowledge', and went on to (possibly) set a new world altitude record for the human body of 24,080 feet on Kabru.[108] Of this ascent, he made the extraordinary assertion, quoted here from the published text of an address to a packed session of the Royal Geographical Society, that 'neither in this nor in any other ascent did we feel any inconvenience in breathing other than the ordinary panting inseparable from any great muscular exertion. Headaches, nausea, bleeding at the nose, temporary loss of sight and hearing, were conspicuous only by their absence'.[109] By way of explanation, he argued that 'unquestionably man's range is increasing. Read any old account of an ascent of Mont Blanc; it was expected that the climber should suffer every possible inconvenience from rarefied air, and the harrowing details were duly forthcoming' but 'now the ascent is mere child's play, and we hear no more of these agonising horrors'.[110] Graham concluded: 'personally I believe that, supposing the actual natural difficulties to be overcome, the air, or the want of it, will prove no obstacle to

[107] Hooker, *Himalayan Journals*, Vol 2, 413–18.

[108] William Woodman Graham, 'Travel and Ascents in the Himálaya', *Proceedings of the Royal Geographical Society* 6, no. 8 (1884): 429. This record is debated, and there remains considerable doubt as to whether Graham was actually climbing on Kabru or a different mountain altogether. See Unsworth, *Hold the Heights*, 232–39.

[109] Graham, 'Travel and Ascents in the Himálaya', 434.

[110] Graham, 'Travel and Ascents in the Himálaya', 435.

the ascent of the very highest peaks in the world'.[111] While he would ultimately be proven right on the last point, his other assertions simply represent a new set of tropes as mountaineering increasingly became a sphere of imperial competition (and national prestige). Indeed, anticipating Graham, Paul Bert noted that by the 1860s and 1870s, many mountaineers stopped reporting altitude sickness symptoms altogether, because 'they were almost as afraid of being ridiculed for mountain sickness as they were for sea sickness'.[112]

Altitude thus continued to be a cipher, and to offer opportunities for travellers to represent their bodily experiences in self-serving ways. While in the early nineteenth century symptoms might have implied heroic challenge, or bodily suffering that could be deployed to establish authority, by the later part of the century they were again sometimes being left unspoken in popular travel accounts (even as systematic scientific studies of respiratory physiology were beginning to be mounted in parallel). This chapter has examined the politics of bodily comparison around altitude in a particularly acute context, that of the imperially motivated but cross-culturally practised early exploration of the Himalaya. In this context, the omnipresence of South Asian guides and porters became central not only to how explorers understood and represented altitude sickness, but also to the way they experienced its unsettling and ill-defined effects. Writing about the Arctic, Michael Bravo and Sverker Sörlin argue that recovering the presence of indigenous peoples often has not gone far enough in histories of exploration, and sometimes 'even studies of field practices in places and spaces where people are in abundance can be carried out while completely ignoring the human beings who are present in the landscape'.[113] In the high spaces of the Himalaya, under the strain of the insidious and invisible effects of altitude sickness, this was never the case in the first half of the nineteenth century. As the many and often questionable tropes that propagated in the exploration narratives of this period demonstrate, the co-presence of bodies, European and South Asian, lowlander and uplander, were impossible to ignore. As we will see in Chapter 4, these ongoing uncertainties around altitude in the Himalaya had wider repercussions. Indeed, they dictated not only the development of coherent medical topographies, but also understandings of the materiality of the mountains through their geology.

[111] Graham, 'Travel and Ascents in the Himálaya', 435.

[112] Bert, *La pression barométrique*, 128. ['on craint presque le ridicule du mal des montagnes, comme celui du mal de mer'.] See also Felsch, 'Mountains of Sublimity, Mountains of Fatigue', 353–54.

[113] Bravo and Sörlin, *Narrating the Arctic*, 5.

4 Frozen Relics

In the first decades of the nineteenth century, although nobody had actually seen a volcano in the Himalaya, it was simply assumed that they had not yet looked hard enough. Even as more and more travellers visited the mountains and failed to locate active volcanoes, naturalists were far from confident that they would not eventually be discovered. As the first Deputy Commissioner of Kumaon, George Traill (1792–1847), remarked in 1832: 'no volcano is positively known to exist, but there are grounds for suspecting that the *Nanda Deví* peak contains something of the kind'.[1] He continued: 'the *Bhotias* and natives of the neighbouring districts bear unanimous testimony to the occasional appearance of smoke on its summit: this is attributed by them to the actual residence of a deity' and in particular, his *chula*, or kitchen (here the phenomenon of spindrift, or snow blown from summits by high altitude winds, was a key source of confusion among European and South Asian observers alike).[2] Continuing to draw on the Bhotiyas' information, Traill noted that doubts about Nanda Devi's nature were ultimately a moot point, because ascending the mountain to resolve them was out of the question: 'a religious *Mela* is held every twelfth year, at the highest accessible point, which is, however, about a mile from the summit: further progress is rendered impossible by a wall of perpendicular ice'.[3] As he concluded: 'the dangers and difficulties incurred by the pilgrims are represented as most appalling ... [and] under these circumstances, it is scarcely possible that the question of a crater can ever be decided by actual inspection'.[4] Accounting for the existence or non-existence of Himalayan volcanoes was thus bedevilled

[1] George William Traill, 'Statistical Report on the Bhotia Mehals of Kemaon', *Asiatic Researches* 17 (1832): 18.
[2] Traill, 18. On spindrift, see for example William Lloyd and Alexander Gerard, *Narrative of a Journey from Caunpoor to the Boorendo Pass, in the Himalaya Mountains* (London: J. Madden, 1840), Vol 1, 242.
[3] Traill, 'Statistical Report on the Bhotia Mehals of Kemaon', 18.
[4] Traill, 'Statistical Report on the Bhotia Mehals of Kemaon', 18.

by a lack of observations in situ.[5] Echoing debates over instrument usage in the mountains, complaints about limited access became an ongoing refrain. This chapter is thus the story of how overlapping frontiers – topographical and cultural as well as political – circumscribed the investigation of geological phenomena in the Himalaya in the first half of the nineteenth century.

Early forays into the material makeup of the Himalaya occurred at the same time the mountains were finally being acknowledged as the highest in the world. This revision of the vertical globe occurred in tandem with momentous changes in understandings of the earth. Indeed, the reconfiguration of scales to imagine the Himalaya was matched and exceeded only by the need to imagine a new vastness for geological time and for processes that could transform the surface of the globe. Meanwhile, the marine nature of fossils found in hills and mountains far from the sea had long been accepted, but these had not yet been adequately explained. Similarly, while it had long been understood that these and other fossils were the remains of organic beings, by the early nineteenth century it was increasingly apparent – if not uncontroversial – that many of these beings were now extinct.[6] As geology's most prolific interlocutor for this period, Martin Rudwick puts it: 'the shift from "mineralogy" to "geology," as the most usual term for what would now be called the earth sciences, encapsulates the dramatic changes in the culture of inorganic natural history that occurred between the late-eighteenth and the mid-nineteenth centuries'.[7] Even as surveyors in the Himalaya were sometimes still seeking non-existent volcanoes, the study of the earth was thus emerging as a modern discipline. As a result, this chapter spans the time before and after which the term 'geology' can be used without anachronism, and echoes the ambiguities this transformation

[5] Peaks like Nanda Devi and Kanchenjunga continued to be speculated as volcanic throughout the first half of the nineteenth century. See 'Volcano in the Himmalaya', *Gleanings in Science* 1 (1829): 338–39; Walter Sherwill, 'Notes upon Some Atmospherical Phenomena Observed at Darjiling in the Himalayah Mountains, during the Summer of 1852', *Journal of the Royal Asiatic Society of Bengal* 23 (1854): 57. These claims were also repeated in Europe, see for example 'Account of a Volcano in the Himalayah Mountains. Communicated to Dr Brewster by a Correspondent in India', *Edinburgh Journal of Science* 4, no. 2 (1826): 209–11; 'Account of Hot Springs and Volcanic Appearances in the Himalaya Mountains', *Edinburgh Journal of Science* 7, no. 2 (1827): 55–56.

[6] Martin Rudwick, *Bursting the Limits of Time: The Reconstruction of Geohistory in the Age of Revolution* (Chicago: University of Chicago Press, 2005), 63–71.

[7] Martin Rudwick, 'Minerals, Strata and Fossils', in *Cultures of Natural History*, ed. Nicholas Jardine, James A. Secord, and Emma C. Spary (Cambridge: Cambridge University Press, 1996), 266. Here Rudwick evokes another classic work in the history of geology: Rachel Laudan, *From Mineralogy to Geology: The Foundations of a Science, 1650–1830* (Chicago: University of Chicago Press, 1987).

entailed.[8] While I sometimes use the term 'geology' in this chapter after the contemporary actors, I thus do so in its broader sense to refer to a diverse range of practices related to questions about the earth, even as I show how these practices simultaneously contributed to narrower processes of disciplinary formation in both the Himalaya and Europe.

Specifically, this chapter considers the place and practices of East India Company (EIC) employees and travellers on the fringes – both geographically and socially – of rapidly evolving debates in the sciences of the earth. As much as it demonstrates the unevenness with which geological information emerged *out* of the mountains, this chapter is also about the unevenness with which up-to-date information was available to the surveyors and naturalists tasked with going *into* the mountains to map the geology of the largely unknown high places. In particular, there was as a distinct lack of geological training afforded to those seconded to surveys in the mountains, and relevant texts and other resources were only haphazardly available. These limits were compounded by material problems around acquiring and transporting specimens, reflecting both the extreme topography and the politics of operating at and beyond nascent political frontiers. Just as significantly, these imperial and geographical frontiers were further complicated by working in, to use Fa-ti Fan's helpful formulation, 'cultural borderlands' (an idea he initially developed in relation to imperial botanical practices in China). As Fan explains, this concept serves as 'a point of entry for explorations into the day-to-day practices of scientific imperialism' where the 'pattern of power relations was complex, dynamic, and localized. It involved constant negotiations among different parties, and the outcome was not uniformly in favor of the [European] naturalists' (though, as he notes, care must be taken not to 'downplay the reality of power differentials').[9] In this chapter, the concept of 'cultural borderlands' is especially useful for understanding the way that geological specimens – for example the so-called 'lightning bones' – could be acquired via pre-existing networks of trade in ritual objects. However, these attempts to circumvent frontier restrictions were only ever a partial solution. Indeed, the dislocated objects these networks returned were often difficult to place accurately on the vertical globe, and meant that the problematic lack of in situ observations was ongoing.[10]

[8] Rudwick, *Bursting the Limits of Time*, 347–48, 468–69. See also Martin Rudwick, *Worlds Before Adam: The Reconstruction of Geohistory in the Age of Reform* (Chicago: University of Chicago Press, 2008), 2–3.

[9] Fan, 'Science in Cultural Borderlands', 221, 224.

[10] On the importance of verticality to geology and territory, see Braun, 'Producing Vertical Territory'. See also Rudwick, *Bursting the Limits of Time*, 83.

In examining everyday activity in 'cultural borderlands', this chapter also highlights surveyors' dependency on guides to locate key collections (for example the Spiti fossils), and to explain changes in topography over time (such as by drawing on multi-generational oral traditions to understand the movement of glaciers). The surviving archives of Himalayan exploration place some constraints on illuminating these episodes fully, but the presences of Himalayan peoples nevertheless haunt this chapter at every turn. While we will meet certain key individuals – including Pati Ram, Nagu Burha and Ram Singh – a key aim here is also to render visible the sheer scale of contributions silenced in the geological accounts of savants published in Europe in this period. Doing so makes it abundantly clear that the fossils extracted from the Himalaya, and eventually placed in the hands of geologists like William Buckland, would not have gotten there without the active participation, and sometimes active resistance, of Himalayan peoples. Indeed, as well as expertise, this chapter emphasises the enormous quantities of labour required to mount expeditions into the high mountains, and to transport often heavy and bulky samples out for closer inspection in the comfort of the lowlands. In a shift from preceding chapters which focused on imaginative disconnect and the discursive strategies of travellers, this chapter thus foregrounds the physical stuff of the mountains – the rocks and the ice, the soil and the shells – in order to better understand the processes involved in making the Himalaya into a coherent place for both science and empire. In highlighting the overwhelming laboriousness of making the Himalaya globally commensurable, the materiality of specimens thus emerges as a key limiter in a context of extreme topography and fragile labour relations.

In tracing the materials used to overwrite South Asian cosmologies with those of European geology (itself a cosmology in flux), this chapter thus expands the argument for the way the mountains were made and remade in this period through comparison.[11] The first appointee to the Geological Survey of the Himalaya, James Herbert, acknowledged of his survey that 'there are no new rocks likely to occur for as Mr Humboldt has observed these are the same in every quarter of the globe'.[12] However, as he elaborated, if 'tertiary strata have hitherto been found in countries of moderate elevation: it is not unlikely ... that the examination of them at such enormous elevations' would be interesting.[13] Key points of direct

[11] For India-based geologists playing up instances of difference to assert their authority, see also Grout, 'Geology and India', 155.

[12] James Herbert to Charles Lushington, 30 November 1826, British Library, IOR/F/4/957/27123(24), f29.

[13] James Herbert, 'On the Organic Remains Found in the Himmalaya', *Gleanings in Science* 3 (1831): 266.

contrast with other mountain ranges included minerals, fossils, strata and glaciers. As with barometrical scales and the symptoms of altitude sickness, these comparisons nevertheless sometimes led to more rather than less uncertainty. Indeed, the question of volcanoes was compounded by misleading inferences arising from the Andes. As in the stories of instrumental measurement and altitude physiology told in previous chapters, Himalayan travellers drew widely on the work of Alexander von Humboldt, whose prominent theories of mountain formation and ascent of the volcano Chimborazo in 1802 loomed large in their imaginative repertoires. The very definitely volcanic nature of the Andes thus made it harder to believe that a mountain range as stupendous as the Himalaya might have none. However, as James Herbert recorded, there is 'considerable difference of physical aspect between these mountains and the Andes, the chain with which it has been most usual to compare them' and 'the numerous volcanoes, extinct or igneous of the Andes ... belong to a totally different order of things from that which prevails in the Himmalaya'.[14] Given these 'usual' comparisons, accepting that the order of things might be different in the Himalaya took time, and the limits of knowledge could continue to extend to questions as large as the existence or non-existence of volcanoes. This chapter thus presents further evidence for why we need to pay attention to both disconnection and connection in the writing of global histories of science, especially by examining moments when comparisons were unhelpful.

While geological investigations had bearings on questions of scientific and cosmological importance, they also had significant economic dimensions.[15] The search for substances like coal to power steamships and later railways, and for valuable minerals like gold and silver, served as key motivators (even if these searches were only haphazardly executed, and remained largely unfulfilled in the first half of the nineteenth century).[16] Potential economic returns were sometimes exaggerated by those pursuing geology for more abstract reasons, to justify the time and expense their investigations incurred. For example, James Gerard rather fancifully claimed that 'the soil of Cabool teems with mineral riches, and it is far from improbable that we shall yet see "Hindoo Khoosh" rivalling in precious ores the mountains of Peru'.[17] Here

[14] Herbert, 'Report upon the Mineralogical Survey of the Himalayan Mountains', cxlvi–cxlvii.

[15] These became particularly important from the 1830s onwards; see Arnold, *Science, Technology and Medicine*, 45.

[16] Grout, 'Geology and India', 13, 29–30. Though not in large quantities or good quality, coal was identified in the Himalayan foothills in this period: James Herbert, 'Notice on the Occurrence of Coal, within the Indo Gangetic Tract of Mountains', *Asiatic Researches* 16 (1828): 404.

[17] Gerard to [Anon], 12 April 1833, British Library, Mss. Eur. C951.

he too evoked the Andes – and in particular Potosi and its extraordinary deposits of silver – painting an optimistic vision that the Himalaya might soon offer similar riches. Such exuberant claims of the mineral potential of the mountains were not uncommon, and as EIC Surgeon Robert Hamilton Irvine wrote: 'with the exception of the absence of volcanoes, the Himalaya range, as far as known, consists in the main body of the very same mineral matter' as the Andes'.[18] Thus, 'reasoning *à priori*, we may conclude that only want of proper exploration has prevented the discovery of metalliferous veins'.[19] As we will see, however, such a priori reasoning had a tendency to come unstuck in the Himalaya.

The hunt for valuable minerals could also heighten existing political tensions. William Moorcroft, for example, recorded while sneaking beyond the frontier to Lake Manasarovar in 1812 that the ground was scattered with rocks bearing minerals, including possibly 'some veins of silver in strata of quartz'. However, he lamented: 'I had no instruments to break stones with, nor did I see any small fragments which I could with convenience place in my girdle. I was obliged therefore rather to leave this point unsettled, than to expose myself to the suspicion of coming into the country in search of precious metals'.[20] Geological practice in the Himalaya, as instrumental measurement, was continually circumscribed by the limits of imperial mastery around the high frontiers. Indeed, as James Herbert complained in his 'Report upon the Mineralogical Survey', the Himalaya and especially the plateau of Tibet 'considered in its various relations to Asia, I might even say to the Old World ... is undoubtedly the most interesting spot on the surface of the globe' and, as such, was in need of urgent mapping. However, 'unfortunately for science, this task is not likely to be soon effected. The jealousy of the Chinese government, to which the greater part of it belongs, opposing insurmountable obstacles to the progress of investigation and discovery'.[21]

In focusing on the high Himalaya, it is worth noting at the outset that this chapter does not deal, except in passing, with the most famous of the Himalayan (or sub-Himalayan) contributions to geology and palaeontology in this period, namely: the extraordinary fossils excavated from the

[18] Robert Hamilton Irvine, 'A Few Observations on the Probable Results of a Scientific Research after Metalliferous Deposits in the Sub-Himalayan Range around Darjeeling', *Journal of the Asiatic Society of Bengal* 17, part 1 (1848): 138.

[19] Irvine, 'A Few Observations', 139. Andrew Grout argues similarly that South America provided a template for expectations around valuable minerals in the Himalaya. Grout, 'Geology and India', 47.

[20] Moorcroft, 'A Journey to Lake Mánasaróvara', 389.

[21] Herbert, 'Report upon the Mineralogical Survey of the Himalayan Mountains', xiii.

Siwalik hills in the 1830s.[22] These fossils, which came to the attention of European naturalists during attempts to dig the Doab canal, made the names of Scottish surgeon and naturalist Hugh Falconer and English engineer Proby Thomas Cautley, and garnered them international attention and the Wollaston Medal from the Geological Society of London in 1837.[23] Scholars have, with good reason, paid particular attention to the way Falconer drew on Hindu cosmology in interpreting the fossils, most famously by using information from the *Puranas* to try and date the extinction of the enormous *Colossochelys Atlas* tortoise. Here Savithri Preetha Nair has argued that 'Falconer's ability to "translate" the sacredness that the natives attributed to fossils into his own scientific concerns was best reflected in his zoological nomenclature'.[24] Indeed, while the appellation 'Atlas' was intended to invoke the Hindu concept of a turtle that carried the world, it was also, as Pratik Chakrabarti and Joydeep Sen have argued, a rhetorical opportunity to get to this idea via the Greek myth of the demigod who held the world on his back.[25] As Chakrabarti and Sen go on to conclude: 'in the context of the making of geology in colonial India, mythology was not simply a device for the popularization of fossils; rather, it served as a methodological tool for interpreting fossils'.[26]

These engagements with cosmology extended beyond the Siwaliks, but we still know relatively little of the practice of geology in the high Himalaya in this period.[27] Indeed, geology in India remains notably understudied, especially when compared to other sciences like botany and astronomy. For a long time, the most extensive contributions remained those of Andrew Grout, whose important work this chapter

[22] See Kenneth Kennedy, *God-Apes and Fossil Men: Paleoanthropology of South Asia* (Ann Arbor: University of Michigan Press, 2000), 86–120; Pratik Chakrabarti, *Inscriptions of Nature: Geology and the Naturalization of Antiquity* (Baltimore: Johns Hopkins University Press, 2020). Much as for 'geology', this study spans the period before and after which 'palaeontology' can be used without anachronism. See Rudwick, *Worlds Before Adam*, 47–48.

[23] Arnold, *Science, Technology and Medicine*, 44–45. See also Grout, 'Geology and India', 142–43.

[24] Savithri Preetha Nair, '"Eyes and No Eyes": Siwalik Fossil Collecting and the Crafting of Indian Palaeontology (1830–1847)', *Science in Context* 18, no. 3 (2005): 380.

[25] Pratik Chakrabarti and Joydeep Sen, '"The World Rests on the Back of a Tortoise": Science and Mythology in Indian History', *Modern Asian Studies* 50, no. 3 (2016): 829–30.

[26] Chakrabarti and Sen, 'The World Rests on the Back of a Tortoise', 839. Falconer's metropolitan peers were nevertheless not especially open to this mix of geology and mythology. The discovery of primate fossils in the Siwaliks in 1836 has also received considerable attention. See Rudwick, *Worlds Before Adam*, 417–20; Chakrabarti, *Inscriptions of Nature*, 66–82.

[27] Rasoul Sorkhabi provides a useful chronology of specifically Himalayan geology, even if his study is largely synoptic: Rasoul B. Sorkhabi, 'Historical Development of Himalayan Geology', *Journal of the Geological Society of India* 49 (1997): 89–108.

builds on.[28] However, Pratik Chakrabarti has recently made a major contribution, eloquently taking examples from across India and the Himalaya to demonstrate how the deep past of the subcontinent was 'naturalized' in the nineteenth century. He also notes that 'because of the fusion of natural and historical imaginations, the study of Indian antiquity was not often confined to specific disciplinary boundaries', which meant that debates around geology often looked quite different to those taking place simultaneously in Europe.[29] Focusing on altitude, dependency and practice in the upper reaches of the mountains, and especially on the role of the vertical in the interpretation of Himalayan geology, this chapter takes a different approach, even as it similarly argues for paying greater attention to the specific nuances of geological science in South Asia, and for attending to the way overlapping cosmologies complicate any simplistic narratives of disciplinary codification.

The chapter begins with a key group of high-altitude fossils reputed as talismans and collected and sold by Bhotiyas under the name 'lightning bones', which reveal tensions between specimens that were both scientifically and cosmologically significant, and the ongoing problem of attaining in situ observations. This is followed by a close examination of a series of fossils from Spiti, focusing on the limited resources – both material and intellectual – of those tasked with retrieving them. The chapter then considers the roles of pre-existing networks in locating and moving material, as well as the contributions of individual brokers such as Pati Ram, who equipped and advised multiple expeditions. I next turn to a broader assessment of how these material remains fit into discussions about the upheavement of the Himalaya, while continuing to emphasise the sometimes-limited vocabulary available to those operating in the decentred spaces of the mountains. Expanding from fossils, the final section of the chapter examines glaciers, especially debates over their existence in the Himalaya (in a reverse of the debate over volcanoes) and experiments to establish that they represented the same phenomenon as had been observed in the Alps. These forays into glaciology are also used to demonstrate naturalists' reliance on multi-generational oral traditions, which highlight both the way imperial mastery was limited in 'cultural borderlands' and the

[28] Andrew Grout, 'Geology and India, 1775–1805: An Episode in Colonial Science', *South Asia Research* 10, no. 1 (1990): 1–18; Grout, 'Geology and India'; Andrew Grout, 'Possessing the Earth: Geological Collections, Information and Education in India, 1800–1850', in *The Transmission of Knowledge in South Asia: Essays on Education, Religion, History, and Politics*, ed. Nigel Crook (Delhi: Oxford University Press, 1996), 245–79. See also the useful overview in Satpal Sangwan, 'Reordering the Earth: The Emergence of Geology as a Scientific Discipline in Colonial India', *The Indian Economic & Social History Review* 31, no. 3 (1994): 291–310.

[29] Chakrabarti, *Inscriptions of Nature*, 3, 6.

multifaceted roles of Himalayan peoples in geological practice. Together, these episodes serve to demonstrate the compounding limits to engaging with the materiality of the mountains, and ultimately, the sheer laboriousness of making the Himalaya globally commensurable in the first half of the nineteenth century.

'Lightning Bones', Ammonites and the Problem of in situ Observation

Among the most widely discussed early geological collections extracted from the Himalaya were a group of fossils often referred to as the *bijli ki har*. Usually translated as 'lightning' or 'thunder' bones, the *bijli ki har* were reputed to have their origins in lightning strikes (an explanation that may have arisen because they were often exposed by heavy rains and thunderstorms which washed away covering soil).[30] These represented some of the first fossils recovered from the high Himalaya, initially collected by Bengal Infantryman William Webb.[31] They were then passed to Henry Colebrooke in Calcutta, who sent them on to Europe. It was by this means that they eventually found their way into the hands of William Buckland, who referenced them in his major geological work *Reliquiæ Diluvianæ* (1823), suggesting that they provided additional evidence of a deluge that had covered the entire earth, including the Himalaya.[32] Interest in these organic remains was thus magnified by the way they figured in discussions of the upheavement of the mountains. However, the *bijli ki har* (examples of which can be seen in Figure 4.1) had been important to South Asian religion long before they mattered to European science. As a result, they provide an excellent case study for the many roles of Himalayan people in geological practice in the mountains.

[30] See Alexandra van der Geer, Michael Dermitzakis, and John de Vos, 'Fossil Folklore from India: The Siwalik Hills and the Mahâbhârata', *Folklore* 119, no. 1 (2008): 85. The colonial accounts contain multiple spellings and transliterations, including *Bijli ka har*, *Bijlee-ke har* and *Bijli ca har*. For other brief mentions of 'lightning bones', see Grout, 'Geology and India', 135–38; Kenneth Kennedy and Russell Ciochon, 'A Canine Tooth from the Siwaliks: First Recorded Discovery of a Fossil Ape?', *Human Evolution* 14, no. 3 (1999): 233; Kennedy, *God-Apes and Fossil Men*, 38; Adrienne Mayor, *The First Fossil Hunters: Paleontology in Greek and Roman Times* (Princeton: Princeton University Press, 2000), 133; Nair, 'Eyes and No Eyes', 361; van der Geer, Dermitzakis, and de Vos, 'Fossil Folklore from India', 72, 83; Chakrabarti, *Inscriptions of Nature*, 57, 104.

[31] 'Asiatic Society of Calcutta – Physical Class, 8 June 1832', *Asiatic Journal (New Series)* 7 (1832): 118.

[32] William Buckland, *Reliquiae Diluvianae, or Observations on the Organic Remains Contained in Caves, Fissures, and Diluvial Gravel, and on Other Geological Phenomena, Attesting the Action of an Universal Deluge* (London: John Murray, 1823), 221–23.

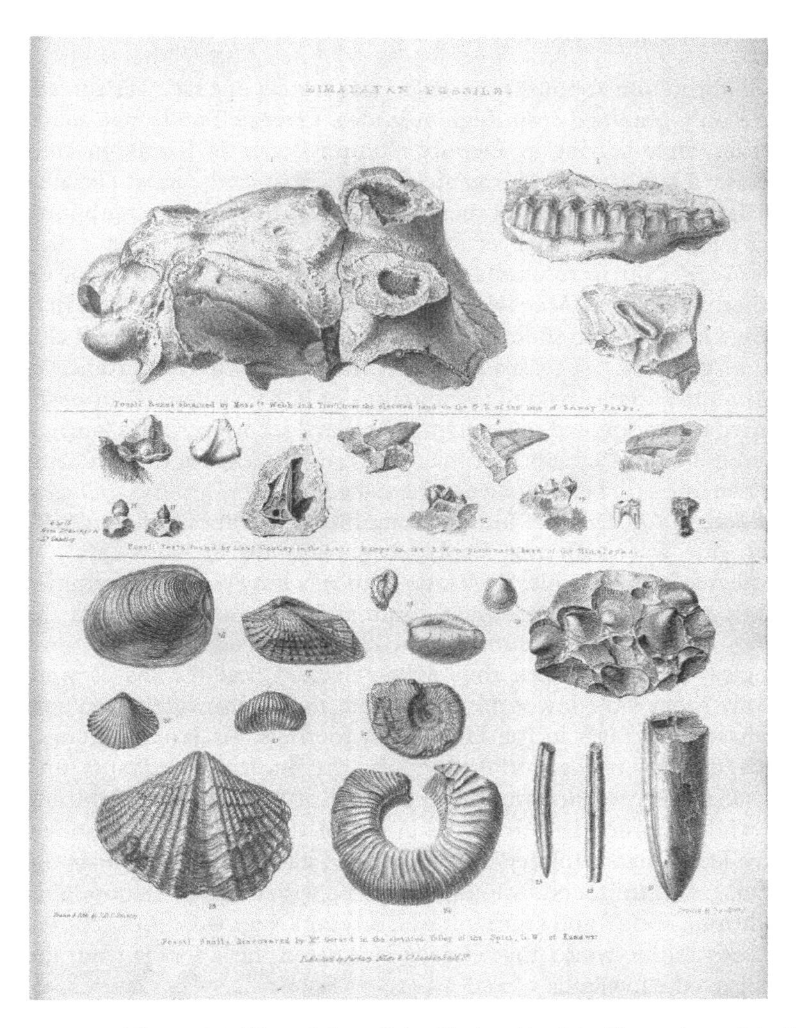

Figure 4.1 Plate 3 from John Forbes Royle's *Illustrations of the Botany and Other Branches of the Natural History of the Himalayan Mountains* (1839).[33] The fossils depicted in the top section are those 'obtained by Messrs Webb and Traill from the elevated land on the N.E. of the line of Snowy Peaks', respectively: '1. Skull of a fossil antelope; 2. Lower jaw of a fossil deer; 3. Fossil tooth of a rhinoceros'. As Royle explained: 'the fossils represented in the upper part of Plate 3 … have also been long known in India by the name of *bijli ke har*, or *Lightning Bones*'.[34]

[33] Royle, *Illustrations*, Vol 2, Plate 3. [34] Royle, *Illustrations*, Vol 1, xxix.

While information about the *bijli ki har* was scarce, one fact was agreed: they were only obtained from high altitudes. George Traill, who made a collection while serving as Deputy Commissioner of Kumaon, confirmed this: 'fossil bones and organic remains exist in the most elevated parts of the Ghats. The former, here called "*Bijlí Hár*" lightning bones, are chiefly found at the crest of the *Niti* pass', while 'the latter, called "*Chakar Patar*" from its resemblance to a wheel, is procured in a ravine on the northern face of the *Mana* pass. In both instances, the elevation may be assumed at seventeen thousand feet above the sea'.[35] Here Traill also refers to a second set of fossils from the mountains with importance in South Asian traditions, namely: ammonites or 'Chakar Patar (Futteer)'. John Batten found some of these on the Niti Pass, recording in his journal: 'ascended the opposite banks [of the Siánkí river], keeping to the north-west for two miles till I came to what was called the fossil ground, (*Chakra patharke makán*)'.[36] Here he found 'ammonites lying about *in hundreds*'. However, he was suffering badly from altitude sickness and as a consequence had to admit: 'I was too ill to stay long picking up ammonites, and, moreover, I can always command a good supply from the Bhotias'.[37] In looking to Himalayan collectors to overcome his own physiological failings, Batten thus acknowledged that the fossils were quite easily procurable lower down. Indeed, they were widely available for purchase at bazaars in the Himalayan foothills. As Hugh Falconer recorded: 'they are brought to Almorah by the Bhoteah merchants, and sold as talismans or charms under the name of "Bijli ki har" lightning bones', while 'ammonites, from the crests of the neighbouring snowy passes, called "Chakar futteer" and venerated all over Hindostan as the sacred Salagram, are generally found mixed up with them'.[38] In obtaining both 'lightning bones' and ammonites, EIC employees were thus tapping into pre-existing networks that already circulated these fossils from the mountains to the lowlands.

These networks existed because the fossils had long-standing uses in Hindu religious practice. As James Herbert reported: 'the first notice of organic remains from the Himmalaya mountains was I believe derived from the fact of the Gunduk river bringing down, with the stones in its

[35] Traill, 'Statistical Report on the Bhotia Mehals of Kemaon', 17.

[36] John Hallett Batten, 'Note of a Visit to the Niti Pass of the Grand Himalayan Chain', *Journal of the Asiatic Society of Bengal* 7, part 1 (1838): 315. For the way that *shaligrams* reveal 'how myths and geohistory became compatible with each other within' a 'naturalistic imagination' of the deep past of India, see also Chakrabarti, *Inscriptions of Nature*, 57–58, 101–9.

[37] Batten, 'Note of a Visit to the Niti Pass', 315.

[38] Hugh Falconer, *Palaeontological Memoirs and Notes of the Late Hugh Falconer*, ed. Charles Murchison (London: Hardwicke, 1868), Vol 1, 173.

bed, specimens of *Ammonites*, the *Saligrami* of the Hindus' but 'as nothing was known at the time of the geology of the mountains, the fact attracted little notice, and indeed was only known perhaps to those who interested themselves in the history and nature of Hindu observances'.[39] Both Falconer and Herbert thus indicate that *shaligrams* were employed in Hindu ritual (where they served as emblems of Vishnu, with their spiral shape thought to resemble his *chakra* wheel). These ammonites were usually gathered along the Gandaki River, especially near the village of Salagrama, from which they took their name. As scholars have noted, 'the ammonites are named after the village, which in turn took its name from the abundant *sala* trees (*Vatica robusta*)'.[40] While both *shaligrams* and *bijli ki har* were used in devotional practice, the 'lightning bones' additionally had a role in medicine. As Herbert continued: 'they were valued, not only as charms, but as medicines; belonging in the latter case to the class of absorbents. As they consist chiefly of carbonate of lime, it appears that they were not unfitted for this office'.[41] Herbert thus indicates that not only did they have a medical application, but they also would be efficacious by the standards of European medicine. John Batten later confirmed this use, seemingly rather shocked at what he saw as the careless destruction of potentially valuable geological specimens, writing that the bones were: 'used as medicine! I am told, in a pounded state'.[42] Batten went on to acknowledge the limitations of these specimens for the advancement of new geological theories, partly because they tended to be in bad shape (even before they were ground up for medicine): 'I have rarely been able to obtain teeth or other *characteristic* specimens'.[43] Hugh Falconer noted similarly that 'judging from the quantities which find their way to Almorah, the fossils are by no means scarce', but 'they are rarely seen entire, consisting generally of fragments'.[44] This mattered, because it made the all-important identification of which animals the bones belonged to difficult. James Herbert later had the opportunity to examine the specimens collected by Traill and passed on to Colebrooke, and gave a detailed analysis of the state they were in: 'they consisted of bones of sizes [sic], including crania or fragments of crania of different animals . . . all these bones were completely mineralised, being converted into carbonate of lime'.[45]

[39] Herbert, 'On the Organic Remains Found in the Himmalaya', 269.
[40] See van der Geer, Dermitzakis, and de Vos, 'Fossil Folklore from India', 72–74.
[41] Herbert, 'On the Organic Remains Found in the Himmalaya', 269.
[42] Manson, 'Capt. Manson's Journal', 1167.
[43] Manson, 'Capt. Manson's Journal', 1167.
[44] Falconer, *Palaeontological Memoirs*, Vol 1, 177.
[45] Herbert, 'On the Organic Remains Found in the Himmalaya', 270. For experiments on the mineral composition of the bones, see also 'Proceedings of the Asiatic Society, Physical Class, 8 February 1832', *Journal of the Asiatic Society of Bengal* 1 (1832): 77.

Along with the bones' poor state of preservation was another and even more serious problem for geology: they had not been observed in situ, and there remained considerable doubts about where they had come from. As James Herbert wrote: 'hitherto, they have been collected only by natives, whose reports—never very precise as to particulars the value of which they do not appreciate—can scarcely be allowed to settle a point of this interest'.[46] He recorded elsewhere that 'from the same people from whom these bones were obtained, great numbers of *Ammonites* and of *Belemnites* were obtained' but 'concerning the locality of these or of the *Belemnites*, I never could get any clear information beyond the fact of their being found North of the range before-mentioned, which, as it is the boundary of the Honorable Company's territory, was likewise that of my investigations'.[47] Here, again, the frontier circumscribed the availability of knowledge, something collectors could never really mitigate, because where fossils were 'brought from beyond our frontier by natives ... neither the distance or the elevation are precisely known'.[48] Indeed, even if these frontiers were considerably more porous for Bhotiyas and Tartars than they were for European surveyors, the problem of verifiable locality data persisted. James Manson, for example, investigated bones he found for sale in the bazaars, and raised some doubt that they were even from the Himalaya: 'the bones which are brought by the Bhoteahs for sale at the fair held at Bageswur, it appears they purchase at Gurtope' and as a result, 'they are not found amongst the Himalaya, which had formerly given an interest to these productions, but which must now, if the above account prove true, cease altogether'.[49] Manson was likely being overcautious, but such doubts remained an issue. As Falconer noted in 1839, 'no competent European observer has as yet seen them in situ', even if – as he went on to acknowledge – the weight of evidence (including the testimony of Bhotiya merchants) suggested that the point should be satisfactorily settled.[50]

However, even at the mid-century, Richard Strachey indicated that the problematic lack of direct observations was still unresolved. He also speculated that in some cases fossils might have been acquired from roadside and mountaintop shrines or cairns (much like the *shughars* which Alexander Gerard thought were Tartar border guards). Strachey noted that these were constructed by 'the superstitious of so many

[46] Herbert, 'Report upon the Mineralogical Survey of the Himalayan Mountains', cxlix.
[47] Herbert, 'On the Organic Remains Found in the Himmalaya', 270.
[48] Herbert, 'Report upon the Mineralogical Survey of the Himalayan Mountains', cxlix.
[49] Manson, 'Capt. Manson's Journal', 1167.
[50] The date of 1839 is that suggested by Falconer's editor, Charles Murchison (although the essay itself was not published until 1868). Falconer, *Palaeontological Memoirs*, Vol 1, 174.

nations, including the Hindus and Tibetans of the Himalayan regions' and 'accumulations of all sorts of oddities are often found in these piles, among which [are] horns and skulls of wild animals, fossil shells, bits of crystal, or eccentric-looking stones' that 'invite the sacrilegious attacks of European travellers, and many of my specimens of ammonites are spoils of this description'.[51] Material specimens revered in one tradition were thus purloined to serve another, though Strachey suggested that the 'Bhotiyas have no scruple in assisting in such proceedings' and 'they generally appear to care but little, unless impelled by considerations of temporal expediency, for the superstitious practices of their Tibetan or Hindu neighbours'.[52] Strachey went on to note that the nature of these fossils, found 'at an elevation of from 14,000 to 16,000 feet above the sea', pointed to some central questions in geology. However, because 'we were altogether ignorant of the precise locality whence they came', no 'conclusions could be formed as to their geological import'. As he continued, 'the Niti Pass, from which it was said that the bones had been brought, was not the place where they were found, but one of the routes only by which they came across the great Himalayan chain from unknown regions beyond'.[53] These doubts about the *bijli ki har* and *shaligrams* thus demonstrate the crucial role of pre-existing Himalayan networks in the acquisition of material specimens, but also a debilitating lack of trust in critical associated information about locality and altitude.[54] This lack of in situ observations nevertheless mattered, because of 'the very important inferences connected with these remains in regard to the elevation of the Himalayahs'.[55] Before examining these debates around upheavement though, it is first necessary to examine a second key set of fossils, and the successive and vexed attempts to extract them from Spiti.

The Spiti Fossils

At the same time that 'lightning bones' were being chased in bazaars, an important and related series of fossils were being acquired by James Gerard, at the 'remarkable' elevation of 16,000 feet (see Figure 4.2),

[51] Strachey, 'Narrative of a Journey', 262–63. [52] Strachey, 'Narrative of a Journey', 263.
[53] Richard Strachey, 'On the Geology of Part of the Himalaya Mountains and Tibet', *Quarterly Journal of the Geological Society of London* 7, no. 1–2 (1851): 306–7.
[54] There continued to be occasional mentions of *Bijli ki har* in the second half of the nineteenth century, including by the so-called 'Pundits' who explored the Himalaya in the 1860s. See Thomas Montgomerie, 'Report of a Route-Survey Made by Pundit, from Nepal to Lhasa, and Thence Through the Upper Valley of the Brahmaputra to Its Source', *Journal of the Royal Geographical Society of London* 38 (1868): 151–52.
[55] Falconer, *Palaeontological Memoirs*, Vol 1, 5.

Figure 4.2 'Matrix of Speetee shell limestone from 16,000 feet elevation'. This specimen was located by James Gerard with the assistance of his Tartar guides, before being laboriously carried down from the mountains and examined at the Asiatic Society in Calcutta. It was subsequently shipped to London, and eventually found its way into the collection of William Buckland. © Oxford University Museum of Natural History.

proving that 'the waters of the ocean had, at some former period, covered these mountains'.[56] For his part, Gerard waived responsibility for identification, citing his limited training: 'if the observation of shells and mountain strata of organic remains at such an altitude be worthy of attention to the geologist, I am happy in having enjoyed the opportunity of verifying the fact' while 'leaving to more experienced hands the recognition of the species and the age of the fossils, the classification of the strata in which they are imbedded, and the theory of their being raised to their present elevation'.[57] Even if Gerard is here leery of philosophising – and given his broader oeuvre, he is perhaps being disingenuous – it is clear that he was aware that these fossils had a bearing on important geological questions. This is further evident in his report on the expedition to Spiti, where he

[56] 'Biographical Sketch of the Late James Gilbert Gerard', *Asiatic Journal (Third Series)* 4 (1844): 68.

[57] Gerard, 'Observations on the Spiti Valley', 277.

wrote that 'the accounts of the *Lamas* confirm the report of calcareous deposits ... wherein shells and various organic remains, with petrified bones, are found' and that 'there is every probability that the whole country lying at the back of the *Himálaya* ... abound with fossil relics, the living prototypes of which have disappeared from the earth'.[58] In this instance, he indicates an acceptance of the still controversial idea of the extinction of species, and a distinct awareness of the place his fossils might have in wider debates.

Moments like this also point to how at every step of the way in acquiring these fossils – from locating to transporting to interpreting – James Gerard relied on the knowledge of Himalayan peoples. His Tartar guides told him of the existence of – and helped him locate – the fossil beds, and his discussions with them led him to speculate that there would be more organic remains found on the Tibetan plateau (even if these remained beyond the frontier and politically inaccessible). Indeed, Tartars seconded by the Qing made it a complex business firstly to get into the region where the fossils were found, and secondly to get specimens out. As Gerard wrote of his experiences at 16,000 feet in Spiti: 'illness, and the languor produced by such an attenuated atmosphere, prevented my taking every advantage of my visit to this interesting region, and my journey was terminated by the limits of the British territory'. As he further lamented, 'just before crossing the boundary of *Ladák* into *Basáhír*, I was gratified by the discovery of a bed of marine fossil shells resembling oysters', but 'the suspicions of the Chinese prevented my bringing away many specimens'.[59] In this instance, as in many others detailed in this book, the material and physiological difficulties of working in these high mountain spaces were thus compounded by the cultural and political.

On one occasion Gerard did manage to obtain a shell from beyond the frontier, traded from a Chinese officer after a night of drinking. Gerard recorded that the unnamed official 'had heard of my searching for fossils and curiosities, and presented me with a petrifaction from Lake Mansarawur; it seems a species of Medusa'.[60] Whatever the potential geological implications of organic remains from the extraordinarily elevated Tibetan plateau, Gerard noted the limited usefulness of obtaining specimens this way, echoing the problems with the *bijli ki har*: 'the very few shells which have thus come to light are chiefly interesting as insulated specimens of the varied resources of the country' because 'being from their unknown situs and position deprived of their value to the geologist,

[58] Gerard, 'Observations on the Spiti Valley', 264–65.
[59] Gerard, 'Observations on the Spiti Valley', 276.
[60] James Gilbert Gerard, 'Letter from a Correspondent in the Himalaya', *Gleanings in Science* 1 (1829): 110.

though still identifying the continuity of character, and pointing out an intimate analogy with the fossil geology of opposite regions of the globe'.[61] Even if problematic, Gerard here also indicates the necessity of comparing the fossils of the high Himalaya globally. Such a comparative outlook is evident in his later evoking of the romantic implications of his discoveries. As he wrote while in Afghanistan with his health irretrievably deteriorating: 'I am quite sick of my situation & would exchange it if I could (at a single leap) alight like the Condor, upon the highest summit of the Himalaya & find myself amongst antediluvian reliquiae, or the fossil bones of those monsters of antiquity, the mammoth & mastodon' though he would 'even be contented with a deposit of oysters or muscles'.[62] In this statement, he thus combines allusions to the sublime with knowledge of key fossil discoveries in other parts of the world, as well as explicitly evoking Humboldt's Andes with the mention of the condor.

James Gerard's report on Spiti and 'a box containing 164 paper parcels of fossils' were later forwarded to the Asiatic Society, where they were discussed at a meeting of the 'Physical Class' in 1832 (for examples of these, see Figure 4.3).[63] EIC Chaplain Robert Everest (younger brother to surveyor George for whom the world's highest mountain would later be named) then had the opportunity of examining, sketching and publishing a brief notice in the *Asiatic Researches* (1833), where several of the Spiti shells were detailed (see Figure 4.3). Everest's sketches were also published in the India-based journal *Gleanings in Science*, where it was noted that 'these organic relics are generally in so mutilated a state that few of the characteristic types are discernible, and the difficulty of naming them is increased by the want of works of reference on fossil conchology'.[64] As was further explained, 'it is hoped by circulating the figures in the GLEANINGS to elicit further opinions on the subject from those who make conchology their peculiar study' and moreover, 'those also who reside among the hills may, by seeing what species the cabinet of the Society possesses, be better able to select fresh varieties, and complete in time this interesting series of Himmalayan fossils'.[65] In this instance, the editors thus specifically argued for circulating information about the fossils back into the mountains, in the hope of filling in gaps in the knowledge.

[61] Gerard, 'Observations on the Spiti Valley', 265.
[62] Gerard to [Anon], 26 March 1833, British Library, Mss. Eur. C951.
[63] 'Proceedings of the Asiatic Society, Physical Class, Wednesday, 15th August, 1832', *Journal of the Asiatic Society of Bengal* 1 (1832): 363.
[64] 'Note by the Secretary' in Herbert, 'On the Organic Remains Found in the Himmalaya', 271.
[65] 'Note by the Secretary' in Herbert, 271.

Figure 4.3 'Himalayan Fossil Shells'. This image showcases a selection of Gerard's Spiti fossils, as depicted by Robert Everest, including several ammonites and belemnites.[66] At the time ammonites, seemingly of marine origin, were highly sought after and discussed in geological circles because they appeared to have no living analogue in the present, clearly suggesting the existence of a former and now extinct world.[67] (Some of the Spiti fossils can also be seen in Royle's collection depicted in Figure 4.1.[68])

[66] Robert Everest, 'Memorandum on the Fossil Shells Discovered in the Himalayan Mountains', *Asiatic Researches* 18, no. 2 (1833): 107–14. See also Grout, 'Geology and India', 127–28, 135–36; Grout, 'Possessing the Earth', 251.

[67] Rudwick, *Worlds Before Adam*, 49; Laudan, *From Mineralogy to Geology*, 83–84, 170–72.

[68] See Royle, *Illustrations*, Vol 1, xxix.

The travails involved in identifying these fossils also points to the ongoing limits of resources. In an 1835 report, the EIC doctor and naturalist John McClelland (1805–1875) outlined the advantages and disadvantages of pursuing geology in India in general, and the mountains in particular. In presenting 'the result of a temporary residence in Kumaon, while on the regular tour of duty with my regiment in that province', he highlighted the unevenness with which journals and new findings circulated: 'India, however, is not a country in which new publications are advertised in every village; and such are the disadvantages of private individuals in remote districts that they often remain ignorant of the existence of the most important works, until some peculiar circumstance or accident presents them to notice'. He went on to explain: 'I mention this, as an apology for not having availed myself of some interesting papers that have recently appeared in India on Geological subjects, and which I had no opportunity of seeing, till my return to Calcutta'.[69] In a period when geological theories were rapidly appearing and disappearing in savant circles in Europe, the availability of up-to-date information mattered. Moments like this are a reminder of why we need to sometimes treat scientific practice in places like Kumaon differently to 'centres in the periphery' like Calcutta. Indeed, while advocating for a new Geological Society in Calcutta (and sub-branches in other parts of the country) to complement the more general work of the Asiatic Society, McClelland argued: 'it is painful to reflect on the number of years the immense empire of Hindustan has been in our possession; and that to this day we should remain as ignorant of its physical structure as we are of that of China, or the interior of Africa'.[70] In chiding the Company (and probably eyeing with some jealously the more expansive approach to scientific funding in the rival French Empire), he remarked that advances were chiefly provided by interested amateurs combining duties, and 'in the few instances in which British governments have patronized the travels of scientific men, the motives have been rather the extension of commerce than the promotion of science'.[71]

Returning to the specific case of the Spiti fossils, James Gerard's physiological and political difficulties mandated the need for return visits. The first of these was carried out not by another EIC surgeon or surveyor, but rather by the French naturalist Victor Jacquemont, who visited to examine the fossils beds in 1830. While much geological practice in the Himalaya in this period continued to be itinerant and opportunistic, carried out by army officers and surgeons like James Gerard, those with

[69] John McClelland, *Some Inquiries in the Province of Kemaon in India, Relative to Geology* (Calcutta: Baptist Mission Press, 1835), vi.

[70] McClelland, 2. See also Grout, 'Geology and India', 210–13.

[71] McClelland, *Some Inquiries*, 3.

better training (and renumerated explicitly for their scientific efforts) did occasionally get opportunities to enter the high Himalaya, and here Jacquemont is a key example. Indeed, he visited under the auspices of the French *Jardin des plantes* between 1829 and 1832, and made significant collections, both botanical and geological. However, he quickly found that his researches too were circumscribed by the frontier, even if while in Tartary he sometimes had 'good luck' in finding 'Chinese vigilance at fault'.[72] For his part, Jacquemont was sceptical about the quality of the geological knowledge being produced by the EIC employees who preceded him. As he wrote: 'these wilds have been travelled over by a good many English, and I have reason to believe that their Flora is sufficiently well known' but in the case of the materials of the earth 'they had all learned geology from books and in India, and I have no faith in their decisions'.[73]

As well as James Gerard, Jacquemont was likely thinking of James Herbert, who was appointed by the EIC government in Calcutta to the 'Geological Survey of the Himalaya' in 1823.[74] While geology remained largely itinerant in the first half of the nineteenth century, Herbert's survey represented an exception, and the first significant attempt at a systematic accounting. This survey nevertheless pushed Herbert to – and beyond – his capacities as a geological observer. Even as his efforts 'rendered a most essential service to the cause of geological science, in giving to the world a connected Geological Map of this part of our great mountain barrier', his observations had critical limitations.[75] Indeed,

[72] Jacquemont, *Letters from India*, Vol 1, xxvi.

[73] Jacquemont, *Letters from India*, Vol 1, 127.

[74] See 'Papers regarding the Bengal Survey Department, 1823', British Library, IOR/F/4/750/20517. The survey itself was abolished for financial reasons in 1829, reflective of the unfulfilled and perhaps exaggerated anticipation of the mineral wealth of the mountains. Frederick Dangerfield and Alexander Laidlaw had been appointed in similar roles previously, but for reasons of competence and health failed to produce results. See Grout, 'Geology and India', 172–77. When Herbert's findings eventually appeared posthumously, as the 'Report of the Mineralogical Survey of the Himmalaya Mountains lying between the Rivers Sutlej and Kalee' (1842), they were distributed gratis as a special issue of the *Journal of the Asiatic Society of Bengal* (a detailed geological map was also published in 1844). See Herbert, 'Report upon the Mineralogical Survey of the Himalayan Mountains'; 'Geological Map of Captain Herbert's Himalaya Survey', *Journal of the Asiatic Society of Bengal* 13, part 1 (1844): 170. A short section was published in 1833 as James Herbert, 'On the Mineral Productions of That Part of the Himalaya Mountains, Lying between the Satlaj and the Kali, (Gagra) Rivers', *Asiatic Researches* 18 (1833): 227–58. Several manuscript versions of this report also exist in the British library: Add. MS. 14381; Mss. Eur. E96; and IOR/F/4/957/27123(24). In this chapter, I quote from the prefatory matter in IOR/F/4/957, and from the published version for the body of the report, specifying in each case.

[75] 'Report of the Curator, Museum of Economic Geology and Geological and Mineralogical Departments, March 1844', *Journal of the Asiatic Society of Bengal* 13, part 1 (1844): xxvi.

Hugh Falconer noted after Herbert's death that he had 'investigated with great zeal', but 'unfortunately, Captain Herbert was a self-taught and book geologist, and he was called upon to describe the geology of an unknown field – a subject new to him, at the very time when he was acquiring his first knowledge of geological science' with the consequence being 'that his labours have been less valuable than they otherwise would have been from his talents and general scientific acquirements with longer study'.[76] While in one sense a rhetorical strategy – pointing out the limitations of a predecessor to elevate one's own contribution to science – such descriptions nevertheless reflected the challenges of seconding surveyors to scientific projects.

EIC secretary Henry Prinsep acknowledged the limitations of these early efforts, while not dismissing their value: 'the scientific geologist and naturalist will perhaps at first regret that they do not find more details falling in with their studies', however, 'we must beg of them to reflect that the writers and editors of such papers, though they may lay no claim to scientific qualifications (so difficult to acquire in India), are nevertheless rendering a service of first rate importance to the cause of science; and this is the important service of *pioneering*'. Indeed, he goes on to note that 'few remember, and many keep out of sight, what they owe to the humble and often forgotten labours' and 'what will not some future Humboldt, with guides like these ready to mark out his path, be able to accomplish amongst the yet hidden wonders of the stupendous mountains of India?'[77] For his part, Herbert was far from unaware of the limitations of his geological knowledge and his status as not a 'Humboldt'. As he wrote in the introduction to his report: 'geology, as a science, has not yet attracted in India that attention which its importance merits, and it would be futile in me to deny that, till selected for this duty, I had but a slender acquaintance with the subject'. As he explained, however, 'while exploring the local phenomena of this tract I have been in reality studying the principles of the science; an advantage in so far as I may hope to have escaped the trammels of system'.[78] Herbert thus tries to spin his ignorance as a positive, claiming that his lack of prior knowledge would prevent him from prejudice towards particular interpretations, and thereby improving him as an impartial observer. Of course, the ongoing role of eclectic amateurism and debates over disciplinary respectability were also playing out in Britain, France and other imperial domains across the globe at this time.[79] Indeed, these dilemmas were far from

[76] Falconer, *Palaeontological Memoirs*, Vol 1, 10.
[77] Editorial note in Manson, 'Capt. Manson's Journal', 1158.
[78] Herbert, 'Report upon the Mineralogical Survey of the Himalayan Mountains', ii.
[79] See especially Endersby, *Imperial Nature*.

unique to the Himalaya (or high mountains), even as naturalists sought to emphasise and exploit their opportunities to try and claim authority from locality and environmental idiosyncrasy.

Returning to Jacquemont, we see that in remaining ostentatiously unconvinced by his predecessors, he sought to frame himself as a different category of scientific investigator. Enjoying the luxury of travelling solely for scientific purposes, thanks to the largesse of the French government (albeit in his opinion insufficiently generously), he saw himself more in the mould of a travelling naturalist *à la* the archetypal Humboldt, and pressed his credentials as such. This form of state sponsorship presents a notable contrast with the sometimes fickle and ad hoc funding of science by the EIC – though also, it must be said, to the activities of the gentlemanly and independently wealthy Humboldt – and echoes wider contrasts with French institutional funding of science in this period (with perhaps the most well-known example being Napoleon's underwriting of the monumental *Description de l'Égypte*).[80] However, these advantages did not make Jacquemont immune to the challenges with the availability of up-to-date information in remote locations, and as he continued, 'my professional friends are urgent that I should send them from time to time a scientific paper which they might publish as a certificate of my existence' but 'if I wished to write some . . . pages which I should not regret at any future time having written, I immediately feel the want of books, which are not at hand' and decided he would 'rather pass for dead than for dying, which might be concluded from feeble and neglected works. I cannot flatter myself that I shall bring home from my journey materials enough to *live upon* India for a score and a half of years, as M. de Humboldt has done on his concerning America'.[81] In this last, he was tragically correct, and his contribution to science – and any aspirations to being the Humboldt of the Himalaya – was blunted by the state of his papers and collections at the time of his death in Bombay at the age of thirty-one (as a result of complications arising from Cholera). Like Herbert, Jacquemont's premature death both limited and delayed the impact of his collections. This was far from uncommon, and the majority of James Gerard's discoveries remained similarly untouched for decades (and were not compared to Jacquemont's and Strachey's collections until 1863). By this time, they were in poor condition – mixed up and lacking labels having been stashed

[80] See Maya Jasanoff, *Edge of Empire: Conquest and Collecting in the East, 1750–1850* (London: Harper Perennial, 2006).
[81] Jacquemont, *Letters from India*, Vol 1, 311.

indiscriminately in a cupboard – a rather ironic fate for a collection so laboriously extracted from the highest mountains in the world.[82]

This perhaps explains why Bengal Infantry officer and geologist Thomas Hutton (1807 – 1874) remarked in 1837 that, even including Jacquemont, little follow-up work had been done on Gerard's discoveries in Spiti, and positioned himself to be the one to do it. He argued that the Siwaliks had drawn the attention of metropolitan savants to the Himalaya more generally, and 'at a time when the attention of the Scientific bodies of Europe is turned to the valuable discoveries of our fossilists in the Sub-Himálayan ranges', this might be used to advantage for promoting 'farther and more complete research' on Spiti.[83] Hutton's reasoning was also couched in language playing up the fear of being scooped by foreign travellers, with particular reference to Jacquemont: 'subsequent to Dr. GERARD's discovery—and wholly dependent on that gentleman for his information—M. JACQUEMONT I believe visited the valley of the *Spiti*'. As he continued, 'shall we, however, allow the riches of our dominions to be brought to light and reaped by Foreign Societies? They send out travellers to glean in the cause of science, through every clime, while we alone, the richest nation of them all, sit idly by and watch their progress'.[84] This was a relatively rare instance of explicitly national chauvinism around scientific practice in the Himalaya in this period, and indeed Hutton's rhetoric perhaps reflected personal vindictiveness or outrage over the Frenchman's indelicate and widely circulated remarks about the EIC's handling of their Indian Empire.[85] In the account of an earlier journey to the Burendo Pass, Hutton also went quite far out of his way to cast aspersions on Jacquemont's intentions, suggesting that to 'some valuable discovery, made near the *Gangtung Pass* on the road from *Dabling* to *Bekhur* on the confines of Chinese Tartary, the hints dropped on his return' presumably referred, otherwise 'why else should

[82] See Henry Francis Blanford, 'On Dr. Gerard's Collection of Fossils from the Spiti Valley, in the Asiatic Society's Museum', *Journal of the Asiatic Society of Bengal* 32 (1863): 124–38. See also John William Salter and Henry Francis Blanford, *Palaeontology of Niti in the Northern Himalaya: Being Descriptions and Figures of the Palaeozoic and Secondary Fossils Collected by Colonel Richard Strachey* (Calcutta: O. T. Cutter, 1865), 110. See also Baud, Forêt, and Gorshenina, *La Haute-Asie telle qu'ils l'ont vue*, 36. On the similar fate of William Griffith's collections from Afghanistan, see also Lachlan Fleetwood, 'Science and War at the Limit of Empire: William Griffith with the Army of the Indus', *Notes and Records: The Royal Society Journal of the History of Science* 75, no. 3 (2021): 285–310.

[83] Thomas Hutton, 'Read the Following Letter from Lieut. Thomas Hutton, 37th N.I. Dated Simla, 27th August, 1837', *Journal of the Asiatic Society of Bengal* 6, part 2 (1837): 897–98.

[84] Hutton, 'Read the Following Letter from Lieut. Thomas Hutton', 898.

[85] Andrew Grout argues that antipathy towards foreigners was more widespread. See Grout, 'Geology and India', 125–31.

he have evinced so much anxiety to prevent any European from visiting that quarter, until he should be able to make known his discovery to the French government and return under their auspices to avail himself of it?'[86] Hutton went on to speculate that gold being 'the discovery hinted at is neither impossible nor improbable. It is certain that none but the precious metals would have been worth the notice of the French government', though it is perhaps telling that there is little indication that the Chinese thought similarly about the supposed potential for geological riches.[87]

Whether or not the Asiatic Society took Hutton's rhetoric seriously, his request to revisit Spiti to follow up on the discoveries of Gerard and Jacquemont was ultimately successful, and they voted 1,000 rupees towards his efforts, 'on the conditions suggested by himself' that the fossils he collected 'be deposited in the Society's Museum'.[88] Hutton set out in 1838, but quickly had political troubles of his own to contend with: 'I experienced the greatest difficulty in reaching the fossil ground owing to the want of supplies and the unwillingness of the *Kiladar* at *Dunkur* to allow me to proceed' as 'he had received instructions from *Ludak* to oppose my advance'.[89] Hutton made violent threats and did eventually cajole his way to the fossil beds, but only managed to acquire a small and disappointing collection, 'certainly not worth one quarter of the trouble they have occasioned'.[90] Hutton thus found his agency limited in the 'cultural borderlands' of the high mountains, struggling to negotiate the long-standing networks of labour and expertise essential to operating in the mountains.

Pre-existing Networks, or the Many Roles of Pati Ram

Thomas Hutton, like James Gerard and Victor Jacquemont before him, relied heavily on Himalayan informants to locate his fossils. In fact, he

[86] Thomas Hutton, 'Journal of a Trip to Burenda Pass in 1836' *Journal of the Asiatic Society of Bengal* 6, part 2 (1837): 924.
[87] Hutton, 'Journal of a Trip to Burenda Pass in 1836', 925.
[88] Hutton, 'Read the Following Letter from Lieut. Thomas Hutton', 898.
[89] Thomas Hutton, 'Extract of a Letter from Thomas Hutton, Soongum, 5 July 1838', *Journal of the Asiatic Society of Bengal* 7, part 2 (1838): 668. See also Thomas Hutton, 'Journal of a Trip through Kunawur, Hungrung, and Spiti, Undertaken in the Year 1838 (Part I)', *Journal of the Asiatic Society of Bengal* 8 (1839): 901–50; Thomas Hutton, 'Journal of a Trip through Kunawur, Hungrung, and Spiti, Undertaken in the Year 1838 (Part II)', *Journal of the Asiatic Society of Bengal* 9, part 1 (1840): 489–513; Thomas Hutton, 'Journal of a Trip through Kunawur, Hungrung, and Spiti, Undertaken in the Year 1838 (Part III)', *Journal of the Asiatic Society of Bengal* 9, part 1 (1840): 555–81. See also Grout, 'Possessing the Earth', 250.
[90] Hutton, 'Extract of a Letter from Thomas Hutton, Soongum, 5 July 1838', 668.

even worked with the very same man as his predecessors: a Bhotiya trader from Sungnam known as 'Puttee Ram', whose services had been indispensable to both Gerard and Jacquemont during their time in Spiti.[91] (As well as to Hutton, Jacquemont and James Gerard, Pati Ram also provided support to James Baillie Fraser in 1815, and Alexander Gerard in 1818 and again in 1821.[92]) Pati Ram was himself very well-travelled, not only 'through the upper hills' but also further afield to 'Kurnal, Delhi, Hansi and Hardwar'. His business interests extended even further still, not only to Delhi, but also to Lahore and Kashmir, and he was thus used to negotiating and advising across a range of contexts.[93] The way Pati Ram cuts across and connects these fossil-hunting expeditions, and his multiple roles as an informant, broker and guide are thus worth reflecting on at some length. Because Pati Ram had previously worked with multiple travellers, Thomas Hutton knew of his value well before he set out for the mountains (through both personal recommendations and references in published travelogues). Having arrived in Sungnam, a village which represented a key staging post for forays into the high mountains, Hutton recorded that 'shortly afterwards the vuzeer himself paid me a visit, and proved to be no less a person than the frank and honest Puttee Ram' the same 'friend of Dr. Gerard, and the source from whence he derived much of his information regarding the higher portions of the hills towards Ladak and Chinese Tartary'.[94] This meeting occurred in 1838, and Hutton continued to record that Pati Ram was 'now grey and bent with age' but 'entered at once into a history of his acquaintance with Dr. Gerard and Mr. Fraser, and talked with pride over the dangers he had encountered with the former in their rambles through Spiti and its neighbourhood'.

[91] 'Puttee Ram' was a title, and he was also referred to as '*the* Puttee Ram'. According to Alexander Gerard, he was 'better known by Lahoureepung, the name of his house'. Hutton spells Puttee with a double 't', Gerard with only one. Gerard, *Account of Koonawur*, 77. Jacquemont refers to him as 'Pattiranme'. See Jacquemont, *Voyage dans l'Inde*, Vol 2, 265. For brief mentions, see also Jahoda, *Socio-Economic Organisation in a Border Area of Tibetan Culture*, 93, 101, 108.

[92] Fraser used the information he gleaned to produce a 'Route given by Puttee Ram from Serān, in Bischur, to Gara, along the Course of the Sutlej'. See Fraser, *Journal of a Tour*, 289–90, 299, 301–11. For Alexander Gerard's interactions, see Lloyd and Gerard, *Narrative of a Journey*, Vol 2, 214, 216, 219, 225, 227–30, 233, 246; Gerard, *Account of Koonawur*, 77, 109, 111–14, 120, 137, 140, 149–50.

[93] Hutton, 'Journal of a Trip through Kunawur (Part I)', 935; Jacquemont, *Voyage dans l'Inde*, Vol 2, 34.

[94] Hutton, 'Journal of a Trip through Kunawur (Part I)', 935. For other references see Hutton, 'Journal of a Trip through Kunawur (Part I)', 906, 935–38; Hutton, 'Journal of a Trip through Kunawur (Part II)', 506, 509–10; Hutton, 'Journal of a Trip through Kunawur (Part III)', 570–71; Thomas Hutton, 'Geological Report on the Valley of the Spiti', *Journal of the Asiatic Society of Bengal* 10, part 1 (1841): 215.

Pati Ram then went on to ask 'if I had ever heard his name before, and the old man's eyes actually sparkled with delight, when pointing to an account of one of Gerard's trips, I told him his name was printed there'.[95] While this presents an arresting image of accounts circulating back into the mountains, such glimpses of Pati Ram through Hutton's telling are undoubtedly highly romanticised. That his services were indispensable is unequivocal; what he really thought of his involvement in these Himalayan expeditions remains elusive. Pati Ram nevertheless emerges as one of the most visible brokers in the colonial archives of early Himalayan exploration, by virtue of both his relatively high status and because he assisted at least five separate explorers over multiple expeditions. Reconstructing Pati Ram's multiple roles in expeditionary practice thus helps us think about Himalayan dependency differently, and in terms that invert traditional accounts of exploration, by placing him at the centre of a multi-generational network of travellers.

Recounting his own trip to the fossil beds of Spiti, Thomas Hutton described in detail the services that Pati Ram provided to the expedition, recording that 'from him I obtained a man who understood the Tartar language, to accompany me through Spiti, and he assured me I should experience no difficulties, as there was now a road across some parts of the mountains where, as in the days when Gerard first visited those parts, there was none at all'.[96] As well as providing access to interpreters and guides, Pati Ram thus also offered information about changes in travel conditions, helpfully showing that James Gerard's travelogue – a key source of information for Hutton – was now outdated. Pati Ram also warned Hutton he was likely to be disappointed when it came to the collection of specimens: 'he said that Spiti produced but few; chiefly ammonites (Salick ram) which were found near Dunkur, but that the best place to procure them was on the Gungtang pass' though 'the Chinese were so jealous of strangers looking at their country that if I went there I should not be allowed to bring anything away' and 'besides this, the pass was at the present season impassable, and from the lateness and quantity of the snow which had fallen, it could not be open before the middle of August'.[97] In this instance, we see how Pati Ram provided information about the quality of the fossil hunting, as well as seasonal information about the state of the route, and an update on the politics of the frontier (earlier, as well as answering 'some questions relative to the physical geography of the interior', he had supplied letters of introduction

[95] Hutton, 'Journal of a Trip through Kunawur (Part I)', 935.
[96] Hutton, 'Journal of a Trip through Kunawur (Part I)', 935.
[97] Hutton, 'Journal of a Trip through Kunawur (Part I)', 935–36.

to Alexander Gerard and made some 'intercessions for [Gerard's] friendly reception at the frontier').[98] In locating and recovering the Spiti fossils, Gerard, Jacquemont and Hutton thus relied on Pati Ram not only for his knowledge of Himalayan geography, but also for the networks he could help mobilise on their behalf.

While unavoidable, this reliance on local networks nevertheless often resulted in tensions. In particular, difficulties recurred around transporting heavy geological specimens out of the mountains. As Alexander Gerard complained, echoing concerns with transporting surveying instruments: 'the carriage of the minerals upon men's backs (most of them for 650 miles through the hills) also involved a great deal of expense, and limited my travels although my inclination for exploring the Himalaya was greatly encreased [sic] much still remains to be done in the vast & interesting field of almost untrodden ground'.[99] These problems could go far beyond expense, however, and as Gerard recorded in his fieldbook while travelling in Kinnaur: 'the specimens picked up are numbered 86, but a great many were lost by the people throwing them away on their return to Camp'.[100] On another occasion, he similarly had to admit that 'the minerals picked up on the way from Shulpeea were as far as I remember similar to those numbered 1&2, & there was some gneiss but the whole of them were thrown away by the guides'.[101] A bit of context explains, perhaps, why the guides were disinclined to haul Gerard's lumps of rock out of the mountains: 'travelling was rendered laborious from our sinking 1½ or 2 feet, the fissures were just beginning to appear [in the snow]'.[102] Indeed, Gerard continued: 'we thought it prudent to order a speedy retreat, especially as the guides were greatly alarmed & strongly remonstrated against our proceeding farther as the snow would sink the whole way & we would certainly fall into some deep chasm'.[103] Given the very real possibility of vanishing into a crevasse, the heavy and seemingly arbitrary rock samples may not have seemed especially important to the guides. As in the previous chapter, such moments could thus be read as acts of resistance to sometimes deeply unpleasant labour conditions. Returning to Thomas Hutton, we find that he, too,

[98] Lloyd and Gerard, *Narrative of a Journey*, Vol 2, 214, 216.

[99] Alexander Gerard, 'Memoir of the Construction of a Map of Koonawur', 1826, British Library, Mss. Eur. D137, f205-6.

[100] Alexander Gerard, 'Remarks Regarding the Geological Specimens, Collected in 1821 by J.G. Gerard and A. Gerard', British Library, Mss. Eur. D137, f42. In the manuscript this remark is annotated in pencil with brackets and an 'X', marking it for deletion.

[101] Gerard, 'Remarks Regarding the Geological Specimens', f48. This remark is also annotated in pencil with brackets and an 'X'.

[102] Gerard, 'Remarks Regarding the Geological Specimens', f42-3.

[103] Gerard, 'Remarks Regarding the Geological Specimens', f43.

had problems exerting his supposed authority as expedition leader, complaining about the disappearance of a set of horns he had obtained, and noting that 'this theft however, was the least of the evil, for the rascally Tartar, thinking his load too heavy, had thrown away a number of valuable rock specimens also'.[104] In moments like this, the overlap of social and environmental factors compounded the difficulties of transporting specimens out of the mountains.

Such episodes also raise the question, however difficult to answer, of what the Bhotiyas and Tartars thought of the laborious process of collecting rock specimens and hauling them down to the lowlands. Edward Madden explicitly speculated on what his guides might have thought he was doing, stating that they were 'very curious to know what the "Sahib-log" did with the sacks and boxes of stones which they carry down to the plains with them!' and claimed they thought 'they must surely contain gold, silver, precious jewels, or very probably the Philosopher's stone, in the reality of which they implicitly believe, may be amongst them!'[105] Here though, his report conveys little of or nothing of his guides' actual perspectives, and is instead couched in a superstitious anecdote which Madden uses to assert his epistemological superiority. Other moments suggest more complex interactions, however, such as the fate of a series of specimens Thomas Thomson gathered while deputed to the ultimately farcical Tibetan Boundary Commission in the late 1840s. This commission had been intended to clear up lingering uncertainty as to where part of the frontier – between Gulab Singh's Kashmir, recently established as a vassal state, and Tibet – actually was. However, it ended up being a unilateral endeavour, with the Chinese ignoring the suggestion to send commissioners to fix the border. Administrators in Calcutta were therefore forced to concede that the status quo was perhaps adequate, and 'that there is nothing which requires adjustment and that we may safely leave matters as they are under their present indefinite form', which was somewhat wryly noted, 'was also the opinion of the Chinese, before the mission started'.[106] The fossils Thomson collected while waiting for the non-existent Chinese commissioners were thus a consolation prize. Most reached Calcutta safely in his possession but some were 'despatched by what was considered an exceedingly safe opportunity; from Hanle to

[104] Hutton, 'Journal of a Trip through Kunawur (Part III)', 557–58.

[105] Edward Madden, 'Notes of an Excursion to the Pindree Glacier, in September 1846', *Journal of the Asiatic Society of Bengal* 16, part 1 (1847): 255.

[106] Lord Dalhousie to the Honorable Court of Directors, 31 July 1851, British Library, IOR/F/4/2461/136806, f3. See also Woodman, *Himalayan Frontiers*; Parshotam Mehra, *An 'Agreed' Frontier: Ladakh and India's Northernmost Borders, 1846–1947* (Delhi: Oxford University Press, 1992); Gardner, 'Moving Watersheds'.

Simla' though they 'never reached the latter place they are supposed to have been plundered, on account of the ammonites (considered sacred by Hindoos) which they contained'.[107] Here Thomson blamed the loss not on apathy, but rather on deliberate theft (given Richard Strachey's plundering of shrines at around the same time, this comes across as rather ironic). Such episodes reveal explorers' sometimes fragile authority over their expeditions and collections. Moreover, they further highlight the issues that could arise in 'cultural borderlands' over fossils which amounted to both scientific specimens and religious objects. Indeed, these remains formed the material foundations of overlapping – though as the next section demonstrates, not always irreconcilable – cosmologies for explaining the origins of the mountains, and this meant tensions were inevitable.

Organic Remains and the Upheavement of the Mountains

Debates around the original locations of fossils, as well as illustrating the laboriousness of scientific practice, provide a window into contemporary discussions about mountain formation and how the Himalaya fit into this unfolding global picture. Specimens of marine fossils like those from Spiti were important, while the *bijli ki har* were of arguably even greater interest, being of large animals. As Hugh Falconer explained, there must have been 'a great upheavement of the Himalayahs, extending to many thousand feet, and equal to the elevation of a tract which formerly bore a tropical fauna, up to a height which now causes a climate of nearly arctic severity' and indeed 'remains of rhinoceros, antelope, hyena, horse, large ruminants, &c., [are] found at 16,000 feet above the sea'.[108] As Falconer continued, the discovery of these remains at such altitudes was notable given the lack of vegetation and 'polar' climate, and 'involves important considerations regarding the physical changes which must have taken place in this part of the Himalayahs since the Rhinoceros remains were entombed in the stratum where they are now met with' (see Figure 4.1).[109] In other words, these animals could not survive in the present climate of the elevations their remains were found at, and this required explanation. The *bijli ki har* thus had to be accounted for in order to make the Himalaya into a coherent region that could be compared globally, but their place in the overall picture was perplexing.

[107] Thomas Thomson, 'Notes on the Geological Structure of Western Tibet', British Library, IOR/F/4/2461/136806, f7.
[108] Falconer, *Palaeontological Memoirs*, Vol 1, 28.
[109] Falconer, *Palaeontological Memoirs*, Vol 1, 173.

Underlying these debates were cosmologically charged questions over whether high-altitude fossils were explained by a deluge – whether merely 'geological' or literally 'biblical' – which had lifted the fossils to their present location, or whether these sites were in fact upraised former ocean beds. Either would explain the deposit of large mammals or marine remains around the world, even in the highest mountains like the Himalaya. Around the time of Herbert's and Gerard's fossil hunting, a diluvial explanation was widely considered likely, even if the connection of this with a very recent biblical flood – most prominently by William Buckland in *Reliquiæ Diluvianæ* (1823) – was no longer considered plausible by most (and indeed, it has been convincingly shown that the supposed schism in this period between 'geology and Genesis' is often exaggerated, and that many geologists managed to practise their science without undue conflict with either their peers or their own worldviews).[110] Others, including notably Georges Cuvier, took a more encompassing approach, pointing to multicultural stories of floods as evidence of a global event. This was echoed by Hugh Falconer, who pointed to Hindu traditions of a deluge, such as the Noah-figure Manu recorded in the *Puranas*, while Joseph Hooker noted that his Lepcha guides from Sikkim also had 'a traditionary deluge'.[111] Indeed, whether through contemplation of material evidence or multicultural cosmologies, most thought by this time that any diluvial event – or 'geological' deluge – must have occurred much earlier than that allowed in purely Biblical terms.[112] In the Himalaya, both European geology and South Asian cosmology could thus sometimes point to the same conclusion.

The Himalaya nevertheless offered a frequently bemusing comparative context, and those operating there seized on opportunities to highlight contradictory claims by savants in Europe. This is evident, for example, in James Gerard's assessment of William Buckland's relation between his famous Kirkdale Cave and fossils like the *bijli ki har*, as evidence for a recent, Biblical and universal deluge. As Buckland had put it in *Reliquiæ Diluvianæ* (1823), 'the occurrence of these bones at such an enormous elevation in the regions of eternal snow, and consequently in

[110] Rudwick, *Bursting the Limits of Time*, 104–5, 115–19, 130–31, 600–2, 620–22, 637–38; Rudwick, *Worlds Before Adam*, 5, 14, 59–61, 80–87, 190–98, 206–7, 346, 424–27, 435–36. See also Laudan, *From Mineralogy to Geology*, 38.

[111] Joseph Hooker to Francis Palgrave, 17 March 1849, Kew Archives, JDH/1/10, f139-140. For Falconer's wider theories of Himalayan formation and the ongoing tension between 'sacred' and 'secular' geography in this period, see also Chakrabarti, *Inscriptions of Nature*, 58–66.

[112] Chakrabarti and Sen, 'The World Rests on the Back of a Tortoise', 10–15. Indeed, the short timescale of the earth implied by a literal reading of Genesis had by this time been, for the most part quietly, put aside. See Rudwick, *Bursting the Limits of Time*, 130.

a spot now unfrequented by such animals as the horse and deer, can, I think' ultimately, 'be explained only by supposing them to be of antediluvian origin, and that the carcases of the animals were drifted to their present place, and lodged in sand, by the diluvial waters'.[113] However, using his Himalayan vantage point, James Gerard was more than happy to refute one of the most eminent savants in Europe:

> Dr. Buckland, theorising from the system of European physics, drawing inferences from the phenomena of the Andes, and conclusions from the empirical formula ... actually appealed to the lofty Himalaya for verification of the agent of those fossil remains which he found in the debris of the Kirkdale caves, because some petrified bones were alleged to have come from the back of Kylas (this word means the skies or heaven), at an estimated altitude of 16,000 feet, consequently, says Dr. Buckland, from the regions of eternal congelation, unvisited and unaccessible by man or animals, therefore the deposits of the flood. Had the Professor known nothing, or known more, he would have arrived at a more rational finale.[114]

Among other errors, Gerard suggests that it was global comparisons with the Andes that had led the metropolitan savant astray, and does little to pull his punches. Indeed, Gerard continued in a similarly sarcastic vein that 'nobody, except very clever people, doubts the deluvian event; but if such futile and preposterous means are used to verify the fact, our credence may indeed be staggered. Dr. Buckland should keep to his caves and the mud, for the Himalaya are beyond the pale of his object'.[115]

By the 1830s, most Himalayan travellers and metropolitan commentators (and, it must be said, Buckland himself by 1836) had moved well beyond any correlation with the Biblical flood.[116] However, some 'Mosaic' or 'scriptural' writers in India, as in Europe, continued to associate the Himalayan fossils with Genesis. Notable is Thomas Hutton, whose important collections of material fossils often had to be detached from controversial theorising. He argued, for example, that the mountains had been upheaved violently to their present height and that 'the fall of man is the true period to which the loss of the fossil marine *Mollusca* of the Spiti and Subathoo fields is to be referred' when 'the increasing depravity of the human race once more called down the vengeance of an offended God, and brought about the second and last grand

[113] Buckland, *Reliquiae Diluvianae*, 223.
[114] [James Gilbert Gerard], 'The Himalaya Country', *Asiatic Journal (New Series)* 6, no. 2 (1831): 111. For the source of the original confusion, see Webb, 'Extract of a Letter from Captain William Spencer Webb, 29th March, 1819', 68.
[115] [James Gilbert Gerard], 'The Himalaya Country', 111.
[116] Buckland acknowledged the weight of evidence and repudiated any connection in his *Bridgewater Treatise* of 1836. See Rudwick, *Worlds before Adam*, 427, 435–36.

revolution which the earth has experienced, namely, the Mosaic *deluge*'.[117] Hutton's views were nevertheless not widely shared, as his editor's rather damning preface makes clear in having the 'great pleasure in giving publicity to this paper, for the views contained in which the author is alone answerable'.[118] In instances like this, it is apparent that divorcing important material specimens from superstitious accompanying explanations was not a process only applied to South Asian cosmology.

That the Himalaya had been upheaved at some point in the past was thus no longer in any real doubt by the 1830s. However, the timescale on which this unfolded remained contested. Increasingly, it appeared to have occurred (relatively) recently in geological time. As Hugh Falconer continued: 'there are unquestionable proofs on the southern side of the chain that important elevations have taken place within a very late period, geologically speaking' and the equivalence of the fossils with existing species 'would show the upheavement, beyond all question, to date, geologically speaking, *since the commencement of* the present order of things'.[119] Victor Jacquemont was largely in agreement: 'my observations on the skirts of the Himalaya, along the plains of Hindostan, are quite confirmatory of my friend M. Elie de Beaumont's views respecting the late period at which that mighty range sprung from the earth'.[120] Here he was applying the widely discussed theory extended by de Beaumont on the origins of mountain ranges, which argued that they came about through successive periods of violent crustal upheavement (revolutions), which he referred to as 'epochs of elevation', separated by longer intervening periods of accumulation.[121] Jacquemont nevertheless suggested, if in somewhat bombastic fashion, that this view was contentious in India. As he wrote to de Beaumont in 1830, 'if M. Pentland has found in Peru some mountains higher than those of the Himalaya, I would not advise him to come to India; and as it is generally admitted that this mighty range, before which the Andes sink into inferiority, is the eldest-born of the creation', so similarly your work 'will in India be considered a personal insult by the geologists of Calcutta, their wives, their children, and their children's dolls. At Bombay I shall take care not to say I am a friend of yours ... to touch the antiquity of the Himalaya is no less a sacrilege in

[117] Hutton, 'Geological Report on the Valley of the Spiti', 227. On 'Mosaic' writers, see Rudwick, *Worlds Before Adam*, 424.
[118] Hutton, 'Geological Report on the Valley of the Spiti', 198.
[119] Falconer, *Palaeontological Memoirs*, Vol 1, 181.
[120] Jacquemont, *Letters from India*, Vol 1, xxvi.
[121] Rudwick, *Worlds before Adam*, 129–33, 334–36, 486–93, 514–16. See also Laudan, *From Mineralogy to Geology*, 197–200.

India'.[122] Here Jacquemont, tongue-rather-firmly-in-cheek, points to ongoing rivalry in global comparison (while also rhetorically provincialising Calcutta as a site of science). Such ridicule was nevertheless both indicative of and complicated by the way that the elevational superiority of particular mountain peaks and ranges, much as universal theories of their upheavement, were prone to dramatic revision in this period.

Just as the timescale for upheavement was debated, so too was the mechanism, and by the 1840s this reflected the broader geological debate often simplified as 'catastrophism vs uniformitarianism'. This amounted to the question of whether changes in the earth's surface (such as the elevation of mountain ranges) were the result of sudden violent revolutions, perhaps involving megaearthquakes or megatsunamis unexperienced in human history, or whether 'actual' or observed causes were adequate to explain gradual changes over a long time frame.[123] Either explanation had to account for mountains as stupendously elevated as the Himalaya. At the mid-century, Richard Strachey was hedging his bets: 'in the present state of our knowledge but little evidence could be adduced as to the degree of rapidity with which all these changes have taken place', but went on to echo de Beaumont: 'I shall only express my own opinion, that though the great regularity of structure and comparative uniformity of upheavement seem to show that the general movement has been quite gradual, yet that there have certainly been well-marked epochs of special activity'.[124] Writing around the same time, Thomas Thomson drew on his experience and observations waiting for the non-existent Chinese boundary commissioners to show up, and was more willing to take a stand, coming down on the side of Charles Lyell's eventually ascendant 'uniformitarian' argument: 'the conclusion has been forced upon me that these mountains have emerged extremely gradually from an ocean'.[125] Even at the mid-century, fuller answers to these questions thus continued to be circumscribed by frontiers, both social and geographical.

Glaciers, Global Comparison and Himalayan Oral Traditions

In 1847, Richard Strachey prefaced an account of 'two most decided Glaciers, which I have just visited' with the perhaps exaggerated claim

[122] Jacquemont, *Letters from India*, Vol 1, 289.
[123] Rudwick, *Worlds before Adam*, 356–61, 393–97, 550–52; Laudan, *From Mineralogy to Geology*, 204–28.
[124] Richard Strachey, 'Physical Geography of the Himalaya (Unbound Proof Copy), c.1851', British Library, Mss. Eur. F127/200, f67.
[125] Thomson, *Western Himalaya and Tibet*, 27.

that a description might be interesting given 'the existence of Glaciers in the Himalayas . . . [is] apparently still considered a matter of doubt by the Natural Philosophers of Europe'.[126] As he continued, 'in all parts of the mountains covered by perpetual snow glaciers abound, and some of them are of great magnitude' and 'the fact that until within the last few years their existence in the Himalaya was doubted, shows, in a manner that needs no comment, what sort of examination this country, perhaps the most remarkable in the world, has received during more than thirty years of British rule'.[127] (Here the uncertainties around glaciers were effectively a reverse of those around volcanoes.) As well as making a political point about the limited resources available to those tasked with surveying the Himalaya, Strachey also tried to offer a scientific explanation. Indeed, he pointed to ongoing confusion around the line of perpetual snow and suggested that complications arose 'from the circumstances of our older surveyors not having a distinct idea of what glaciers were, in consequence of which they invariably talk of them as great snow-beds'.[128] In particular, Strachey singled out William Webb and John Hodgson, noting: 'neither of them knew what a glacier was. Capt. Webb, as we have seen, talks of the Gori emerging from the snow, when we know that in reality it rises from a glacier. Capt. Hodgson falls into a similar error in his description of the source of the Ganges'.[129] Strachey went on to voice scepticism about the observations of the Gerard brothers, ultimately suggesting that credit should go to botanist Edward Madden for 'having first removed all doubts on the subject'.[130] Indicative of how rapidly understandings of the mountains shifted across this period, by the mid-century that there had ever been doubt was met with incredulity in some quarters. As Joseph Hooker wrote to his friend Charles Darwin in 1848: 'Why glaciers were denied to the Himal. I cannot conceive, nor anyone else, there are plenty'.[131]

[126] Richard Strachey, 'A Description of the Glaciers of the Pindur and Kuphinee Rivers in the Kumaon Himálaya', *Journal of the Asiatic Society of Bengal* 16, part 2 (1847): 794.

[127] Richard Strachey, 'On the Physical Geography of the Provinces of Kumáon and Garhwál in the Himálaya Mountains, and of the Adjoining Parts of Tibet', *The Journal of the Royal Geographical Society of London* 21 (1851): 71.

[128] Richard Strachey, 'Physical Geography of the Himalaya (Unbound Proof Copy), c.1851', British Library, Mss. Eur. F127/200, f237-8.

[129] Strachey, 'On the Snow-Line in the Himalaya', 297. Thomas Hutton thought Strachey was being disingenuous in his assessment of his predecessors: see Thomas Hutton, 'Remarks on the Snow Line in the Himalaya', *Journal of the Asiatic Society of Bengal* 18, part 2 (1849): 954–55.

[130] Richard Strachey, 'Physical Geography of the Himalaya (Unbound Proof Copy), c.1851', British Library, Mss. Eur. F127/200, f237.

[131] Hooker to Charles Darwin, 13 October 1848, Kew Archives, JDH/1/10/112–114.

Even once resolved, these existential doubts were followed by questions over whether glaciers in the Himalaya presented exactly the same phenomenon as that which had recently been described in Europe. Edward Madden, as Richard Strachey was at pains to point out, not only 'prominently directed attention to the fact of the existence of glaciers in these mountains', but he also noted they were 'in all points identical with those of the Alps.*'[132] Adding weight to this was that Madden, having travelled and botanised extensively in the Himalaya, was on his return from India able to visit the Alps personally.[133] Victor Jacquemont, meanwhile, was one of the very few Himalayan travellers in this period to make observations the other way around; that is, he had seen the glaciers of the Alps first, before venturing into the Himalaya. In comparing, he remarked that Himalayan glaciers were found higher and not always where he expected based on his experiences in Europe. Describing the region around Jumnotri, he noted that 'the thickness of the snow is more than twelve metres in some places. I do not know if it melts entirely in summer. In the Alps, at a much lower elevation, it would form a glacier'.[134]

A decade or so later, Thomas Thomson expanded on these investigations into the universality of glaciers, noting that 'the general appearance of an Indian glacier seems in every respect to accord with those of Switzerland and of other parts of the temperate zone'.[135] Thomson also suggested that if confusions had arisen, then this was excusable because glaciers and their appearance and motion were only relatively recently understood anywhere, let alone in the remote corners of India.[136] While acknowledging a fundamental equivalency, he nevertheless went on to indicate there were some differences in situation when compared globally: 'in every part of the Himalaya, and of Western Tibet, wherever the mountains attain a sufficient elevation to be covered with perpetual snow, glaciers are to be found . . . it may be laid down as a general law, that every glacier has its origin in perpetual snow' but 'the converse of this

[132] Richard Strachey, 'Physical Geography of the Himalaya (Unbound Proof Copy), c.1851', British Library, Mss. Eur. F127/200, f237-8. The editor suggested in the footnote that '*Mr Vigne had incidentally alluded to an unmistakable glacier in Western Tibet, before the publication of Major Madden's paper'. Here he was referring to Godfrey Thomas Vigne, *Travels in Kashmir, Ladak, Iskardo, the Countries Adjoining the Mountain-Course of the Indus, and the Himalaya, North of the Panjab* (London: Henry Colburn, 1842), Vol 2, 285–86.

[133] Madden, 'Notes of an Excursion to the Pindree Glacier', 257–58.

[134] Jacquemont, *Voyage dans l'Inde*, Vol 2, 88. ['L'épaisseur de ces neiges est de plus de 12m en certains lieux. J'ignore si l'été les fond entièrement. Dans les Alpes, à une élévation bien moindre, elles formeraient un glacier'.]

[135] Thomson, *Western Himalaya and Tibet*, 475.

[136] Thomson, *Western Himalaya and Tibet*, 475.

proposition does not seem to be so universal. We have the high authority of Humboldt for the fact that no glaciers occur in the Andes of tropical America, from the equator to 19° north latitude'.[137] In concluding, Thomson also suggested that the latitude of the Himalaya meant that expectations skewed thinking on the discovery of glaciers: 'it has also, singularly enough, long been the custom to look upon the Himalaya as a tropical range of mountains, in which it was, as a matter of course, regarded as impossible that glaciers could exist'.[138]

While demonstrating 'that these phenomena exist in the Himalaya, under forms apparently identical with those observed in the Alps', Richard Strachey noted the caveat that 'as these are the first Glaciers that I have ever seen, it is right to add that I am only acquainted with those of the Alps, through the medium of Professor Forbes's accounts, and that as I lay no claim to originality'.[139] In displaced locations, global comparisons often occurred through a limited lens – or at least surveyors insisted as much to excuse failings. Strachey also blamed the topography and politics for circumscribing his observations: 'to guard against mistakes I would also mention, that these Glaciers were selected for examination only on account of their accessibility, and that consequently no inferences should be drawn from them, of the general extent of Glaciers in the Himalaya'.[140] In making comparisons between European and Himalayan glaciers, he later laid out his findings in the form of diagrams (see Figure 4.4). Elaborating on these images, Strachey wrote that 'the glaciers of the valley of Chamonix are not by any means the largest in Switzerland' but 'the valley of Chamonix is so well known, that the comparison with its glaciers will probably be more appreciated'.[141] Such compromises reflected a need to make his comparisons intelligible to as wide an audience as possible, and also why well-known mountains like Mont Blanc or Snowdon were so often used as touchstones.

As well as observing and comparing glaciers, Strachey conducted some of the most extensive experiments in this period to measure their motion in the Himalaya (see Figure 4.5). Strachey recognised the importance of these instrumental measurements to questions about both the nature of glaciers and their global commensurability, noting that he returned to the Pindari glacier in 1848 'chiefly with the intention of making an accurate

[137] Thomson, *Western Himalaya and Tibet*, 474.
[138] Thomson, *Western Himalaya and Tibet*, 475–76.
[139] Strachey, 'A Description of the Glaciers', 795.
[140] Strachey, 'A Description of the Glaciers', 795.
[141] Strachey, 'Narrative of a Journey', 411.

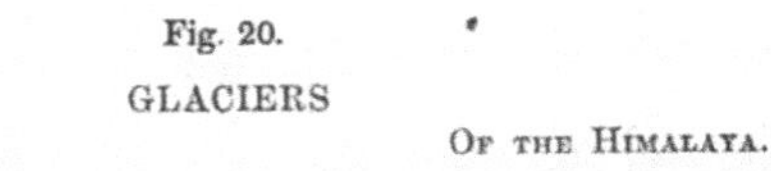

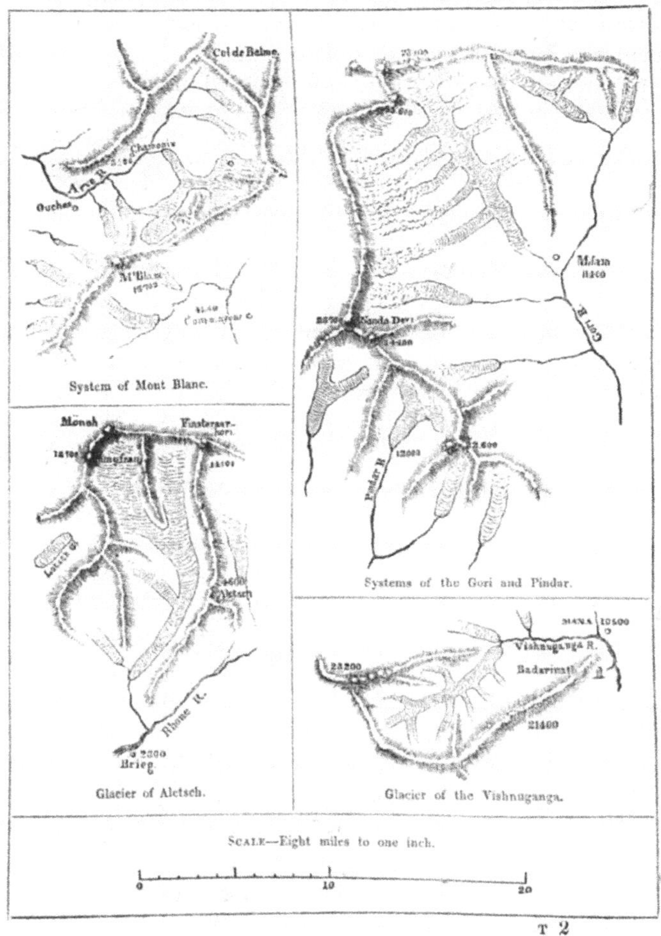

Figure 4.4 Comparison of glaciers of the Alps and in the Himalaya. From the proofs of Richard Strachey's unpublished 'Physical geography of the Himalaya' (c.1851). He noted that 'the annexed woodcut . . . will convey an idea of the size of the Gori and Vishnuganga glaciers, of which we have rough plans, as compared to some of the best-known glaciers of the Alps'.[142] Courtesy of The Society of Authors as agents of the Strachey Trust.

[142] Richard Strachey, 'Physical Geography of the Himalaya (Unbound Proof Copy), c.1851', British Library, Mss. Eur. F127/200, f239. A version of this image was eventually published in Strachey, 'Narrative of a Journey', 412.

Figure 4.5 'Glacier of the Pindar, 12,000 feet'. Plate from Richard Strachey, 'The Physical Geography of the Himalaya' (1854). Richard Strachey is here portrayed as an intrepid observer, surveying the motion of the glacier with a theodolite. His camp and several assistants are also depicted, hinting at the significant labour requirements of the expedition.
Source: British Library, Mss. Eur. F127/202. Courtesy of The Society of Authors as agents of the Strachey Trust.

measurement of its motion'.[143] He described this operation in detail: 'a stake was driven into the moraine, at its highest point, close to the rock on the line between the two crosses, and a Theodolite was set up over it. Five other marks were also made on the glacier, at intervals along the same line' and 'on the following day . . . the distance between the centre of the stick and that of the fixed mark was then measured, which evidently showed the downward progress of the ice at that point of the glacier since the mark was made the day before'.[144] Strachey went on to summarise his measurements by showing that while Himalayan glaciers were relatively slow-moving, 'these rates do not differ in an important degree

[143] Richard Strachey, 'Note on the Motion of the Glacier of the Pindur in Kumaon', *Journal of the Asiatic Society of Bengal* 17, part 2 (1848): 203.
[144] Strachey, 'Note on the Motion of the Glacier', 203. For these calculations, see also Richard Strachey, 'Pindari Glacier', British Library, WD2332, 38b.

from those commonly observed in the summer months on glaciers in the Alps, which lie between 9 and 27 inches in 24 hours'.[145]

As is tacitly apparent in Strachey's descriptions and depictions of his measuring process, these observations were heavily dependent on local labour. But like the case of Hutton's and Gerard's fossil hunting, these relationships also went beyond mere muscle to include the provision of critical knowledge. Whatever the uncertainties among European surveyors, Himalayan peoples had long understood the nature of glaciers. Indeed, Richard Strachey's brother Henry noted that 'glaciers are well known to the Tibetans under the name of *Kangri*, i.e. *Iceberg*, which is also loosely used to denote any high mountain covered with perpetual snow or *nevée*', but 'their true character is sufficiently attested by the Tibetan inhabitants, who (incurious as they generally are in such matters) have observed their progressive motion and *éboulement*, as evidenced by fragments of ice falling to the foot of the mountain'.[146] As Henry Strachey continued, this was not only true for the Tibetans, and 'these glaciers are well known to the Bhotias, under the term *Gal*'.[147] Himalayan peoples were thus crucial for locating glaciers, much as they had been for fossils. For example, on visiting the Pindari Glacier in 1846, Edward Madden wrote: 'we were accompanied here from Khathee by Ram Singh, the accredited guide to the glacier'.[148] Guides like Singh provided not only knowledge of routes and safe travel practices; they also gave access to Himalayan oral traditions which were used to map changes in glaciers over time. Information about changing routes of trade and pilgrimage through the mountains made it clear that glaciers were not static, even in the short term. This knowledge could also be used to consider changes over multiple generations. As Madden recorded elsewhere: 'the Bhotiahs of Milum affirm that their glacier has receded from the village two or three miles to its present site, and Ramsingh assured me that the same is true, in a less degree, at Pinduree'.[149] In a similar case, Joseph Alexander Weller, at the time Junior Assistant Commissioner of Kumaon, relied on his guides Nagu Burha and Dhun Singh while investigating the Gori Glacier on a shooting expedition. He recorded: 'Nagoo Boorha tells me that his father (who lived to 98 years) remembered the source of the Goree nearly opposite Milum, and Nagoo himself has seen the recession of the snow-bed some 3 or 400 yards in the course of 40 years'.[150]

[145] Strachey, 'Narrative of a Journey', 156.
[146] Strachey, 'Physical Geography of Western Tibet', 52.
[147] Strachey, 'Narrative of a Journey to Cho Lagan', 107.
[148] Madden, 'Notes of an Excursion to the Pindree Glacier', 249.
[149] Madden, 'Notes of an Excursion to the Pindree Glacier', 258.
[150] Cited in Manson, 'Capt. Manson's Journal', 1166.

To European naturalists, much as to Ram Singh or Nagu Burha, it was readily apparent that the extent of the glaciers of the Himalaya had changed considerably, even if the cause remained debated. As Richard Strachey wrote: 'in Kumaon the shepherds . . . have no idea of any actual motion in the whole bulk of the glacier, though they everywhere suppose the ends of them to be gradually receding'.[151] He considered this information to be of uneven quality and 'very vague'; however, 'in the case of the glacier of the Gori, the termination of which is within a mile of the village of Milam, these stories of the gradual decrease of the glaciers become more trustworthy' and indeed 'the people point out certain places to which the ice was known to have extended in their fathers' time, and there is every reason to think their statements correct'.[152] Strachey went on to place these concerns in global context, comparing evidence from the Himalaya with theories from the Alps, to note that it was possible that 'here too there has been a period of cold, a glacial epoch, similar to that now generally admitted to have occurred over the area of Europe'. He continued to suggest that this was more likely 'only the result of a change of climate consequent on the upheaval of the great plains of Northern India', before ultimately acknowledging that a fuller determination of such climatic fluctuations was beyond his resources.[153]

Conclusion

This chapter has examined the way political frontiers, topographical barriers and 'cultural borderlands' circumscribed the everyday practice of geological activities in the Himalaya in the first half of the nineteenth century. Here political boundaries limited access to important fossils and necessitated turning to existing networks to obtain key specimens (especially from Tibet). However, acquiring *bijli ki har* and *shaligrams* in this fashion, and forcing them to answer to European science rather than South Asian religion, was a fraught process. In particular, these networks exacerbated the existing problem of a lack of in situ observations, whether as a result of recalcitrance on political frontiers or the topographically unreachable and ice-entombed summits of mountains. Indeed, without being securely fixed on the vertical globe, specimens acquired via bazaars and brokers had only limited ability to answer what were increasingly pressing questions about universal deluges and the upheavement of mountain ranges. In tracing the existing networks that circulated ritual

[151] Richard Strachey, 'Physical Geography of the Himalaya (Unbound Proof Copy), c.1851', British Library, Mss. Eur. F127/200, f245.
[152] Strachey, 'Physical Geography of the Himalaya (Unbound Proof Copy), c.1851', f245.
[153] Strachey, 'Physical Geography of the Himalaya (Unbound Proof Copy), c.1851', f244.

objects from the uplands to the lowlands, this chapter has demonstrated that operating in 'cultural borderlands' meant South Asian cosmology could inflect scientific practice in this period, as theories rose and fell with dizzying regularity, and 'geology' began to emerge as a recognisably modern discipline. Such rapid changes in understandings of the earth were not always easy to assimilate and engage with in remote locations, which were slow to receive up-to-date journals and books. At the same time, as is evident in the case of James Gerard, limited training and resources did not stop contradictions – especially those that emerged through global comparison – from being seized on and exploited by EIC employees in the mountains.

Ultimately, compounding limits to geological practice meant that volcanoes could be speculated into existence, and the enormous masses of moving ice that constituted glaciers could be doubted. More broadly, this chapter advances the overall claims of the book by emphasising the material dimension of the mountains, and the extraordinarily laborious nature of scientific practice in this period. Making glaciers knowable and fossils move so that they could be compared globally required engagement with pre-existing networks (of both muscular power and multi-generational expertise), and this dependency only exacerbated existing challenges to imperial mastery in the mountains. In considering the way agency could be subverted and resisted in 'cultural borderlands', this chapter thus demonstrates that as much as the history of Himalayan geology cannot be told without the names Richard Strachey, Thomas Hutton and Victor Jacquemont, it also cannot be written while omitting the names Pati Ram, Nagu Burha and Ram Singh. As the next and penultimate chapter details, similar questions of labour and expertise run through the story of the founding of the botanical gardens at Saharanpur and Mussoorie, and understandings of their ambiguous positions on the vertical globe, straddling the highlands and the lowlands.

5 Higher Gardens

In the descriptions she revised to accompany George Francis White's *Views in India* (1838), English travel writer Emma Roberts (c.1794–1840) suggested that 'the city of Saharunpore is of very ancient date, but possesses few or no remains of interest . . . with the exception of the botanical garden, which forms, indeed, its principal attraction'.[1] Roberts went on to note that before being appropriated by the East India Company (EIC) as a scientific establishment in 1817, the Saharanpur garden had a precolonial history, and 'common report states, that this useful and ornamental work owes its existence to the family of Zahita Khan, a former chief', though she hastened to add that 'it must have undergone great changes since its early formation [which] . . . render it truly English in its aspect'.[2] Hinting at the limited resources available to those operating in remote stations, Roberts also remarked that 'though not so great a pet of the government as the Calcutta establishment, the garden at Saharunpore is kept in excellent order, the most being made of the comparatively small sum allowed for its maintenance'.[3] When it came to engaging with the Himalayan mountains in the early decades of the nineteenth century, the limitations of the EIC's flagship garden at Calcutta, with its tropical climate and vast distance from the Himalaya, had quickly became apparent. If the EIC – or rather, an eclectic cast of its employees – were to exploit the botanical opportunities offered by expanding frontiers in the mountains, then they quickly realised they would need a 'northern' garden to do so.[4] In this period, Saharanpur thus became the interface – if unevenly and at times

[1] White, *Views in India*, 20. Emma Roberts, one of the most prolific travel writers in nineteenth-century India, rewrote and greatly expanded the original text which White had composed to accompany his images. She did so based on 'descriptions drawn from the notes of several tourists, and her own experience of Indian life'. See 'Lieutenant White's Views in India, Edited by Emma Roberts', *The Spectator*, 16 December 1837. These additions and revisions were so extensive that throughout this chapter I have chosen to refer to Roberts as the author rather than White.

[2] White, *Views in India*, 20. [3] White, *Views in India*, 20.

[4] Royle, *Illustrations*, Vol 1, 2.

haphazardly – between the hugely diverse and potentially highly valuable botany of the Himalaya and the rest of the globe.

Long in the shadow of the Calcutta Botanic Garden in both contemporary and modern scholarship, the garden at Saharanpur played a central role in the early scientific and imaginative appropriation of the Himalaya. However, while eminently suitable for growing a different set of plants to the Calcutta garden, at only 1,000 feet of elevation Saharanpur was never sufficiently temperate nor 'mountainous' enough in its own right to extract all the botanical riches the Himalaya seemed to offer. As Roberts continued: 'Saharunpore may be called the threshold of the hill districts' but 'the plants are generally cultivated in the first instance at Mussooree, a station in the hills, and the experiments made at Saharunpore'.[5] Established in 1826 at an elevation of 6,500 feet, the 'experimental nursery in the hills' at Mussoorie had a symbiotic relationship with Saharanpur. Each garden compensated for the other's seasonal variations in temperature, allowing them to encompass the flora of the mountains, and indeed much of the globe. The development of these two gardens, envisioned in terms of a vertical scale, is the subject of this chapter. In what follows, I demonstrate that their complementary positions on the vertical globe were essential to understanding the mountains through their botany. However, I also show that the location of these gardens was ambiguous, and that they represented liminal spaces between the uplands and lowlands, sometimes awkwardly combining characteristics of both.

As previous chapters have demonstrated, growing recognition of the commensurability of mountain environments saw the necessity of reframing scientific phenomena in three dimensions. This was never truer than in the case of plants. Mountains and botany were increasingly interconnected, both conceptually and practically. For example, John Forbes Royle, who became the second superintendent of Saharanpur in 1823, argued that 'as mountains situated in hot countries embrace every variety of climate, they are capable of producing the plants of both temperate and polar regions' and 'to succeed in the cultivation, therefore, of the plants of any particular country, it is first requisite to determine the height to which it is necessary to ascend a mountain, to obtain a similar climate'.[6] The

[5] White, *Views in India*, 21.

[6] John Forbes Royle, 'Extract of a Report on the Medicinal Garden at Musoorea', *Transactions of the Medical and Physical Society of Calcutta* 4 (1829): 412. For more on Royle's 'ecological' approach, see Satpal Sangwan, 'From Gentlemen Amateurs to Professionals: Reassessing the Natural Science Tradition in Colonial India 1780–1840', in *Nature and the Orient: The Environmental History of South and Southeast Asia*, ed. Richard Grove, Vinita Damodaran, and Satpal Sangwan (Delhi: Oxford University Press, 1998), 219; Arnold, *The Tropics and the Traveling Gaze*, 164.

recognition that the Himalaya were, by some margin, the highest mountains in the world gave them elevated importance in understanding the geographical distribution of plants in the vertical realm. This chapter and the next tell this botanical story. In this chapter, I address practices around the acclimatisation of botanical specimens, the role of Indian gardeners and collectors, and the institutional challenges in operating from 'mountain' gardens with limited resources, displaced from both Calcutta and London. By focusing on institutions rather than itinerant expeditions, this chapter also broadens and adds another dimension to the key themes of this book, that is, the necessity of further decentring the spaces of scientific practice, paying closer attention to everyday relationships between Europeans and South Asians in scientific knowledge-making and tracing the way the Himalaya were remade through comparison with the Alps and the Andes. The final chapter then deals with questions of vertical plant geography and the imagining and tracing of elevational limits. Together, these two chapters thus tell the story of the remaking of the mountains through botanical appropriation and an often-laborious process of global comparison.

If the mountains remained largely 'blank space' for European science at the turn of the nineteenth century, the excitement around their botanical potential was nevertheless becoming palpable. As Scottish surgeon, and the first superintendent of the Saharanpur Garden, George Govan wrote: it was in 'vain to attempt describing the enthusiasm and delight experienced by the admirers of nature on first entering these districts with the invading army' during the Anglo-Gurkha War beginning in 1814.[7] Indeed, botanical knowledge and military expansion were integrated, as the EIC's acquisition of Kumaon and Garhwal meant naturalists gained more reliable access to a much higher altitudinal cross-section of the Himalaya. Even where the mountain frontiers remained closed to Europeans – such as in the newly diminished but still staunchly independent kingdom of Nepal – South Asian plant collectors were deployed to bring back plants and seeds of all kinds. However, as Chapter 4 detailed in the case of fossils, specimens obtained in this way brought about particular challenges for knowledge production, having not been observed in situ, and this is why this chapter follows a diverse cast of both European and South Asian actors. Indeed, while following the first three EIC superintendents at the Saharanpur Garden – George Govan, John Forbes Royle and Hugh Falconer – I simultaneously demonstrate that scientific practice at Saharanpur and Mussoorie was always, to a greater or lesser extent, a cross-cultural, negotiated affair. As we saw

[7] Govan, 'On the Natural History and Physical Geography', 19.

in Emma Roberts's description, Saharanpur had a precolonial history, and more important even than the carryover of infrastructure, the legacy of this older garden was embodied in the transfer of personnel, most notably Hari Singh, who was employed as head gardener by both iterations. Meanwhile, the hill nursery at Mussoorie, however much it was intended to advance European scientific and imperial ends, was often left entirely in the hands of Indian gardeners. Reflecting the sheer scale of these contributions in both labour and expertise, this chapter thus foregrounds the eclectic networks of – usually silenced and unnamed – plant collectors, bazaar druggists, gardeners and artists who were fundamental to the operation of European botanical sciences in and of the Himalaya. As in this book more broadly, by focusing on practical engagements between Europeans and South Asians in remote locations as much as questions of epistemology and ontology, I aim to demonstrate the role of everyday sociability in the scientific and imaginative constitution of the mountains.

After a presentation that John Forbes Royle gave to the Asiatic Society in Calcutta in 1832, a member of the audience shared the 'pleasure and surprize of finding in the Botanic Garden, at Sahára̍npúr, the English daisy, looking up from the plain of India to the lofty snows of the Himalaya'.[8] Of course, British interest in the mountains, botanical and otherwise, was never this whimsical, naïve or innocent. The foundational purpose of Saharanpur, as with the numerous colonial gardens that dotted the globe by the early nineteenth century, was both implicitly and explicitly imperial, and its maintenance was couched in the language of utility, 'improvement' and empire.[9] This included the avoidance of famine (a legacy of widespread recent disasters in Bengal), but also the production of medicines and materials like fibres and timber (the latter often explicitly intended for naval ends), as well as staples and luxury foodstuffs, and even ornamental garden plants, which had an increasingly valuable market in both India and Europe.[10] Perhaps no plant was more

[8] 'Proceedings of Societies – Asiatic Society, Wednesday, 7 March 1832', *Journal of the Asiatic Society of Bengal* 1 (1832): 117.

[9] Londa L. Schiebinger and Claudia Swan, *Colonial Botany: Science, Commerce, and Politics in the Early Modern World* (Philadelphia: University of Pennsylvania Press, 2005), 13. See also Richard Drayton, *Nature's Government: Science, Imperial Britain, and the 'Improvement' of the World* (New Haven: Yale University Press, 2000); Jayeeta Sharma, *Empire's Garden: Assam and the Making of India* (Durham: Duke University Press, 2011).

[10] On famines, see Vinita Damodaran, 'The East India Company, Famine and Ecological Conditions in Eighteenth-Century Bengal', in *The East India Company and the Natural World*, ed. Vinita Damodaran, Anna Winterbottom, and Alan Lester (Basingstoke: Palgrave Macmillan, 2015), 80–101. The list of plants that the Saharanpur superintendents mooted or trialled was long. In terms of Himalayan species, medicinal plants like rhubarb and ginseng were prominent, as were materials like cedar and hemp, and spices

famously experimented with than *camellia*, or the tea plant, with which Royle, as well as Saharanpur's third and fourth superintendents – Hugh Falconer and fellow Scotsman William Jameson (1815–1882) – were all enthusiastically involved. Indeed, Saharanpur was pivotal to the introduction and ultimate success of Chinese tea in India, though this now well-known story (if the role of Saharanpur perhaps less so) is not my focus in this chapter.[11] As well as commerce, the rhetoric of military utility was prevalent in justifying expenditures on the garden. Royle, for example, sometimes emphasised that Saharanpur had critical imperial and strategic dimensions, downplaying his own more abstract personal interest in scientific botany. As he argued, the garden was 'admirably adapted for enabling an observer to obtain a knowledge of the Flora of the plains of Northern India, as well as of the Himalayan Mountains', all 'within thirty miles of the commencement of the successive ranges which form that great barrier between the dominions of the British and the territories of the Chinese'.[12] The result was that sometimes irreconcilable tensions could emerge between prioritising the acquisition and acclimatisation of commercially, medically and ornamentally interesting plants and the advancement of philosophical botany, all of which nevertheless shaped emerging perceptions of the distinction between uplands and lowlands.

The history of colonial botanic gardens, and especially their entangled and sometimes insidious relationships to empire, has borne much fruit in recent years. It is not my intention in this chapter to recapitulate the now well-trodden ground around 'colonial' science and the development of Indian botany or, except where necessary, to detail the circulation of specimens to Europe.[13] Examining the role of Saharanpur and

like saffron. Attempted introductions via global interchange included medicines, most notably (if initially unsuccessfully) cinchona, as well as food staples like rice, maize and potatoes, and other valuable cash crops like cotton, sugar, tobacco and coffee. See Royle, *Illustrations*; John Forbes Royle, *Essay on the Productive Resources of India* (London: W.H. Allen, 1840).

[11] See among many Hugh Falconer, 'On the Aptitude of the Himalayan Range for the Culture of the Tea Plant', *Journal of the Asiatic Society of Bengal* 3 (1834): 178–88; John Forbes Royle, 'Report on the Progress of the Cultivation of the China Tea Plant in the Himalayas, from 1835 to 1847', *Journal of the Royal Asiatic Society of Great Britain and Ireland* 12 (1849): 125–52; Griffith, *Journals of Travels*. See also Zaheer Baber, 'The Plants of Empire: Botanic Gardens, Colonial Power and Botanical Knowledge', *Journal of Contemporary Asia* 46, no. 4 (2016): 659–79.

[12] Royle, *Illustrations*, Vol 1, 1.

[13] The classic study is Lucile H. Brockway, *Science and Colonial Expansion: The Role of the British Royal Botanic Gardens* (New York: Academic Press, 1979). See also Grove, *Green Imperialism*; Drayton, *Nature's Government*. For colonial botany more broadly, see also Londa L. Schiebinger, *Plants and Empire: Colonial Bioprospecting in the Atlantic World* (Cambridge, MA: Harvard University Press, 2004).

Mussoorie in the constitution of both the Himalayan mountains and the vertical globe is nevertheless only possible by building on the excellent previous scholarship on colonial gardens, in particular Calcutta.[14] Here both the city of Calcutta and its botanic garden have been read by scholars as both 'centres of calculation' and 'contact zones'.[15] Given its prominence in global botany, it is perhaps unsurprising that a majority of scholarly attention on Indian gardens has been directed towards Calcutta. However, this also highlights the historiographical and archival draw of 'centres in the periphery', which while successfully making important reorientations away from the metropolitan centres have also had the side effect of obscuring other important sites and spaces of scientific practice in India. Eugenia Herbert touches on this issue, noting that 'with the exception of Saharanpur, a former Mughal garden, these [gardens] were colonial creations in the capitals of the three presidencies' and argues that 'Calcutta remained to Indian gardens what Kew was to the empire as a whole'.[16] While true, this is nevertheless only part of the story. By privileging 'centres in the periphery' in our histories, we ultimately risk replicating diffusionist, metropole-centric histories of 'colonial' science.

In thinking about Saharanpur and Mussoorie, a useful counter-approach emerges from the work of Sujit Sivasundaram who, in his study of the Peradeniya garden in Sri Lanka, argues that things look different when we make this garden the focus instead of Kew. In so doing, he shows that 'the fragility of Kew's reach is more apparent' and thus 'by looking at the island and from the island, it is possible to take locality seriously in the history of science on the global stage'.[17] Following

[14] Marika Vicziany, 'Imperialism, Botany and Statistics in Early Nineteenth-Century India: The Surveys of Francis Buchanan (1762–1829)', *Modern Asian Studies* 20, no. 4 (1986): 625–60; Satpal Sangwan, 'The Strength of a Scientific Culture: Interpreting Disorder in Colonial Science', *The Indian Economic and Social History Review* 34, no. 2 (1997): 217–50; Arnold, *The Tropics and the Traveling Gaze*; Richard Axelby, 'Calcutta Botanic Garden and the Colonial Re-Ordering of the Indian Environment', *Archives of Natural History* 35, no. 1 (2008): 150–63; Mark Harrison, 'The Calcutta Botanic Garden and the Wider World, 1817-46', in *Science and Modern India: An Institutional History, c.1784–1947*, ed. Uma Das Gupta (Delhi: Pearson Education India, 2011), 235–53; Khyati Nagar, 'Between Calcutta and Kew: The Divergent Circulation and Production of Hortus Bengalensis and Flora Indica', in *The Circulation of Knowledge between Britain, India and China: The Early-Modern World to the Twentieth Century*, ed. Bernard Lightman, Gordon McOuat, and Larry Stewart (Leiden: Brill, 2013), 153–78; Adrian Thomas, 'Calcutta Botanic Garden: Knowledge Formation and the Expectations of Botany in a Colonial Context, 1833–1914' (PhD Dissertation, King's College London, 2016).

[15] Latour, *Science in Action*; Raj, *Relocating Modern Science*.

[16] Eugenia W. Herbert, *Flora's Empire: British Gardens in India* (Philadelphia: University of Pennsylvania Press, 2011), 146.

[17] Sivasundaram, 'Islanded: Natural History in the British Colonization of Ceylon', 143. Sivasundaram also argues that the site of the garden came to be seen as 'intermediate'

from this, taking the mountains from the mountains, things also look different, especially when considering Saharanpur as removed from both Calcutta and Kew, and Mussoorie at another remove again. Writing in 1825, John Forbes Royle remarked that Saharanpur had been envisioned 'as a collateral branch of the institution at Calcutta', though it remained in many ways independent.[18] Of course, the garden operated as something of a scientific centre in its own right. This was not limited to botany, and Saharanpur became, if briefly, a staging ground for operations in the high mountains, and a clearing house for geological, zoological and palaeontological specimens (not least the extraordinary fossils from the nearby Siwalik Hills), as well as a station for long-term meteorological observations, and a key node in the early trigonometrical survey operations into the mountains.[19] Studying Saharanpur nevertheless offers the opportunity to further decentre the spaces of science by considering the resources available to those at another level of displacement from London than Calcutta (and Mussoorie as displaced even further). As we move from London to Calcutta, Calcutta to Saharanpur and Saharanpur to Mussoorie – closer to the mountains and higher up the vertical globe – scientific practice begins to look different (or at least, those operating there insisted that it looked different). In examining these differences – often expressed in terms of limited resources, time and distance – it is thus important to consider not only circulatory successes, but also the many ways that these networks could fail and become confused. Across the period of this study, we will see that the Saharanpur superintendents sometimes – intentionally or otherwise – exaggerated the botanical possibilities of the plants they extracted from the Himalaya, and overestimated their ability to circulate them. Emphasising Saharanpur's disconnect from Calcutta, let alone London, I argue that for remote institutions like Saharanpur and Mussoorie – as much as for expeditions in the high mountains – we need to pay particular attention to the moments when things did not travel, and practices broke down.

Historical scholarship on the Saharanpur and Mussoorie gardens is considerably more limited than it is for the EIC's garden at Calcutta.[20]

between highlands and lowlands, an argument I draw on here. Sivasundaram, *Islanded*, 182.

[18] Royle to Nathaniel Wallich, 1 June 1825, British Library, IOR/F/4/955/27123(2), f150. Initially, as established under Govan, the garden was independent (although Govan was expected to cooperate with Nathaniel Wallich in Calcutta). However, following Govan, it took on officially subordinate status, even if this did not always translate into practice.

[19] See John Forbes Royle, 'Account of the Honourable Company's Botanic Garden at Saharanpur', *Journal of the Asiatic Society of Bengal* 1 (1832): 56.

[20] H. Montgomery Hyde's early 1960s biographical article on George Govan remains the most substantial piece of scholarship for the early days of the garden. H. Montgomery

As part of his broader work on landscape and tropicalisation in India, David Arnold is right to argue that 'Saharanpur, a thousand miles to the northwest [of Calcutta], had an equally important role in the development of India's botany and in the increasing delineation of its tropical, as opposed to temperate, climatic zones and flora', even if he does not go on to make a sustained examination of the garden.[21] More recently meanwhile, Pratik Chakrabarti has examined Saharanpur's important role in geological practice, but not its involvement in imperial botany.[22] The limits of the existing scholarship not notwithstanding, this chapter is not intended as a comprehensive institutional history of Saharanpur and Mussoorie. Instead, what follows is the story of the 'northern' gardens' roles in the making of the Himalaya in European scientific and imperial imaginations. It begins with the geography of Saharanpur and the modification of the existing garden for the purposes of scientific botany. Next, I consider debates around the need for a higher garden and the foundation of the Mussoorie hill nursery. This is followed by a discussion of the centrality of two Indian gardeners – Hari Singh and Murdan Ali – to the functioning of the gardens, and an examination of the role of collectors in the botanical delineation of the highest mountains in the world. The final section then draws these threads together to consider the problems of distance and limited resources in the making of Himalayan botany as both globally commensurable and imperially valuable.

The Geography of Saharanpur and the Appropriation of a Mughal Garden

Writing in 1832, John Forbes Royle acknowledged that 'it is singular, and at the same time most fortunate, that nearly at the most northern limit of the British territories, and in one of the most eligible situations for the purpose, a public garden should have been established by the native Governments which preceded the British'.[23] His predecessor, George Govan, also explained the eligibility of Saharanpur's geography in some detail: 'the proximity of the situation to the base of one of the noblest

Hyde, 'Dr. George Govan and the Saharanpur Botanical Gardens', *Journal of The Royal Central Asian Society* 49, no. 1 (1962): 47–57. For brief notices, see Isaac Henry Burkill, *Chapters on the History of Botany in India* (Calcutta: Botanical Survey of India, 1965), 30–36, 78–83; Ray Desmond, *The European Discovery of the Indian Flora* (Oxford: Oxford University Press, 1992); Rajesh Kochhar, 'Natural History in India during the 18th and 19th Centuries', *Journal of Biosciences* 38, no. 2 (2013): 206–7.

[21] Arnold, *The Tropics and the Traveling Gaze*, 163. See also Arnold, *Science, Technology and Medicine*, 48–56.

[22] Chakrabarti, *Inscriptions of Nature*, 55–58.

[23] Royle, 'Account of the Honourable Company's Botanic Garden at Saharanpur', 43.

mountain ranges in the world, and of which the natural history is among the desiderata of men of science', made it the ideal location 'to attempt the naturalisation either from the Himmalaya into Europe or vice versa'.[24] As this indicates, Saharanpur was always intended to facilitate not only the acclimatisation and extraction of Himalayan plants, but also the naturalisation of those of Europe and the rest of the globe into India and the mountains. As Royle noted, this was allowed by the climate of Saharanpur, 'which is described as being tropical at one season, and partially European at another, and as having, in consequence, an equally varied cultivation'.[25] He went on to elaborate that Saharanpur's latitude 'embraces in its course a greater variety of interesting country than perhaps any other' because 'the above parallel, or that of 30°, leaving India, passes through Persia, Arabia and Egypt, and over the southern boundaries of Libya, Barbary and Morocco' then 'across the Atlantic, through New Orleans, between Old and New Mexico, and passing the Pacific Ocean, crosses the very centre of China and Tibet'.[26] Royle later showed that these theoretical comparisons and equivalencies translated into practice, and 'taking the Saharunpore garden as an example, we have collected in one place, and naturalized in the open air, the various fruit-trees of very different countries, as of India and China, Caubul, Europe, and America'.[27] While no doubt reflecting some of the exaggerated rhetoric of a superintendent defending his contributions and expenditures, it is nevertheless apparent that, a thousand miles to the northwest of Calcutta, the garden at Saharanpur was excellently situated to become the necessary 'northern garden' in the EIC's nexus of colonial botanical institutions.[28]

This existing Mughal garden – sometimes referred to as the 'Farhat Baksh [delight giver]' – had originally been instituted in 1779 by the orders of the Rohilla chief Zabita Khan. Its founding decree included the revenue of seven villages for the garden's upkeep, which continued after the garden fell to the Marathas in 1788. This was reduced to two villages in 1801, an arrangement that carried through into the garden's EIC period.[29] This section traces the sometimes exaggerated and never entirely complete transition of Saharanpur from a Mughal Garden to an EIC scientific establishment. In particular, I examine the physical transformations of the spaces of the garden (the

[24] Govan to James Hare, 17 April 1816, British Library, IOR/F/4/587/14218, f41; Govan to Charles Lushington, 20 January 1821, British Library, IOR/F/4/660/18324, f73.
[25] Royle, *Illustrations*, Vol 1, xxxiii. See also Arnold, *The Tropics and the Traveling Gaze*, 164.
[26] Royle, 'Account of the Honourable Company's Botanic Garden at Saharanpur', 43.
[27] Royle, *Illustrations*, Vol 1, 10.
[28] Royle to J.C. Melvill, 31 December 1838, British Library, IOR/F/4/1828/75444, f12v.
[29] Neil B. Edmonstone and George Dowdeswell to the Court of Directors, 28 October 1817, British Library, IOR/F/4/587/14218, f1-2. This name is the same as that given to the more famous 'Farhat Baksh' in Lucknow.

transfer of personnel is taken up later), showing that these changes were often as insisted as much as they were real. Here the co-opting of older, precolonial gardens into the service of empire was far from unprecedented, and served as a means of asserting imperial control through continuity and change.[30] While referenced frequently in colonial sources (including both administrative records and widely-read travel narratives), the generic way in which Saharanpur was usually described as a 'Mughal Garden' nevertheless elides the diversity of earlier Persianate institutions and botanical traditions. Indeed, Pratik Chakrabarti suggests that the continuities in the case of Saharanpur were overwhelmingly superficial, and prefers to emphasise the disjuncture brought about by the British naturalisation of antiquity arising from excavations of the nearby Doab canal and in the Siwalik hills.[31] While I would agree that the continuities largely consisted of superficial posturing, this was not so clear cut in the case of botany, and the EIC was happy, at times, to leverage this previous history, even while denigrating the garden's earlier management. At the same time, it is worth noting that it chose to perform its overwriting of the previous 'pleasure garden' with the principles of scientific botany in different ways for different audiences – whether residents of the district, colonial visitors, or consumers in Europe – emphasising at various times utility and improvement, continuity and change, ornament and science.

Although nominally in British hands since 1803, it was not until 1817 that, after visiting the garden and acceding to the suggestions of George Govan and others, the Governor-General of India, Francis Rawdon-Hastings decided 'that which was intended only for the gratification of an Asiatic sensualist, should contribute to the advancement of science' and decreed it be formally established as a botanical garden.[32] Despite recognising the benefits of adapting an existing garden, the EIC employees who were placed in charge of Saharanpur were eager to emphasise what they saw as the poor condition of the institution, and thereby elevate their contributions to improving and remaking it. Indeed, there was an insistence on 'the Garden having gone into a state of rapid decay whilst under the immediate charge of the Natives'.[33] This, as was argued by the Magistrate of Saharanpur in 1814, was also central to the justification for the expense of re-establishing and maintaining the garden, which would be of great benefit to the 'native subjects in this part of the country, who fondly cherishing the recollection

[30] See Sivasundaram, 'Islanded: Natural History in the British Colonization of Ceylon', 130.
[31] Chakrabarti, *Inscriptions of Nature*, 33–34, 58. Here he is pushing back against the argument in Grove, *Green Imperialism*, 409–12.
[32] Royle, 'Account of the Honourable Company's Botanic Garden at Saharanpur', 43.
[33] 'Extract Public Letter to Bengal', 28 June 1820, British Library, IOR/F/4/660/18324, f6.

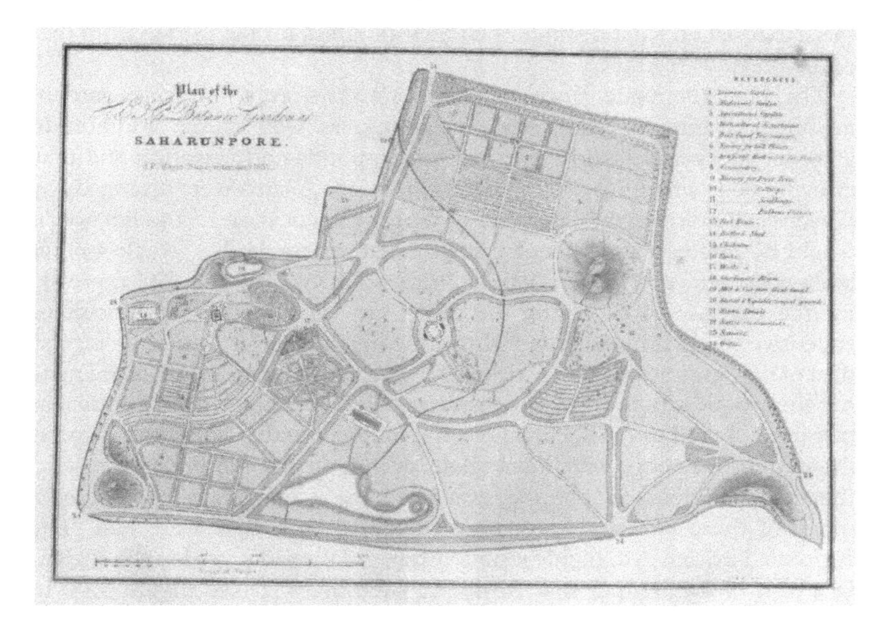

Figure 5.1 Map of the Saharanpur Botanic Garden produced under Royle's direction in 1831.[36] This map was produced in the context of Governor-General Bentinck's visit the same year, and concern that the garden would be abolished as part of a general reduction in EIC spending on economic botany. This context perhaps resulted in an exaggerated order, making the garden appear grander and more scientific than it actually was. This map was later published by Royle as the frontispiece to his *Illustrations of the Botany and Other Branches of the Natural History of the Himalayan Mountains* (1839).

of what it once resembled . . . will, in as much as they now regret the state it lies in, gratefully admire the liberality that shall cause it once more to rise in all and more than former grandeur'.[34] Govan's and Royle's reports indicate that in transforming the garden, they focused especially on removing what they saw as an 'indiscriminate mixture of saleable produce' and 'superfluous and common plants' which were 'incompatible with the improvement and Botanical objects of the garden'.[35] Indeed, this emphasis on science allowed

[34] 'Extract from a Report from the Magistrate of Seharunpore to the Governor General', 31 December 1814, British Library, IOR/P/9/8, f22r-v.
[35] James Hare to C.M. Ricketts, 21 April 1816, British Library, IOR/F/4/587/14218, f30; Royle to Nathaniel Wallich, 1 June 1825, British Library, IOR/F/4/955/27123(2), f152.
[36] Royle, *Illustrations*, Vol 2, frontispiece.

them to insist on what distinguished their program for the garden from that of the previous regimes.

The changes made since Saharanpur's days as a 'pleasure garden' included clearing, levelling and replanting, as well as the addition of a conservatory and seed house, new roads, ponds and irrigation and new growing beds where Royle recorded that 'in one, plants were arranged [by Govan] according to the Linnaean system of classification . . . though now it would be preferable to change it for the natural method'.[37] Royle's other main addition was a 'physic' garden, ordered in 1825 by the EIC expressly for the purpose of supplying the dispensary with medicines difficult or expensive to obtain otherwise.[38] In summing up, Royle wrote that 'a good deal of new ground has been enclosed, and many alterations made in laying out the grounds. In these the English style of gardening has been as much as possible adhered to'.[39] The qualification inherent in this statement suggests these transformations were incomplete, even if the superintendents insisted that they were doing the best they could with the resources available. Indeed, many of these erasures were more rhetorical than they were real. The continued existence of highly visible vestiges of the older garden are evident in a plan of Saharanpur that Royle prepared in 1831 (see Figure 5.1). In layering these additions on top of the older establishment, the garden never became – whatever Emma Roberts' insistence – a wholly European space, in either design or aspect. As the map indicates, a 'Chabutra' (tower) was retained at the centre of the garden, and a much older 'Hindu Temple' continued to be a key node, visible from many parts of the garden. Hugh Falconer commented on the former, noting that 'an old native Chubootra [sic] occupies nearly the centre' and from this several 'broad roads lead off in serpentine curves, intersecting the garden in all directions'.[40] Whatever the ostensible reorganisation of the garden beds at ground level along the lines of scientific botany, the tower thus remained the garden's focal point and a prominent landmark for visitors – as indeed it does to this day.

Mussoorie and the Need for a Higher Garden

For all its advantages over the Calcutta Botanic Garden when it came to the appropriation of Himalayan flora and the naturalisation of European

[37] Royle, 'Account of the Honourable Company's Botanic Garden at Saharanpur', 47. See also Burkill, *Chapters on the History of Botany in India*, 33.
[38] Royle to Nathaniel Wallich, 1 June 1825, British Library, IOR/F/4/955/27123(2), f157-8.
[39] Royle, 'Account of the Honourable Company's Botanic Garden at Saharanpur', 46.
[40] Falconer to Currie, 11 April 1839, British Library, IOR/F/4/1828/75444, f57r. See also David Arnold, 'India's Place in the Tropical World, 1770–1930', *The Journal of Imperial and Commonwealth History* 26, no. 1 (1998): 10.

and North American plants, Saharanpur was only ever a partial solution. Indeed, it was never high enough on the vertical globe to encompass the entirety of the botanical offerings of the Himalaya in its own right. The debates about the need for a higher garden are thus revealing of the haphazardness of the early imaginative and scientific constitution of the Himalaya, and the haziness around emerging distinctions between uplands and lowlands. George Govan was aware of the limitations of Saharanpur as early as 1820: 'of those [plants] brought from elevations of 8,000 and 12,000 feet above the level of the sea scarce one plant in 100 can be preserved during the first hot season and rains, although subsequently they become more hardy' but 'success may in future be expected ... in collecting them in a depôt in the hills during the rains and hot weather, to dispatch them to Saharunpore only during the cold season'.[41] Though a higher garden even than Saharanpur was quickly recognised as essential, Mussoorie was not the only potential location, and Govan had initially settled on Nahan. However, it was only slightly higher in elevation than Saharanpur, and did not offer enough of a climate differential to be a long-term solution. Royle suggested Mussoorie as an alternative, noting that it was located at approximately 30.5° of latitude and 'attains an elevation of about 5,500 feet', going on to explain that 'as Decandolle states that 500 feet of elevation are deemed equal to one degree of latitude, it will be necessary to add 11 degrees to the latitude of the place and we have 41½ as the latitude to the climate of which that of this place should correspond'.[42] As Royle continued, 'it may therefore be safely inferred from the above facts that the climate is analogous to that of the temperate parts of Europe'.[43] In choosing Mussoorie, Royle also explicitly placed the nursery in a global context, and indicated the climatic preferences of the species he was hoping to accommodate: 'the latitude being nearly that of Shiraz and Cairo, and of the southern boundaries of Barbary and Morocco, the solar rays must of course be powerful' though 'as its elevation is also considerable, the causes which produce a reduction of temperature as we ascend in the atmosphere reduce its temperature to that of the southern parts of Europe'.[44] Here Royle was grappling with ongoing and unresolved questions around the relationship between latitude and

[41] George Govan to [Anon], 8 July 1820, British Library, IOR/F/4/660/18324, f52-3.
[42] Royle to [Nathaniel Wallich], 7 April 1826, British Library, IOR/F/4/955/27123(2), f115-6.
[43] Royle to [Nathaniel Wallich], 7 April 1826, British Library, IOR/F/4/955/27123(2), f118.
[44] Royle, 'Extract of a Report on the Medicinal Garden at Musoorea', 407.

altitude, even while evoking global comparisons to indicate his ambitions for the garden. The 'hill nursery' at Mussoorie was thus intended to complement, and even compensate, for the weaknesses of Saharanpur (which was, in its turn, compensating for Calcutta). Royle reflected that 'by taking advantage of the different months adapted for cultivation in the hills and in the plains, a complete year of moderate climate may be obtained for the germination of the seeds, and for the growth of the plants of the temperate climates of every part of the globe', and summed up that 'many plants have actually been thus introduced and preserved, which if confined to either would, while young, have been destroyed by the hot winds of the plains, or killed at Masúrí by the frosts of winter'.[45] While undoubtedly a clever solution, this is nevertheless also a reminder of the ambiguous position of these two gardens on the vertical globe, sometimes awkwardly straddling the uplands and lowlands.

Suitability of climate and appropriate elevation were important, but the 'hill nursery' would never have succeeded without attention to accessibility. Mussoorie, around 50 miles to the north, was 'well adapted for the purpose and has the advantage of being at an available distance from Saharunpore'.[46] The establishment of the hill nursery at Mussoorie also paralleled the town's inauguration as a colonial 'hill station', and it would go on to be one of the most popular of these alongside Shimla and Darjeeling.[47] Similarly, in the 1830s, Surveyor General George Everest chose a location a little further down the ridge – and only a short walk from the Mussoorie nursery – for 'Hathipaon', his house and base while overseeing the Survey of India (though not the survey of the mountain that would come to bear his name, which he never saw and was named in his honour by his successor). Mussoorie thus became, for a time, a crossroads for scientific practice and the mapping and cataloguing of the Himalaya, as well as being a popular sanatorium for convalescing soldiers and a social hub for Europeans escaping the plains (see Figure 5.2).

While Royle considered that Mussoorie admirably fulfilled the purpose he had chosen it for, the 'hill nursery' was nevertheless temporarily defunded in 1831, as part of a general albeit passing scepticism on the Company's part around the extent of expenditures on economic

[45] Royle, 'Account of the Honourable Company's Botanic Garden at Saharanpur', 46.
[46] Royle to [Nathaniel Wallich], 7 April 1826, British Library, IOR/F/4/955/27123(2), f115.
[47] See Kennedy, *Magic Mountains*; Jayeeta Sharma, 'Producing Himalayan Darjeeling: Mobile People and Mountain Encounters', *Himalaya, the Journal of the Association for Nepal and Himalayan Studies* 35, no. 2 (2016): 87–101.

Figure 5.2 'Mussooree and the Dhoon, from Landour' with 'the dim plains of Saharunpore still farther in the distance'.[48] This image was published in George Francis White's *Views in India* (1838), and depicts the hill station in highly romanticised fashion. Emma Roberts, in the description she edited to accompany the image, noted that in the mid-1830s Mussoorie was still a fledgling town, but was 'daily increasing in size'.[49]

botany.[50] Moments like this reflect the occasional fickleness of EIC institutional support, and the sometimes ambivalent attitude of the Company to the scientific enthusiasms of its employees. Writing to the government in Bengal, EIC administrator Henry Thoby Prinsep (1792–1878) suggested that 'the establishment at Mussoorie Teeba' has 'not been successful as a garden'.[51] As part of a reduction in costs which also affected Saharanpur (including the Superintendent's salary), Prinsep recommended that 'the subordinate garden at Mussoorie may be dispensed with altogether – and if the prosecution of Physical Botany be considered essential, the establishment for the purpose can be provided out of the 200 Rs per mensem allowed for the Saharunpore Garden'.[52] Falconer argued that these reductions went too far, especially given

[48] White, *Views in India*, 30, 35. [49] White, *Views in India*, 34.

[50] See Grove, *Green Imperialism*, 414; Drayton, *Nature's Government*, 131.

[51] Henry Thoby Prinsep to G.A. Bushby, 30 March 1831, British Library, IOR/F/4/1828/75444, f43v.

[52] Henry Thoby Prinsep to G.A. Bushby, 30 March 1831, British Library, IOR/F/4/1828/75444, f45r.

Mussoorie's essential role in acclimatising valuable mountain species to send to Europe and vice versa. He put the case as follows: 'a subordinate garden in the Himalayah mountains is a desideratum of vital importance to the Saharunpoor Garden for the reception of the fruit trees and productions of Europe'.[53] As he further explained, 'the experimental nursery which Mr. Royle established at Mussooree was abolished in 1831, but the ground is still available, and it possesses the great advantage of being in a measure stocked with valuable fruit trees and other plants and of being within a short distance of Saharunpoor'. He then went on to 'strongly recommend' its reestablishment, which duly occurred under the patronage of Lord Auckland.[54] Falconer's arguments resonated, and after this temporary shuttering, the 'hill nursery' was never in danger of being closed again before the end of the nineteenth century. The ongoing debates about expenditures nevertheless reflect inherent tensions between the gardens' scientific, ornamental and economic purposes, which were often at odds with the superintendents' interests in the botanical delineation of the mountains.

Hari Singh and Murdan Ali at Saharanpur

Returning now to Saharanpur, it is notable that even more than the spatial organisation or the architecture, the most significant continuity between the Mughal and Company periods was the number of Indians employed by the garden in both guises (most prominently the head gardener Hari Singh). Such carryovers and the appropriation of existing orders was essential to the establishment of Company rule more generally, as Chris Bayly and others have demonstrated.[55] In recent decades, scholars have also been especially interested in the roles of Indians in the development and practice of botany in India, whether as botanists, assistants, painters, gardeners and collectors or, in modern scholarly parlance, informants, intermediaries and go-betweens. Botany, perhaps more than other sciences, has left an archive that allows us, if unsatisfactorily and imperfectly, to trace these intermediaries and glimpse their roles and contributions to the making of knowledge (if rarely to recover their idiosyncrasies and personalities). Here some of the most innovative work comes from anthropologists such as Erik Mueggler, who has carefully and creatively recovered the worlds that natural historical knowledge was made in and moved through in early twentieth-century China, and the complex roles of paper materials, inscriptions and brokers

[53] Falconer to Currie, 11 April 1839, British Library, IOR/F/4/1828/75444, f69r.
[54] Falconer to Currie, 11 April 1839, British Library, IOR/F/4/1828/75444, f69r.
[55] Bayly, *Empire & Information*.

in navigating these. In focusing on everyday practice, I thus follow Mueggler and others in emphasising the importance of this sort of historical recovery work, and in tracing the development out of 'arduous experience a comprehensive vision of a region' while acknowledging that such transformations also 'emerge[d] out of the contingent heterogeneity of daily experience'.[56] When considering Saharanpur's role in making the botany and plant geography of the Himalaya, it is worth emphasising, as Falconer noted in his report of 1839, that 'the Superintendent of the Botanical Garden is the only European connected with the Establishment'.[57] The day-to-day operation of the garden, and the everyday practices required for it to function – including but far exceeding mere labour – were conducted almost entirely by Indian employees.[58]

In his report, Falconer went on to detail the organisation of the garden's thirty-three directly employed personnel, listing the names of those in senior roles and their salaries, as well as indicating the dates they were first employed.[59] Among the most interesting names on Falconer's roll is that of Hari Singh, who succeeded his father 'Moze Ram' (who had been charged with the Mughal establishment) as head gardener, indicating a family occupation. Singh was retained in the same role when George Govan became superintendent in 1817, and the longevity of his career is notable.[60] As well as spanning the Mughal/Maratha/Company periods, Singh maintained his position under the first four superintendents of Saharanpur, and was apparently still paying rent on a house attached to the garden in 1851.[61] The archival record for Singh is nevertheless sparse, with his most substantial documentary presence relating to a salary dispute that occurred in the 1830s under Hugh Falconer's tenure as superintendent. In his petition for a pay rise, Singh outlined that when his father had been appointed head gardener by Zabita Khan's government, this had included the support of three villages and 'thus it remained during the whole of the Nawab's possession as also during the time the Mahrattas' had the garden. Singh went on to argue that 'on the Company becoming its possessors Mr Chamberlain deprived me of

[56] Erik Mueggler, *The Paper Road: Archive and Experience in the Botanical Exploration of West China and Tibet* (Berkeley: University of California Press, 2011), 17.

[57] Falconer to Currie, 11 April 1839, British Library, IOR/F/4/1828/75444, f64r.

[58] In common with many other British colonial institutions at this time, the garden also had access to convict labour, in this case thirty daily workers from the Saharanpur jail. Falconer to Currie, 11 April 1839, British Library, IOR/F/4/1828/75444, f57v.

[59] Hugh Falconer, 'Detailed statement of salaries and Establishment of the H.C. Botanic Garden Saharunpoor on the 1st May 1838', British Library, IOR/F/4/1828/75444, f86v.

[60] Falconer to Currie, 11 April 1839, British Library, IOR/F/4/1828/75444, f52r. Hari Singh was initially appointed to joint charge with his brothers 'Bowanee' and 'Dongee' Singh.

[61] William Jameson, 'Statement shewing the quantity of land occupied by the Gardens at Saharanpoor and Mussooree', 26 February 1851, British Library, IOR/F/4/2498/141673, f26v.

the villages and in lieu thereof settled upon me a salary of 26 Rs monthly. Since then eleven having been cut, it has been reduced to fifteen rupees'.[62] Proby Thomas Cautley, perhaps best known for his 'co-discovery' of the Siwalik fossils with Falconer, outlined in a letter of 1838 the evidence he could find for this earlier arrangement. He suggested that Hari Singh's two brothers had also held appointments, the joint value of which was 26 rupees, and that Hari Singh himself had never earned more than 15.[63] Cautley concluded that 'on the grounds put forward by the petitioner therefore for an increase of salary I cannot see that he has any claim whatever', but demurred that 'as an old servant of the Garden and as a man far superior in the botanical acquirements to any native that could in all probability be found to take his place, I would ... recommend him to the consideration of ... a monthly salary of 25 Rs'.[64] Though a backhanded endorsement, this is nevertheless indicative of Hari Singh's importance to the functioning of the garden. Indeed, when Hugh Falconer later weighed in, he agreed with Cautley that while the grounds of the petition were inaccurate, given 'the smallness of that salary for the duties he has to perform', an increase was nevertheless deserved.[65]

As the head gardener at Saharanpur right across the period the garden facilitated Himalayan botany's expansion on a global stage, Hari Singh's contributions to the making of the mountains must have been significant. It is evident that he had a close working relationship with John Forbes Royle in particular, and played a key role in the production of Royle's *Illustrations*. For example, Royle noted in this that the potentially valuable timber tree *pinus longifolia* 'is called *cheer*, *sullah*, and *thansa*, also *surul*; but Huree Singh, the head native in the Saharunpore Botanic Garden, informed me that the last is a variety, if not a distinct species'.[66] More than simply providing local names and uses, the language of species and varieties implies Singh was engaging with a Europeanised approach to botany. It seems Royle was also on occasion happy to use Singh's findings to correct the errors of Europeans savants. As he wrote regarding the way the '*Jatamansi* of the Hindoos has been considered to be the Spikenard of the ancients ... the proofs and reasoning of Sir Wm. Jones appearing to me so satisfactory'. However, 'one day accidentally asking Huree Sing, an intelligent and respectable native at the head of the establishment of the Saharunpore Botanic Garden, whether a plant (*Valeriana villosa*) in the Conservatory was not like the *Jatamansi*'

[62] 'Petition of Huree Singh Chowdree to the Honourable Company's Garden at Seharunpore', British Library, IOR/F/4/1828/75444, f34r-f34v.
[63] Cautley to Thomason, 31 March 1838', British Library, IOR/F/4/1828/75444, f36r.
[64] Cautley to Thomason, 31 March 1838', British Library, IOR/F/4/1828/75444, f36v.
[65] Falconer to Currie, 11 April 1839, British Library, IOR/F/4/1828/75444, 49v.
[66] Royle, *Illustrations*, Vol 1, 349.

he responded 'in the negative, and pointed to a *Plantago*, with lanceolate leaves, as that which most nearly resembled it'.[67] In following up on these discussions of *Jatamansi*, Royle also sought to tap into trade networks and obtain *materia medica* already circulating out of the mountains via fairs and bazaars. As in the case of the fossil specimens discussed in Chapter 4, pre-existing networks were emerging as important sites for tracing the natural history of the high mountains, and individuals like Hari Singh were essential to navigating these. Royle noted that *Jatamansi*, 'better known in Northern India by the name *bal-chur*', was brought down annually from the Himalaya. He procured some fresh roots and found they 'exactly resembled those sold in the Saharunpore bazar as *Jatamansi*'.[68] He concluded that the original error had arisen because 'either by accident or design, a wrong plant was sent from Bootan, and figured and described in the Asiatic Researches, at a time when it was not possible to detect the imposture, as it was long before we had free access to the hills'.[69] Here Royle's suggestion of possibly deliberate deception indicates the way that the frontier continued to circumscribe the limits of scientific practice (as well as potentially providing opportunities to resist the appropriation of botanical knowledge).

Royle also consulted Hari Singh regarding juniper, and remarked of two varieties: 'the former appears to be the plant called *theloo* by the natives, and seen by Huree Sing between Simla and Phagoo ... and by Murdan Aly, a very intelligent plant collector, near Saughee Ke Ghat, a high hill to the southward of Rol'.[70] As well as indicating that Hari Singh travelled in the mountains on behalf of the garden, Royle's note serves to introduce another key figure in this story, namely: the *munshi* Murdan Ali, who was appointed 'Herbarian' at Saharanpur in 1825.[71] Unusually for Indians in this period, Ali came to be considered as a 'botanist' in his own

[67] Royle, *Illustrations*, Vol 1, 243.

[68] Royle, *Illustrations*, Vol 1, 243. For other plants marked 'grown in the Saharanpore Botanic Garden from seed bought in the bazaar', see Royle, *Illustrations*, Vol 1, 184, 197, 199, 312. See also John Forbes Royle, 'List of Articles of Materia Medica, Obtained in the Bazars of the Western and Northern Provinces of India', *Journal of the Asiatic Society of Bengal* 1 (1832): 458–71. On bazar medicines, see Arnold, *The Tropics and the Traveling Gaze*, 181; Herbert, *Flora's Empire: British Gardens in India*, 152–53. See also Bayly, *Empire & Information*, 272–73; Pratik Chakrabarti, 'Medical Marketplaces beyond the West: Bazaar Medicine, Trade and the English Establishment in Eighteenth-Century India', in *Medicine and the Market in England and Its Colonies, c. 1450–c. 1850*, ed. Mark S. R. Jenner and Patrick Wallis (Basingstoke: Palgrave Macmillan, 2007), 216–37. Another important source of Royle's medical and botanical knowledge was 'Sheikh Nam Dar, commonly called Nanoo, the head medical assistant in the Civil Hospital of Saharunpore'. See John Forbes Royle, 'On the Identification of the Mustard Tree of Scripture', *Journal of the Royal Asiatic Society of Great Britain and Ireland* 8 (1846): 114.

[69] Royle, *Illustrations*, Vol 1, 243.

[70] Royle, *Illustrations*, Vol 1, 351.

[71] Hugh Falconer, 'Detailed statement of salaries and establishment of the H.C. Botanic Garden Saharunpoor on the 1st May 1838', British Library, IOR/F/4/1828/75444, f86v.

right, rather than merely a 'collector' or 'gardener'.[72] Like Hari Singh, his importance is revealed as an unintended consequence of the colonial paperwork generated around 'a claim to consideration for augmentation of salary'. This was, Hugh Falconer wrote, 'on the grounds of great merit, long service and inadequate salary at present. He has attained ... a fair knowledge of the natural arrangement of plants, and can distribute new species into their families and genera, with [a] considerable deal of accuracy'. Such was, Falconer continued, 'a rare attainment for a native and his merits have been strongly acknowledged by Mr Royle'.[73] Falconer went on to recommend Ali's salary be increased from 10 to 20 rupees per month, tacit recognition of his importance to the garden.[74] Indeed, Ali's talents as a botanist are here extolled by Falconer, if again in a culturally hierarchical fashion which conveniently serves to minimise the degree of dependency the European superintendent has to admit to.

Irish-born officer-turned botanist Edward Madden also worked with Ali in the 1840s, and provides a little more information about his multiple contributions to Himalayan botany. While travelling in the mountains, for example, Madden noted a plant 'discovered here by Moonshee Murdan Alee* of the Seharunpoor Botanic Garden'.[75] In the footnote he remarked that '*this very intelligent and respectable Syyud, the first of his race, perhaps, who addicted himself to Natural History or any useful knowledge, and in whose honor Dr. Royle established the genus Murdannia, has' while 'under the occasional instruction of Messrs. Royle, Falconer, and Edgeworth, his masters and mine, attained a considerable proficiency in Botany'.[76] Even if this does much to minimise Ali's agency and transfer it to his European instructors (similarly to the case of instrument makers in Chapter 2), this is nevertheless notable for indicating the etymology of *Murdannia* (see Figure 5.3). While naming newly discovered Himalayan species after travellers was common practice in the early nineteenth century, including for relatively lesser

[72] Sometimes also Aly or Alee or Ulee. For brief notices of Ali's accomplishments and the supposed distinction between 'botanists' and 'collectors', see Arnold, *The Tropics and the Traveling Gaze*, 183; Deepak Kumar, 'Botanical Explorations and the East India Company: Revisiting "Plant Colonialism"', in *The East India Company and the Natural World*, ed. Vinita Damodaran, Anna Winterbottom, and Alan Lester (Basingstoke: Palgrave Macmillan, 2015), 28.

[73] Falconer to Currie, 11 April 1839, British Library, IOR/F/4/1828/75444, f50r.

[74] As Falconer also noted, Ali 'was formerly on the same footing with Hurry Sing, but did not participate in the augmentation of 5 rupees per mensem which the chowdry [head gardener] got in 1831'. Falconer to Currie, 11 April 1839, British Library, IOR/F/4/1828/75444, f50r.

[75] Edward Madden, 'The Turaee and Outer Mountains of Kumaoon', *Journal of the Asiatic Society of Bengal* 17; 18 (1848): Vol 17, 416.

[76] Madden, 'The Turaee and Outer Mountains of Kumaoon', Vol 17, 416–17.

Figure 5.3 *Murdannia scapiflora*, named for Murdan Ali, the long-time 'Herbarian' at Saharanpur.[77] In his notes, Royle added that this plant 'has some repute in Hindoo Materia Medica'.[78] This watercolour also serves as a reminder of another critical group of Indian technicians who were essential the development of Himalayan botany at Saharanpur, namely: the so-called Company school artists, whose extraordinary contributions are detailed in the rich and archivally exhaustive studies of Henry Noltie.[79] Among those who worked at Saharanpur were some of the most famous and acclaimed artists of this tradition, including Lakshman Singh and Vishnupersaud, who were briefly seconded there from the Calcutta garden while Nathaniel Wallich was away in Europe.

[77] Royle, *Illustrations*, Vol 2, Plate 95. [78] Royle, *Illustrations*, Vol 1, 403.

[79] See 'Extract Public Letter from Bengal, 1 July 1828, British Library, IOR/F/4/1191/30877, f3r. On the 'Company School Artists' see especially Henry Noltie, *Indian Botanical Drawings, 1793–1868, from the Royal Botanic Garden* (Edinburgh: Royal Botanic Garden Edinburgh, 1999); Henry Noltie, *Robert Wight and the Botanical Drawings of Rungiah and Govindoo*, 3 vols (Edinburgh: Royal Botanic Garden Edinburgh, 2007); Theresa M. Kelley, *Clandestine Marriage: Botany and Romantic Culture* (Baltimore: The Johns Hopkins University Press, 2012), 163. See also Mildred Archer, *Natural History Drawings in the Indian Office Library* (London: Her Majesty's Stationery Office, 1962); Desmond, *The European Discovery of the Indian Flora*.

known figures like Bengal Infantry surveyors William Webb and Alexander Gerard, the naming of species for Indian assistants was highly unusual.[80] Royle's explanation of his dedication is thus especially interesting: 'I have named *Murdannia*, in compliment to Murdan Aly, a plant collector and keeper of the Herbarium at Saharunpore, who collected many of the plants described in this work' and who has 'acquired a remarkable tact and quickness in detecting new plants, as well as in remembering the characters by which genera and families are distinguished, so as to be able at once to arrange a new discovery in its appropriate place'.[81] That such an unusual honour was bestowed on an Indian botanist indicates just how important Ali was to the work that made Royle's name, as indeed does Royle's acknowledgement here.

It was not only Royle and Madden who Ali discussed plants with, and he later gave botanical advice to Richard Strachey, and assisted with specimens (as is evident for example in Figure 5.4). Perhaps even more intriguingly, Edward Madden indicates that Ali had 'compiled a Hindoostanee work' on botany (probably in Urdu), 'containing a general introduction to the study, followed by a detail of the orders and genera, after the Natural System, comprising most of those indigenous to the upper provinces of India and the Himalaya'.[82] He noted that 'the work still languishes in MS, the expenses of printing being beyond the author's means' but 'with some previous supervision, it is deserving the attention and patronage of the Asiatic or any other Society interested in the progress of Botany in India, amongst the Indians'.[83] While this description places Ali's work in a lower tier of practice – Indian rather than European botany – it is nevertheless highly interesting that Ali was producing a work according to the natural system in this period. Royle also recorded that Ali worked on plant catalogues, writing in 1825 that 'I have also been employed in affixing systematic names to the native ones in the two catalogues prepared by Mr Alli [sic]: copies of which I have had the honour of forwarding, the first containing the Hindoostanee and the second the Hill names of plants arranged alphabetically'.[84] Like Hari Singh, Murdan Ali had an exceptionally long career. Indeed, William Jameson's superintendent's report for 1866–7 acknowledges the contributions of Murdan Ali, now the 'the Curator' of Saharanpur's newly formed museum, which aimed to showcase the botany, geology

[80] *Pinus Webbiana* and *Pinus Gerardiana*. [81] Royle, *Illustrations*, Vol 1, 403.
[82] Madden, 'The Turaee and Outer Mountains of Kumaoon', Vol 17, 416–7.
[83] Madden, 'The Turaee and Outer Mountains of Kumaoon', Vol 17, 417.
[84] Royle to Nathaniel Wallich, 1 June 1825, British Library, IOR/F/4/955/27123(2), f160-1. Neither of these appear to have survived.

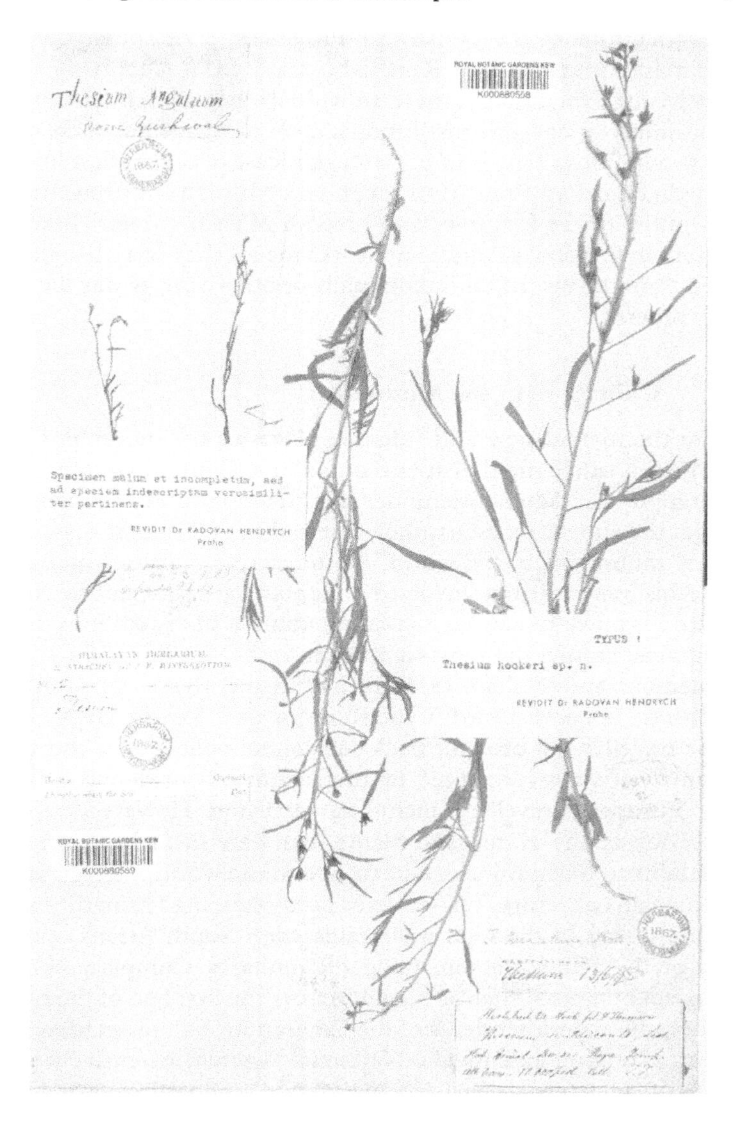

Figure 5.4 A specimen of *Thesium* marked in the top-left corner 'from Murdan Ali'. Other specimens include notes such as 'Royle and his man swear that this grows wild' and 'Royle's man says', though whether these refer to Ali or someone else is unclear. Source: http://specimens.kew.org/herbarium/K000880559 © The Board of Trustees of the Royal Botanic Gardens, Kew.

and agriculture of the Northwest Provinces.[85] Although operating in a different context to Pati Ram (who featured in Chapter 4 equipping and advising expeditions), these individuals thus all demonstrate considerable ability to navigate institutional and patronage networks over multiple generations. This is important, not least because what little can be gleaned about Pati Ram, Hari Singh and Murdan Ali from the archives owes much to the longevity and breadth of their careers. Indeed, their ubiquity in imperial scientific networks meant they were less likely to be elided from the record, unintentionally or otherwise, as was the fate of so many others.

Collectors in the Mountains

Before the myriad new and valuable plants of the mountains could be identified, catalogued, experimented on and disseminated from Saharanpur and Mussoorie under the direction of Hari Singh, Murdan Ali and the superintendents, they first had to be brought down from the highest mountains in the world. As the example of the *Jatamansi* indicates, this was a task that involved both political and scientific challenges. The 1830s onwards saw an increasing number of expeditions in the high mountains, among the most significant for botany being those of Victor Jacquemont and William Griffith, and in the 1840s, those of Thomas Thomson, James Edward Winterbottom and Joseph Dalton Hooker. These travellers all brought back substantial collections, and their herbariums were complemented by the significant incidental collecting of other European travellers, including surveyors. However, a significant proportion of the Himalayan plants that were brought down from the mountains to Saharanpur – and thence to the world – were gathered by South Asian collectors. This reliance partly stemmed from the limitations on EIC power in the high mountains, with South Asians operating in parts of the Himalaya remaining off limits to Company control and European botanists, such as Nepal. Across the first half of the nineteenth century, the superintendents of the Saharanpur botanic garden (much as Calcutta superintendents like Nathaniel Wallich) extensively employed parties of collectors, as well as attempted to co-opt pre-existing networks of trade and pilgrimage to obtain plants and seeds from the high mountains. In this section, we see that being a successful nineteenth-century Himalayan botanist meant, in no small part, negotiating South Asian talent and labour, and directing it towards scientific ends. The most

[85] William Jameson, 'Report on the working of the botanical gardens at Seharunpore, 1866–7', British Library, IOR/V/23/121 Pt 37 Art 4, f73v.

successful botanists were, in other words, those who were best able to tap into networks to bring plants down from the mountains, and most capable of managing the South Asian employees and botanical experts involved at all levels of the process, whether collecting, drying, identifying, arranging or painting.

As Royle's interactions with Singh and Ali and visits to the bazaar indicate, he relied far more on collectors than his own travelling to obtain new plants. Royle's duties – and perhaps temperament – meant he travelled infrequently, and he thus built the important collection that formed the basis of his *Illustrations* around the labour of local collectors. Beginning in 1824, once Saharanpur was functioning to such a standard that he felt he could spare the personnel – 'previous to that the garden itself required the labours of the whole establishment for its internal improvement and management' – Royle had parties of collectors almost constantly employed in the mountains. This included the high mountains, and as Royle continues, 'in 1825, I first endeavoured to get a collection of specimens from Kanáwar, but the gardeners whom I sent, unfortunately, ran away'.[86] Such reporting reveals the limits of Royle's agency when it came to his collectors while they were in the mountains, and whose actions here might also be understood in terms of resistance. From Royle's side, however, denigrating the collectors served as an excuse for the limits of his report. In this way, the presence of collectors could be played up or played down, and their agency deliberately depicted in imperial communications or rendered invisible, depending on needs of the naturalist. This rhetorical massaging was especially necessary given that these were not the only setbacks encountered in employing collectors, and as Royle continued, problems often multiplied:

> The plants from Kashmir were first procured in 1828, by sending two of the gardeners belonging to the Seháranpúr establishment along with the northern merchants who bring down fruit, &c. for sale. In the following year or 1829, the merchants themselves brought me down a number of dried specimens in a book which I had given them for the purpose, but these were generally duplicates of the former year. Last year I again sent two of the establishment, but they brought an indifferent collection in point of numbers, though the specimens were generally large and well dried.[87]

Perhaps most intriguing here is the revelation that, once it was known that Royle wanted particular plants, merchants began taking their own initiative to bring them to him (even if there was apparently some level of misunderstanding over what was actually wanted). Royle went on to note

[86] Royle, 'Account of the Honourable Company's Botanic Garden at Saharanpur', 48.
[87] Royle, 'Account of the Honourable Company's Botanic Garden at Saharanpur', 48.

that some of these collectors were established on a semi-permanent basis and 'men have been stationed in the Hills for the purpose of collecting seeds and drying plants, of the former 4 or 5 parcels have been forwarded monthly to Calcutta'.[88] Similarly, in obtaining rhubarb, Royle used collectors to establish trade with 'Tibet, or Western Mongolia, by means of the Tatars who resort to the Hill Fairs'. As he continued, 'this trade might easily be encouraged by the government purchasing all the Rhubarb it requires, which might thus be employed for hospital use after crossing the frontiers, instead of now, after making a journey of 20,000 miles or nearly the circuit of the globe'.[89] (Though he went on to admit that 'as no naturalist has visited this part [of Tibet], and neither seeds nor plants have been obtained thence, it is as yet unknown what species yields this Rhubarb'.[90])

By the time of Falconer's tenure, beginning in 1832, networks of collectors had been set up to gather plants from the mountains on an annual basis, though this led to compromises over the allocation of labour and resources. As Falconer noted, 'there being no separate establishment of plant collectors attached to the Garden, parties have to be detached from the other departments', and went on to explain that 'they are generally sent off towards the close of the cold weather (or early in March) and return in November; and the workmen employed in the garden are reduced during these months to the extent of from 5 to 6 hands by this arrangement'.[91] The seasons also dictated collecting in the mountains in other ways, and as Falconer recorded: 'I left plant collectors in Cashmeer so that none of the summer flora might be lost during my absence'.[92] Whatever the complaints of Royle and Falconer around the accuracy and diligence of collectors, this system of collecting for the garden was evidently still in place at the mid-century. Jameson, for example, sent collectors all over the mountains 'as far as Niti the frontier town' and 'to the jungles bordering on the snow, to collect all the seeds and plants procurable in these elevated regions'.[93] Collectors were also sometimes employed by European travellers and surveyors who were setting out for the mountains and enthusiastic to help with their botanical appropriation, but who were themselves wanting in botanical talent. For example, when Henry Strachey was presented with the opportunity to visit the frontier as part of the Tibetan Boundary Commission in the

[88] Royle to Nathaniel Wallich, 1 June 1825, British Library, IOR/F/4/955/27123(2), f157.
[89] Royle, *Illustrations*, Vol 1, 315–6. [90] Royle, *Illustrations*, Vol 1, 315–6.
[91] Falconer to Currie, 11 April 1839, British Library, IOR/F/4/1828/75444, f59v.
[92] Falconer to Currie, 18 April 1839, British Library, IOR/F/4/1828/75444, f111r.
[93] William Jameson to J.W. Sherer, 14 August 1852, British Library, IOR/F/4/2528/145895, f5r.

1840s, he confessed in a letter to William Hooker at Kew: 'I am absolutely ignorant of Botany'. In acknowledging this, he suggested that the solution was to 'apply to Dr Jameson (supt. Saharanpore Garden) for a botanical collector', and in so doing outsource his botanical obligations and compensate for his own deficiencies as a polymath.[94]

Employing collectors, especially South Asian collectors, nevertheless created particular problems in terms of the credibility of the knowledge produced about the mountains. Collectors' efforts were often placed lower in hierarchies of knowledge even than the contributions of named assistants like Hari Singh and Murdan Ali. Just as there was a hierarchy of gardens, so too was there a hierarchy of gardeners and collectors. These hierarchies of trust and credibility come through for example in the writings of Edward Madden, and the curious case of *melianthus major*, of which 'we are told that the Doctor's [Royle's] plant collectors obtained a species ... on "the lofty mountains of Kumaoon"' and a specimen of which was growing in the Saharanpur garden.[95] Madden noted that this would make the *melianthus* 'remarkable for being found both at the Cape of Good Hope and in Nepal without any intermediate station', but of this he was suspicious, especially as 'a considerable number of these lofty mountains of Kumaoon have been explored by Lieut. Strachey, Mr. Winterbottom, and myself, and we could scarcely have missed so conspicuous a shrub if it existed in any of the localities visited'.[96] Madden explained this discrepancy by suggesting that '"plant-collectors" are glad enough to load their Herbaria with garden specimens, and are for the most part not enthusiastic at all in exploring "lofty mountains."'[97]

Of course, what Madden saw as laziness might also be seen as resistance, and he seemed to have a particularly antagonistic relationship with his collectors, venting to David Moore at the botanic garden in Dublin that 'there is no trusting them, & so apathetic are they, that sooner than go out & gather me seeds for a good sound reward they will stay here and cut wood at 4 [Rs] a day!'[98] He continued in a similar vein to William Hooker at Kew that 'when the spring comes I will see what can be done about employing collectors – I have myself very little faith in them; they are quite up to the plan of staying at home & saying they have been to such & such a place'.[99] Madden also went out of his way to denigrate the practice of

[94] Henry Strachey to William Hooker, 11 June 1850, Kew Archives, DC/54/485.
[95] Madden, 'The Turaee and Outer Mountains of Kumaoon', Vol 18, 627.
[96] Madden, 'The Turaee and Outer Mountains of Kumaoon', Vol 18, 627.
[97] Madden, 'The Turaee and Outer Mountains of Kumaoon', Vol 18, 627.
[98] Madden to David Moore, 29 October 1845, Kew Archives, DC/54/337.
[99] Madden to William Jackson Hooker, 5 November 1847, Kew Archives, DC/54/336.

using Indian collectors more generally, perhaps in order to assert his own value and contributions: 'having resided for several years in the British portion of the Himalaya Mountains, and more especially in the province of Kemaon, which borders on the Nepalese territories, I possessed opportunities for examining its botany, which up to that period had been investigated by native collectors only'.[100] Ultimately though, Madden's complaints were hollow ones, revealing his limitations and frustrations. Indeed, whatever the caveats and breakdowns endemic in these networks in the mountains, the use of South Asian collectors remained essential across the period of this study, especially in places where the frontiers continued to circumscribe the acquisition of specimens.

The Problem of Distance

The centrality of the EIC's 'northern' gardens to the acclimatisation of the flora of the vertical globe was always contingent on their locality and access to the mountains. However, even while creating privileged opportunities for making botanical knowledge, displacement simultaneously presented challenges for scientific practice (and not only as a result of dependence on South Asian gardeners and collectors). This final section of the chapter examines the tensions around scientific practice in spaces removed not only from London, but also from 'centres in the periphery' like Calcutta, and the challenges – some real, some insisted – described by the EIC surgeons and botanists who operated there. Hugh Falconer, for example, wrote in a letter to William Hooker at Kew that he had managed to 'amass a large collection of plants from Cashmeer all along to the Nepal frontier', but 'everything relating to the described Botany of this part of India has been brought out so much piece meal and detached that it is hard' and 'indeed almost impossible for a person placed as I am – removed from European sources of information – to know how much of what I possess is new and how much already known and described'.[101] Distance was also an issue in other ways, and as Royle noted in 1838: 'numerous useful plants have been introduced into India by the Calcutta Botanic Garden, and others by that of Saharunpore' but 'more might have been introduced into the former from the new world, had there been more frequent direct communication with different parts of South America, Africa and India' and, moreover, 'the Northern Garden might have acclimated many of Europe and North American plants [sic] had it

[100] Edward Madden, 'On the Occurrence of Palms and Bambus, with Pines and Other Forms Considered Northern, at Considerable Elevations in the Himalaya', *Transactions of the Botanical Society of Edinburgh* 4 (1853): 185.

[101] Falconer to William Jackson Hooker, 20 May 1837, Kew Archives, DC/53/68.

not been so remote from both Calcutta and Bombay'.[102] Indeed, writing after he had retired to England to take up a professorship in *materia medica* at King's College London – as one of the few EIC employees featured in this book to successfully transition to a metropolitan scientific career– Royle suggested that more would have been achieved, but as 'Saharunpore is remote from the sea, the means of obtaining European plants are few and difficult, and seeds in a vegetative state arrive but seldom'.[103] Transfer into the mountains could be especially problematic, and circulation in both directions often broke down. Sometimes, this made the introduction of European species prohibitive: 'attempts were also made to obtain some of the fruit trees of England . . . but the distances are so great, and the modes of transport were so little understood, that only one apple tree arrived alive at Saharunpore, and thus cost no less than £70'.[104]

The other prominent issue around which the difficulty of displacement coalesced was the lack of a substantial, let alone comprehensive, library and herbarium, as well as the unreliable receipt of up-to-date European scientific journals. This was not only a problem for itinerant travellers, as we saw in Chapter 4, but also for those operating out of decentred institutions. As Royle wrote: 'there is no library either public or private within many hundred miles of this station' and 'without books it is utterly impossible to make any progress in the study of any of the branches of natural history'.[105] He continued: 'possessed of only a few elementary works ... I felt I could not do justice to the duties of my appointment if I did not possess myself of a library at whatever expense and pecuniary sacrifice this was to be effected'.[106] Royle indicated that he had spent more than 4,000 rupees of his own money, and asked for an extra allowance to purchase what he considered the necessary books. However, while his work was complimented, the request was refused (although a few duplicate books from the Calcutta garden were forwarded).[107]

Without an adequate library and herbarium, the Saharanpur garden – in spite of its privileged access to the Himalaya – risked being reduced to simply a data-gathering station, and its superintendents dismissed as mere collectors rather than 'philosophical' botanists. Certainly, that was

[102] Royle to J.C. Melvill, 31 December 1838, British Library, IOR/F/4/1828/75444, f12v.
[103] Royle, *Essay on the Productive Resources of India*, 217.
[104] Royle, *Essay on the Productive Resources of India*, 226–27.
[105] Royle to Nathaniel Wallich, 1 December 1827, British Library, IOR/P/12/28, f25v.
[106] Royle to Nathaniel Wallich, 1 December 1827, British Library, IOR/P/12/28, f25v-f26r.
[107] Royle to Nathaniel Wallich, 1 December 1827, British Library, IOR/P/12/28, f27v; Nathaniel Wallich to Charles Lushington, 13 December 1825, British Library, IOR/P/11/46, f3r.

their fear.[108] These problems were not necessarily unique to remote mountainsides and the edges of empire. Playing up locality and resource limits was a way of claiming relevance in debates over roles and practices across all levels of a changing discipline, from amateur natural history in Britain itself to an increasingly formalised imperial botany ostensibly governed from Kew. Here Jim Endersby has fruitfully examined the tensions around recognition and exclusion in 'philosophical' botany, especially with regard to the career of Joseph Hooker and his collectors in both India and Europe. Endersby shows how 'Hooker was one of the people who created the modern scientist, not least because he showed how it was possible to earn a living from science without sacrificing one's respectability'.[109] However, for those who needed an income for their science that tied them to India for much of their careers, and who were less well-connected than Hooker, the practical and social challenges to not be relegated to a lower tier of knowledge-making were even greater. Likewise, even if not unique, these problems were very much real and ongoing. Hugh Falconer was still complaining about the lack of a library in 1839, although an extensive herbarium of around 4,000 specimens and 500 drawings had by this time been accumulated by Royle (who left a set at Saharanpur while disseminating duplicates to Calcutta, and later to London).[110] Even while naturalists in remote locations wrestled with the question of what sort of science they could or could not do with the resources available, others were already judging them from afar. Lacking access to books could cement inferiority, and mean that a botanist and their work might not be taken seriously. As English officer William Munro commented of Edward Madden: 'I met him at Simla and did not form any very rich notion of his Botanical acquirements, indeed he had no books'.[111] Given that Madden spent so much time denigrating his collectors and placing them lower than himself in hierarchies of knowledge production, it is worth taking a moment to appreciate the irony that others would, in turn, do the same thing to him.

As apparent in previous chapters, the grafting of scientific interests onto official duties was an ongoing source of tension in this context, and the role of the superintendent of the Saharanpur botanic garden was always

[108] For this tension elsewhere, see Jim Endersby, '"From Having No Herbarium." Local Knowledge versus Metropolitan Expertise: Joseph Hooker's Australasian Correspondence with William Colenso and Ronald Gunn', *Pacific Science* 55, no. 4 (2001): 354.

[109] Endersby, *Imperial Nature*, 5.

[110] Falconer to Currie, 11 April 1839, British Library, IOR/F/4/1828/75444, f64v. See also Royle, *Essay on the Productive Resources of India*, 240–41.

[111] William Munro to William Jackson Hooker, 21 March 1848, Kew Archives, DC/54/352.

combined with that of district surgeon in the first half of the nineteenth century. As Royle lamented in 1825, the issue 'principally tending to obstruct the full and efficient prosecution of my Botanical pursuits is the unceasing attention which the Medical duties of the Station demand', although from the Company's point of view it was for these duties that at least part of his salary was paying.[112] These medical duties were also limiting in that they restricted Royle's ability to conduct what he considered to be essential travel. As he continued, discussing the outpost that Govan had established at Nahan (before it was replaced by Mussoorie), this nursery 'cannot be expected to thrive when left to the exclusive management of the native establishment, but evidently requires the occasional presence of the Superintendent' and similarly 'it is in the Hills also that the richest field for Botanical researches will be found and it is from them that the most rare and valuable plants have been obtained'. Royle continued to explain that 'these Botanical excursions have always indeed been considered as an essential part of the Superintendent's duty and were frequently carried into execution by Dr Govan' but 'instead of occasionally examining the nursery of Hill plants, I have literally never had an opportunity of even seeing it', and he instead had to rely on collectors.[113] These workload issues nevertheless remained unresolved, and Falconer was still making the same arguments to his superiors about the need to split the medical and garden duties in 1839, though with no more success than Royle.[114]

Complaints about excessive duties and problems operating in a remote location like Saharanpur were, however, double-edged. If displacement brought challenges for scientific practice, there was always a tension arising from the opportunities it simultaneously conferred (an important motivation for taking such a post in the first place). Indeed, Falconer went on to acknowledge some of the advantages: 'the Garden at Saharunpoor is happily situated in many important respects' because 'placed almost on the northern limit of the British possessions in Hindoostan, it is adapted to supply the wants of the north western provinces, which the remote position of Calcutta, precludes the noble institution there, from doing with effect'.[115] This distance could also give the superintendents authority from locality. As John Lindley wrote in 1839: 'with regard to the seeds which can be transmitted with the best hopes of success from England to India for distribution to public officers and other residents in the

[112] Royle to Nathaniel Wallich, 1 June 1825, British Library, IOR/F/4/955/27123(2), f165.
[113] Royle to Nathaniel Wallich, 1 June 1825, British Library, IOR/F/4/955/27123(2), f166-8.
[114] Falconer to Currie, 11 April 1839, British Library, IOR/F/4/1828/75444, f50v.
[115] Falconer to Currie, 11 April 1839, British Library, IOR/F/4/1828/75444, f55v.

Himalayan range of mountains', ultimately 'it is impossible for one who is personally unacquainted with the country to advise the Honorable Company with so much confidence as another person might who has himself resided in the provinces referred to'.[116] Royle too, however much he complained about the limitations of being stationed at Saharanpur, was aware that the possibilities it opened up were ultimately responsible for his successful career as a naturalist. The prospectus for his *Illustrations* made this abundantly clear: 'Mr Royle having been for several years Superintendent of the Honorable East India Company's Botanic Garden at Saharunpore, in 30° of latitude, one thousand miles to the north-west of Calcutta, and within thirty miles of the commencement of the ranges of the Himalaya' has necessarily had 'both from his situation and duties considerable opportunities for becoming acquainted with, and making collections of, the natural production of the parts of these mountains, which he had an opportunity of visiting, or could reach be means of his Plant collectors'.[117] Operating out of decentred locations like Saharanpur was thus characterised by attempts to balance often irreconcilable tensions between the scientific, economic, political and ornamental purposes of the gardens. However, this was always a catch-22 for the superintendents (especially Royle and Falconer) because it was their time at Saharanpur that they were able to leverage into positions in the scientific world beyond India, transitioning with some success to metropolitan careers in London.

As the nineteenth century unfolded, new technologies seemed to offer to make the globe smaller, and potentially alleviate some of the problems of remoteness. In particular, the use of steamships was touted by several of the Saharanpur superintendents as a means of diminishing distance.[118] As Royle argued, if 'different ages of the world have been memorable for the different routes of commerce, as well as for the interchange of the useful plants of different countries', then 'so may the present time be distinguished by the more numerous introduction into India of useful plants, in consequence of the facilities afforded by Steam Navigation'.[119] Writing in 1838, Royle elaborated on this in relation to Saharanpur, suggesting that 'in the North of India therefore much may be done, and here steam navigation proceeding from the South of Europe to Bombay and thence over land for seeds or up the Indus for plants affords every

[116] John Lindley to J.C. Melvill, 16 January 1839, British Library, IOR/F/4/1828/75444, f21v.

[117] Royle to William Jackson Hooker, 21 October 1832, Kew Archives, DC/53/102. The printed prospectus was enclosed with this letter.

[118] See Arnold, *Science, Technology and Medicine*, 101–5.

[119] Royle, *Essay on the Productive Resources of India*, 440–41.

desirable facility'.[120] Hugh Falconer, the superintendent at the time, was similarly enthusiastic about sending seeds via this route, as was Lord Auckland, the Governor-General.[121] However, extended discussions of the potential for steam technology to shrink travel times actually serve to highlight that distance was an as yet unresolved concern. Indeed, steamboats never managed to fulfil their promise for the pursuit of Himalayan botany, at least in the first half of the nineteenth century. William Jameson was still hopeful in 1852, writing: 'there are a great many interesting and valuable Himalayan Plants the introduction of which into England has up to this time failed ... [but] the experience that we have now gained in transmitting plants in Wardian cases convinces me that all the plants so desirable to introduce might most advantageously be forwarded ... to Calcutta by the first available steamer' and then onto London, and unless this method of transportation was followed, they 'will never reach England alive or in a germinating condition'.[122] The ongoing imaginative panacea of steamships thus reminds us that this is far from a story of unfettered circulation and appropriation. As well as the issue of distance in transporting viable live specimens and seeds, drawings were lost in the post, labels were mixed up and specimens went mouldy.[123] Indeed, the history of botany has sometimes been told as one of accomplishment and the remaking of the world, and certainly the long-term impacts of the circulation of species on a global scale can hardly be underestimated. However, as this chapter has aimed to demonstrate, paying attention to the disconnects and moments that these practices broke down can be just as enlightening.

Conclusion

This chapter has examined the period in which Saharanpur and Mussoorie became the facilitators of the botanical appropriation of the Himalaya, in a broader context of the scientific imagining of a vertical globe. Though never as lauded as Calcutta, when the eyes of European

[120] Royle to J.C. Melvill, 31 December 1838, British Library, IOR/F/4/1828/75444, f13r-f13v.
[121] Falconer to Currie, 11 April 1839, British Library, IOR/F/4/1828/75444, f70r; Lord Auckland to Court of Directors, 16 August 1838, British Library, IOR/F/4/1732/69955, f2r; Lord Auckland to Court of Directors, 16 August 1838, British Library, IOR/F/4/1732/69955, f3r-f4r.
[122] William Jameson to J.W. Sherer, 14 August 1852, British Library, IOR/F/4/2528/145895, f5v-f6r. There were nevertheless significant overall improvements in success rates in plant transportation across this period. See for example, Hooker, *Himalayan Journals*, Vol 2, 247–48.
[123] See for example, George Govan to Fraser, 14 April 1820, British Library, IOR/P/10/10, f12v-f13r.

savants and gardeners turned to the Himalaya, Saharanpur had its moment as a nexus of global botany.[124] Indeed, as Hugh Falconer wrote in 1839: 'for the investigation of the Botany of the northern part of the plains of India and of the Himalayah mountains . . . the extensive and important contributions which it has made in this department, under the zealous management of Mr Royle have given the Botanic Garden a well-known European celebrity'.[125] Much has been made by scholars of Kew's role as a global garden, but this could also be true of gardens in the peripheries.[126] Saharanpur was from its earliest days a space of global science, oriented both upwards to the peaks and outwards to the rest of the globe, and ultimately serving as the interface between the mountains and the plains. The situations of the 'northern' gardens were globally compared by the superintendents in terms of both altitude and latitude, and in this manner fed into the scientific imagination of mountains more broadly. However, located in the foothills, the gardens of Saharanpur and Mussoorie also indicate the inherent uncertainty in differentiating between lowlands and highlands, existing – sometimes uneasily – in the grey area between the two. Indeed, the need for Mussoorie to compensate for the limits of Saharanpur, which had been founded only a decade earlier to compensate for the limits of Calcutta, is indicative of the haphazardness of early attempts to mark the graduations of the vertical globe, and to formulate the highlands and lowlands as distinctive spaces for thinking about environmental and scientific concerns.

By looking at two interconnected institutions, this chapter has broadened the focus of previous chapters on expeditions, many of which nevertheless relied on places like Saharanpur to facilitate their itinerant scientific and imperial activities in the high mountains. Indeed, while in Kashmir in 1848, Thomas Thomson used Saharanpur as his relay, writing to William Hooker at Kew that 'in future I shall forward [seeds and specimens] through Jameson at Saharanpore addressed to Dr Royle as

[124] Later in the century, especially through the middle and latter parts of William Jameson's long superintendence, Saharanpur became most known for the promotion of tea cultivation, while Mussoorie also continued to operate, but in an increasingly commercial capacity. Jameson's role evolved into that of Superintendent of Botanical Gardens, North-Western Provinces, but by this time Saharanpur was no longer a major scientific clearing house. Indeed, in the second half of the nineteenth century the 'northern' gardens took on a more agrarian and horticultural focus, aimed at supplying seeds for both public and private gardens, and especially to line imperial roads and supply materials for the railways. See Deepak Kumar, *Science and the Raj, 1857–1905* (New York: Oxford University Press, 1995), 80.

[125] Falconer to Currie, 11 April 1839, British Library, IOR/F/4/1828/75444, f56r- f56v.

[126] See especially Richard Drayton, who makes Kew his 'protagonist' for a world history. Drayton, *Nature's Government*.

you recommend'.[127] In this chapter, I have also particularly focused on the way that obtaining, identifying and acclimatising specimens in order locate them on the vertical globe relied heavily on the likes of Hari Singh and Murdan Ali, who played key roles in hierarchical networks of Indian gardeners and collectors. Because the Himalaya were understood in this period in no small part through their botany, this dependency matters. However, as this chapter has simultaneously demonstrated, dependency was only one of many reasons that global sciences were prone to breakdown: whether through the imperfect circulation of material, the limits of the imagination or as a result of EIC politics and economics. The persistence of Saharanpur and Mussoorie, and their central – if uneven – part in the making of Himalayan botany nevertheless played a key role in providing material for the global imagining of mountains in this period, as Chapter 6 on vertical limits details.

[127] Thomas Thomson to William Hooker, 26 April 1848, Kew Archives, DC/54/502.

6 Vertical Limits

In 1837, Robert Boileau Pemberton (1798–1840), at the time special 'Envoy to Bootan', made an extended trip to the country with 'two very excellent barometers' firmly in hand. In an official report submitted to the East India Company and later published, he noted that 'the accurate determination of heights is a point of such vital importance in every investigation relating to the geographical limits of certain descriptions of vegetation' as well as 'the habitats of animals and birds, that many most valuable and extensive collections have been rendered comparatively useless by inattention to it'.[1] William Griffith, who accompanied Pemberton to Bhutan – officially as surgeon, but in practice as expedition naturalist – was similarly blunt: 'the Botanist who travels without the means of determining these points, destroys half the value of his collections'.[2] Pemberton and Griffith were part of a new breed of travellers who were adamant that observations of the natural world were meaningless without being attached to precisely measured and quantified elevations above sea level.[3] This chapter builds on the previous five, arguing that the first decades of the nineteenth century were decisive in the delineation of an imagined globe that was not only round but also vertical. In so doing, it continues to emphasise the unevenness with which information travelled over imperial and scientific networks, and shows that necessary global comparisons could sometimes increase rather than alleviate uncertainties around the vertical organisation of the world. However, it also adds another important facet to this broader argument by focusing on the way that horizontal and lowland norms inflected the imagining of the vertical globe, and added a new dimension to the imperial bounding of global space.

[1] Robert Boileau Pemberton, *Report on Bootan* (Calcutta: Bengal Military Orphan Press, 1839), 138.

[2] Griffith, *Journals of Travels*, 430.

[3] Bourguet, 'Landscape with Numbers', 116–17. See also Feldman, 'Applied Mathematics and the Quantification of Experimental Physics'.

While Chapter 5 considered the mountain gardens of Saharanpur and Mussoorie and the appropriation of Himalayan botany through institutions, we now return to itinerant travellers in the spaces of the high mountains. In what follows, I consider the observation, comparison and visual representation of a range of different (if often interconnected) altitudinal limits and zones in the Himalaya: of plants, of animals, of crops and of human habitation. These limits were addressed by naturalists and travellers through ideas and practices that fed into the modern discipline of biogeography. However, while capturing the same impulse to map flora and fauna through space and time, the term 'biogeography' was not used until the twentieth century, which is why I prefer contemporary terms like 'geography of plants' and 'botanical geography' to describe the practices in this chapter, even while pointing to the role of these mountain activities in the formation of the discipline.[4] Originally applied to the horizontal distribution of plants, the geographical distribution of species was coming to be understood as equally important in the realm of the vertical. Among other aesthetic, imperial and scientific strategies for imagining this, European travellers in the Himalaya took existing horizontal divisions – tropical, temperate and arctic (or polar) – and, as had been done for the Andes, mapped them onto the vertical.[5] In so doing, they analogised and stretched ideas about two-dimensional distribution to encompass the third. As John Forbes Royle wrote in 1839:

> One regrets the poverty of the language at present applied to the geography of plants, as it is impossible to indicate the nature of mountain vegetation by merely using the name of the range; for as we have seen in the case of these mountains, the vegetation varies, and is analogous to that of very different countries, according to the elevation or as peculiarities of local circumstances cause a variation in climate. The inconveniences of this might, it appears, be considerably remedied, if botanical regions on the surface were more circumscribed according to their respective climates, or taking the several zones of latitude.[6]

[4] Analogues in other languages, including 'géographie des plantes' and 'pflanzengeografie', were also widely used. See especially Janet Browne, 'Biogeography and Empire', in *Cultures of Natural History*, ed. Nicholas Jardine, James A. Secord, and Emma C. Spary (Cambridge: Cambridge University Press, 1996), 484. See also Nicolson, 'Alexander von Humboldt, Humboldtian Science and the Origins of the Study of Vegetation'; Pär Eliasson, 'Swedish Natural History Travel', in *Narrating the Arctic: A Cultural History of Nordic Scientific Practices*, ed. Michael Bravo and Sverker Sörlin (Canton, MA: Science History Publications, 2002), 125–54.

[5] For a study that considers places of high altitude and high latitude together, see Cosgrove and Della Dora, *High Places*.

[6] Royle, *Illustrations*, Vol 1, 310. See also Arnold, *Science, Technology and Medicine*, 52–53; Arnold, *The Tropics and the Traveling Gaze*.

Accounting for a globe that was vertical as well as round was thus initially achieved by borrowing from the language of latitude, and adapting it to explain variations in the realm of up and down. This had both short- and long-term implications for understandings of mountain environments and mountain peoples. Applying existing horizontal divisions meant simultaneously overwriting long-standing local cosmologies, and broader South Asian imaginings of the Himalaya. Much as inscribing 'blank spaces' onto maps erased indigenous presences, so too did dividing up the mountains according to European norms derived from the horizontal. This is precisely why it is necessary to historicise the reconfiguration of the vertical globe in this period, and ultimately to recognise the imperial utility of notions of a commensurable 'globe' more widely.

It was only occasionally acknowledged that these vertical zones were being written over existing indigenous topographies. To take a rare example, polymathic Himalayan scholar and diplomat Brian Houghton Hodgson remarked that 'the third region of Nepal is the juxta-Himalayan, called by [Francis] Buchanan the Alpine, and by the natives denominated the Kachár'.[7] Edward Madden, meanwhile, remarked that 'the vegetation of Cheenur and Nynee Talthus presents some difficult problems, which the natives resolve at once by the assertion that the Oak, Cypress, Limonia, Colquhounia, &c, were imported from the snowy range and planted here by Devee herself'.[8] While framed as the dismissal of indigenous superstition and myth, Madden acknowledges that these stories pointed to some fundamental questions about the distribution of plants in high mountains. To be clear, this chapter is primarily concerned with the European imagining, representation and appropriation of the Himalaya. It is not intended to reconstruct – as has been ably done by other scholars – the cosmologies that these scientific imaginings were imposed on top of, even while it pays careful attention to the vestiges of older notions of space that informed European visions of the mountains.[9] Instead, it demonstrates that imagining the vertical globe through the language of latitude was implicitly a form of imperialism, and allowed for the subsuming and appropriation of the Himalayan landscape, flora, fauna and peoples into a framework that was explicable and therefore exploitable by European science and empire. In this chapter, I thus build on the work of Janet Browne, who in characterising plant geography more broadly notes that 'the nature of the assumptions and explanations put

[7] Brian Houghton Hodgson, 'On the Mammalia of Nepal', *Journal of the Asiatic Society of Bengal* 1 (1832): 338.

[8] Madden, 'The Turaee and Outer Mountains of Kumaoon', Vol 17, 361.

[9] See, for example, Mathur, 'Naturalizing the Himalaya-as-Border in Uttarakhand'; McKay, *Kailas Histories*.

forward to account for biogeographical regions can be attributed to the overriding ethos of colonization . . . it provided metaphors and a rationale; the raw materials and a way to understand'.[10]

Overwriting pre-existing spatial arrangements and organising vertical zones according to horizontal sensibilities was thus essential to establishing the Himalaya as a coherent space that could be used to make long-distance comparisons. As we have seen, references were most frequently made to the Alps and the Andes, though Tenerife, North America, China and even Siberia also sometimes featured.[11] As with altitude physiology and geology discussed in previous chapters, the imagining of plant geography in the Himalaya inevitably played out through global comparisons. This chapter thus serves as the culmination of my argument for the inherently comparative nature of mountain sciences in this period, and the dawning of a global consciousness of verticality.[12] I have argued elsewhere (particularly in Chapters 4 and 5) that this globality had a material dimension, in the circulation and movement of enormous quantities of things – specimens, inscriptions, drawings and personnel. This chapter, however, especially expands on the imaginative dimension of global comparison, and considers the way surveyors and naturalists envisioned their science on a global scale.

While the commensurate nature of mountain environments was increasingly recognised, comparisons nevertheless sometimes continued to reveal significant differences. For example, as James Gerard wrote in a letter published in the first volume of India-based scientific journal *Gleanings in Science* (1829): 'I came upon a village at a height of 14,700 feet; are you not surprised that human beings could exist at such an elevation?'[13] The question mark here is explained by the global comparative context, one that was both contradictory and surprising: 'men, animals, and vegetable productions succeed better here than in the valley below, all thriving profusely in a zone that contracts and terminates every trace of plants in the Andes under the Equator'.[14] Comparisons made it clear that certain phenomena – such as a line of perpetual snow – occurred in high places everywhere, but inconsistences in the heights at which these appeared could thus lead initially to more confusion rather than coherence. These issues were sometimes magnified, as we have seen, by an

[10] Browne, 'Biogeography and Empire', 320.

[11] See, for example, John Forbes Royle, *An Essay on the Antiquity of Hindoo Medicine* (London: W. H. Allen, 1837), 16.

[12] I take inspiration here from Michael Reidy's use of the term 'vertical consciousness'. See Reidy, *Tides of History*, 280.

[13] Gerard, 'Letter from a Correspondent in the Himalaya', 109.

[14] Gerard, 'Observations on the Spiti Valley', 244.

overreliance on Alexander von Humboldt and the better-known norms of the Andes, which could cause interpretive problems when it came to addressing the Himalaya. (Aside from volcanoes and glaciers, other difficulties when comparing with the Andes stemmed from the way South America's largest range runs mostly north–south – that is, perpendicular to lines of latitude – while the Himalaya run mainly east–west and broadly parallel to their median latitude.)

Unravelling contradictory comparisons was also exacerbated by the difficulties inherent in operating in locations displaced not only from London, but also from 'centres in the periphery' like Calcutta. While John Forbes Royle, William Griffith, Victor Jacquemont, Thomas Thomson and Joseph Dalton Hooker – all of whom were immediately or eventually recognised by their peers as botanists – play key roles in this chapter, it is nevertheless also significant that observations were often made by those with limited or no botanical training. This was especially so in the case of officers and surgeons seconded from the Bengal Infantry to work on surveys in the high mountains. As James Herbert instructed a fellow Bengal Infantry officer who was about take over his survey: 'the elevation of the different zones of the vegetable kingdom, including its highest limit, (though strictly speaking not within the line of duty), yet may be deemed for their interest, not unworthy a little time and attention'.[15] As in early measurements of the heights of the Himalaya (and their superiority over the Andes), the social status of the India-based East India Company employees sometimes led to tensions. For example, James Gerard, confessing his lack of botanical expertise, nevertheless argued that 'the excessive cold that reigns at the highest cultivable levels of the *Intra Himalayan* regions during the greater part of the year in no way cramps the progress of vegetation, since this is effected by the necessary quantity of heat during the appropriate season' and indeed 'the solar rays of this parallel of latitude, in so thin and transparent an atmosphere, are infinitely more powerful' but 'these facts, and their effects upon the constitution of men, animals, and vegetation, are not properly understood in Europe, or if known, are explained upon theoretical assumptions which have no grounds of existence in nature'.[16] Gerard is here leveraging his authority from locality to elevate his contributions to the debate. Based on his own observations from within the displaced spaces of the high mountains, he was able to argue that received theories (developed based especially on the Alps and the Andes) could founder when it came to encompassing the Himalaya.

[15] Herbert to Thomas Oliver, October 1821, NAI/SOI/DDn. 152, f225.
[16] Gerard, 'Observations on the Spiti Valley', 251.

This chapter investigates these debates in five sections. The first examines the absolute limits of vegetation, sublime responses to the end of the 'habitable world', and attempts to divide up the highest mountain range on the globe using a vocabulary of verticality borrowed from the horizontal. The second section extends these debates to animals. The third examines the temporal dimensions of mountain plant geography, debates over the 'tropicality' of the Himalaya and inconsistences in the line of perpetual snow. The fourth section considers the altitude limits of cultivation, firewood and human habitation and the ways these circumscribed life and movement in the mountains. The fifth and final section considers attempts to represent and understand these altitude limits visually. Throughout, I demonstrate that as much as from abstract natural historical interests, observations of altitude thresholds were wrapped up with the concerns of empire. Understandings of verticality informed discussions of the potential for European colonisation in the mountains, and the question of how high was habitable was simultaneously a question of how secure the frontiers were. As Peter Bishop has eloquently argued, in the early nineteenth century, European travellers and administrators looking up to the Himalaya struggled to impose imaginative coherence on their newly acquired frontiers.[17] Organising the mountains according to the language of latitude, and from norms derived from the Alps and the Andes, thus represented a key – if never wholly successful – strategy for addressing this imaginative incoherence.

Limits, Zones and the Language of Verticality

As they gained more reliable access to parts of the high Himalaya in the first decades of the nineteenth century, European travellers were keenly interested in the absolute upper limit at which plants could exist. The highest altitude reached in this period was probably that attained by the Gerard brothers on Reo Purgyil, as discussed in Chapter 2. While painstakingly measuring this high point, Alexander Gerard also took note of the vegetation: 'we reached the elevation of 19,411 feet, and found mosses'.[18] Travelling two decades later, Bengal Infantryman Henry Strachey found 'a few minute plants chiefly mosses [and] lichens ... at heights of 18,800 feet and the most stunted forms of lichens that gave a color rather than a coating to the surface of stones, seemed to extend one or two hundred feet higher'.[19] Observations and specimens thus served to

[17] Bishop, *The Myth of Shangri-La*, 89.
[18] Gerard, *Account of Koonawur*, appendix, xxv.
[19] Henry Strachey, 'Tibetan Commission Reports', British Library, IOR/F/4/2461/136807, f560.

fix the upper ceiling at around 19,000 feet, which coincided with the limits of the high passes and pre-existing routes of travel through the mountains. Though impressed by the ability of certain species to survive in an extreme environment, there was nevertheless often a sense of melancholy at reaching the end of vegetation. Henry Strachey's brother Richard wrote of the 'cessation of life as we ascend in elevation', while Victor Jacquemont sighed as, at 11,500 feet on the side of the peak of Kedarkantha, 'the forest expired'.[20] In a similar vein, James Gerard wrote of making 'preparations for ascending the parent chain, and I may say, to take leave of the world for some time'.[21] Reaching and recognising these thresholds thus often saw recourse to the sublime. This was never entirely melancholy, and elsewhere in the same letter, penned while wracked with altitude sickness in a cramped tent some 13,800 feet above sea level, Gerard remarked that 'the idea that we are beyond the habitable world makes us catch eagerly the stir of the wind, the flutter of an insect, or the noise of some rock in its fall; and although we feel an emotion that we cannot describe, the mind still partakes of the serenity of the region around'.[22] These romantic, picturesque and sublime reactions were intimately linked with the scientific constitution of the mountains. Where the 'habitable world' ended, for what and for whom was nevertheless a subjective judgement, and to some extent always relative. Brian Hodgson, for example, suggested that the central band of mountains were 'splendidly wooded', and 'this generally from 10 to 16,000 feet above the sea, up to the limit of habitability; where, of course, I stop'.[23]

As well as invoking the sublime, travellers operating in the Himalaya frequently speculated on the underlying causes of these vegetational changes and their ultimate ceiling. Beyond the obvious factor of decrease in temperature, they also considered precipitation, solar radiation, lack of soil and the rarity of the air. William Griffith suggested that temperature was crucial, using examples of microclimates he encountered: 'I had here an opportunity of observing the curious effect of a patch of snow in retarding vegetation, all the plants about, being as it were a spring flora, even such as at similar elevations elsewhere, were all past seed'.[24] However, he acknowledged that moisture was perhaps equally important, and Thomas Thomson agreed, singling out 'the change produced in the

[20] Strachey, 'On the Physical Geography', 78; Jacquemont, *Voyage dans l'Inde*, Vol 2, 127. ['la forêt expira']
[21] Gerard, 'A Letter from the Late Mr J.G. Gerard', 293.
[22] Gerard, 'A Letter from the Late Mr J.G. Gerard', 298.
[23] Hodgson, 'On the Mammalia of Nepal', 338. For more on Hodgson, see Waterhouse, *The Origins of Himalayan Studies*.
[24] Griffith, *Journals of Travels*, 396.

vegetation in the temperate and subalpine zones as we advance towards the interior of the mountains, in consequence of the diminution in the amount of rain'.[25] The brutal debility experienced by human and non-human bodies as a result of the rarity of the air, even if still little understood, was nevertheless generally ruled out as a principal cause for the vegetational ceiling. As Joseph Hooker put it at the mid-century: 'it has long been surmised that an alpine vegetation may owe some of its peculiarities to the diminished atmospheric pressure' but 'I know of no foundation for this hypothesis; many plants, natives of the level of the sea in other parts of the world, and some even of the hot plains of Bengal, ascend to 12,000 and even 15,000 feet on the Himalaya, unaffected by the diminished pressure'.[26] Even if change in pressure could be dismissed, other environmental and climatic factors could not be easily isolated. Holistic explanations were thus required, indicating a shift towards an ecological way of thinking *avant la lettre*.

If the limits of knowledge meant that theories were in constant danger of contradiction, the idea that laws might be established was not in doubt. A logical extension of the discussion of altitude limits was to use them in a predictive capacity, and to estimate heights based on species that occurred. For example, Victor Jacquemont, relying on Kashmiri traders as plant collectors, wrote that vegetation, when 'taking into account the law according to which the temperature decreases from the equator to the pole', could 'speak so precise a language to one who can interpret it, concerning the height of places' and that 'in the complete ignorance which existed before my journey of the level of this celebrated valley [Kashmir], I had fixed it at between five and six thousand English feet, from a small number of plants which I had seen brought by merchants'.[27] When he later visited Kashmir, Jacquemont could smugly recount: 'now, my own observations make it about five thousand one hundred and fifty feet. It was with the most lively satisfaction, that I saw the final logarithm of my calculation transform itself into this number'.[28] Much as human bodies might be calibrated to estimate height, so too it seemed that plants might become botanical barometers, even for places not yet visited by European naturalists.[29]

[25] William Griffith, *Itinerary Notes of Plants Collected in the Khasyah and Bootan Mountains, 1837–8 and in Affghanistan and Neighbouring Countries, 1839–41*, ed. John McClelland (Calcutta: J.P. Bellamy, 1848), 395–96; Thomas Thomson, 'Sketch of the Climate and Vegetation of the Himalaya', *Proceedings of the Philosophical Society of Glasgow* 3 (1855): 203.

[26] Hooker, *Himalayan Journals*, Vol 2, 415.

[27] Jacquemont, *Letters from India*, Vol 2, 76–7.

[28] Jacquemont, *Letters from India*, Vol 2, 77.

[29] The term 'botanical barometer' is adopted from Bourguet, 'Landscape with Numbers', 102, 106.

While searching out and trying to explain absolute limits, naturalists also developed and applied various schemes for dividing up the Himalaya into altitudinal zones. Usually borrowed from the horizontal, these could help both imagine and explain the vertical realm. John Forbes Royle offered one of the most comprehensive of these schemata. Adopting the natural system of classification – 'the only method which enables us to treat systematically of their Geographical distribution'– he argued that a new vocabulary was needed.[30] Here he proposed that the language of latitude might encompass the vertical realm of the Himalaya:

> Mountains might be similarly divided into zones or belts, according as elevation, climate, and vegetation, displayed sufficient differences to warrant the distinction. We have frequently seen, that according as we observe [sic] the natural phenomena, at the base or towards the apex of these mountains, the correspondence is either with tropical, European, or polar regions. This might be indicated by a word compounded of that of the mountain range, and of the zone to which the belt corresponded, as Tropico-Himalayan, Arcto-Himalayan, &c., which would sufficiently indicate the nature of the vegetation at different elevations, as well as the geographical situation.[31]

Royle was aware that these ideas were not necessarily his own or novel, but pointed out that they had never been systematically applied in the Himalaya. He also acknowledged that the 'bounds are in a great measure arbitrary to which each of these belts have been restricted, for the changes, both in temperature and vegetation, are so gradual'.[32]

Two decades later, Thomas Thomson was similarly conscious of the fundamental subjectivity of dividing up the range into belts given the way they bled into one another. He nevertheless argued that 'some mode of subdivision is quite necessary for the purpose of description, as otherwise the mind would be puzzled by the multitude of facts' and of course the simpler 'the mode of division is, the more intelligible it will be'. He therefore suggested that it was 'quite sufficient to refer the forms of vegetation to three groups, similar to the three zones interposed between the equator and the pole, namely, tropical, temperate, and arctic' (or in the latter case 'to use the term more commonly applied in the case of mountains, *alpine* vegetation').[33] More so than some of his predecessors, Thomson was also interested in the significant differences in vegetation that occurred laterally across the vast span of the Himalaya and its associated ranges. He remarked that describing elevational changes would be relatively easy if vegetation was uniform across the range, but

[30] Royle, *Illustrations*, Vol 1, 443. [31] Royle, *Illustrations*, 311.
[32] Royle, *Illustrations*, 16.
[33] Thomson, 'Sketch of the Climate and Vegetation of the Himalaya', 200.

'few indeed of the plants of the eastern extremity of the Himalaya ... [are] identical with those which occur in the far west. In general terms, it may be said, that to the eastward the vegetation is very much more luxuriant and tropical'.[34] Indeed, despite the homogenising tropes that remade the Himalaya into a globally commensurable range, they remained hugely diverse in terms of topography, and their western and eastern reaches presented very different pictures in terms of plant geography. This variety is thus a further reminder of the need to pay attention to the unevenness of exploration and access, especially given the outsized role places like Kumaon and Garhwal came to play in imperial imaginative repertories.

Aside from the language of latitude, another key criterion for divisions, and an analogical language for imagining, understanding and appropriating the Himalaya, was the appearance of so-called 'European' forms. These were associated with the temperate bands of the Himalaya, or the middle zone in Royle's triptych between 'tropical, European, or polar regions'. Comparisons with Europe could nevertheless result in contradictions, and Victor Jacquemont wrote of several species on Kedarkantha which were 'all European forms' but 'found similarly associated in the Alps at less than half the height'.[35] (William Griffith also added a somewhat wry caveat to his own observations: 'European vegetation continues, so far as such a statement is assumable by one who never was beyond Paris'.[36]) Royle was particularly enamoured with the similarities to home, stating that a European, 'on his first arrival in a tropical country, is struck by the magnificent peculiarities of its vegetation' though 'to one who has long resided in such a clime these become familiar, and his attention is more quickly excited by the re-appearance of forms with which he was familiar in his youth, and which characterize the more humble and verdant, but not less beautiful Flora of temperate climates'.[37] Royle thus found himself nostalgic for the 'familiar' flora of the mountains, as opposed to the 'exotic' flora of the plains. As he continued: 'in proportion as we ascend these mountains, the plants of India disappear, and we are delighted at finding the increase in number and variety of those belonging to European genera' though 'at first we see only a few straggling towards the plains, which in a more temperate climate would be their favourite resort' and indeed 'it is not until we have attained a considerable elevation that, having apparently lost all traces of tropical vegetation, we enter a forest of pines or oaks, and lofty

[34] Thomson, 'Sketch of the Climate and Vegetation of the Himalaya', 196–97.

[35] Jacquemont, *Voyage dans l'Inde*, Vol 2, 124. ['tous genres européens' but 'trouve pareillement associés dans les Alpes à une hauteur de moitié moindre']

[36] Griffith to William Jackson Hooker, 6 August 1840, Kew Archives, DC/54/231.

[37] Royle, *Illustrations*, Vol 1, 15.

rhododendrons, where none but European forms are recognizable'.[38] Royle's characterisation of 'European' vegetation 'straggling towards the plains' might easily be read as a metaphor for European bodies. Indeed, David Arnold notes that insecurities about mortality in the tropical lowlands 'gave added poignancy to this discovery of "European" flora in upland India'.[39]

These geographical imaginings thus fed into highly fraught debates about the long-term colonial possibilities for India, which hinged on the potential for European bodies to adapt and acclimatise (see also Chapter 3).[40] The analogical language of 'European forms' went hand-in-glove with the construction and popularisation of colonial hill stations at places like Mussoorie, Shimla and Darjeeling, beginning in earnest in the 1830s.[41] As Victor Jacquemont wrote when he first arrived in the Himalayan foothills, 'the English are so rich that no obstacle can stop them. I shall find them everywhere on the first and second *stories* of the mountains'.[42] In terms of the vertical globe, these intermediate elevations – or 'first and second *stories*' – came to be seen as the subcontinent's goldilocks zone for European bodies. At moderate altitudes they might thrive, much as 'European' plant species appeared to. Lower down in the hot plains, pestilence abounded, while climbing too high into the 'arctic', 'polar' or 'alpine' zone meant returning to a state where habitability was again questionable (albeit for very different reasons).

Brian Houghton Hodgson explicitly linked this vertical zone of colonisation potential to plant geography: 'the small or hill species of bamboo, which prevail from 4,000 to 10,000 of elevation, mark with wonderful precision the limits of the central healthful and normal region of the Himalaya'.[43] He also put forward some of the most formulated – if fanciful – thoughts on settling the mountains in this period, in an essay titled 'On the Colonization of the Himalaya by Europeans' (1856). He argued that this was a political imperative and the 'eminent fitness for European colonization having once been taken up, will never be dropped till colonization is a "*fait accompli*"' such that 'the accomplishment of this greatest, surest, soundest, and simplest of all political measures for the stabilitation [sic] of the British power in India, may adorn the annals of

[38] Royle, *Illustrations*, Vol 1, 15.
[39] Arnold, 'India's Place in the Tropical World, 1770-1930', 12–13.
[40] Harrison, *Climate and Constitutions*; Arnold, 'Race, Place and Bodily Difference'.
[41] See Kennedy, *Magic Mountains*. [42] Jacquemont, *Letters from India*, Vol 1, 184.
[43] Brian Houghton Hodgson, *Essays on the Languages, Literature, and Religion of Nepal and Tibet, Together with Further Papers on the Geography, Ethnology, and Commerce of Those Countries* (London: Trübner, 1874), part II, 22. See also Arnold, 'Hodgson, Hooker and the Himalayan Frontier, 1848–1850', 199.

the present Viceroy's administration'.[44] Hodgson did acknowledge that this would have to be a long-term process but clarified that he did 'not mean wholesale and instantaneous colonization, for any such I regard as simply impossible' given the 'distance and unpopularity of India', and especially its reputation for disease-ridden mortality.[45] He thus thought it necessary to demonstrate that 'in regard to the Himálaya, the vulgar dread of Indian diseases is wholly baseless – to show also that its infinite variety of *juxtaposed* elevations, with correspondent differences of climate', ultimately 'offer peculiar and almost unique advantages (not a fiftieth part of the surface being now occupied) to the colonist … to cultivate a wonderful variety of products ranging from the tropical nearly to the European'.[46] Hodgson presented a rather rosy view, concluding that 'there is, in fact, no end [to] the mineral and vegetable wealth of the Himalaya'.[47]

Writing in 1848, however, Edward Madden had been far less optimistic, arguing that 'data, fortified by experience, will enable us to rate at its proper worth the colonization cant which so often fills the gazettes, combined with the most exaggerated pictures of Himalayan resources, and the most chimerical schemes for railways, in a country where we are only too happy to find any roads at all'. As he continued, 'in sober truth, the resources of the mountains are not many, and are already as much developed as the nature of the country will admit of'. In particular, he remarked that 'the soil, except in the low vallies where the European colonist cannot exist, is generally poor, besides being pre-occupied, and often exhausted, by the aboriginal population', while 'the fine tracts of rich meadow, which flank the Snowy Range, are too remote for settlers, and are too high and too cold to ripen grain'.[48] Madden thus recognised the significant differences between the lower valleys and the high meadows, even while arguing that both offered insurmountable challenges. He concluded his pessimistic picture by reflecting on what he saw as the implications of this for the security of the empire: 'if the above be a true view of the case, it appears chimerical to hope that the Himalaya can ever maintain an independent body of colonists, such as might supersede the necessity of drawing recruits from Europe, or such as, on any emergency, could be brought down to act in the defence of the Lower Empire'. As he

[44] Hodgson, *Essays on the Languages, Literature, and Religion of Nepal and Tibet*, part II, 88.
[45] Hodgson, *Essays on the Languages, Literature, and Religion of Nepal and Tibet*, part II, 88.
[46] Hodgson, *Essays on the Languages, Literature, and Religion of Nepal and Tibet*, part II, 88.
[47] Hodgson, *Essays on the Languages, Literature, and Religion of Nepal and Tibet*, part II, 85.
[48] Madden, 'The Turaee and Outer Mountains of Kumaoon', Vol 17, 420–1. See also Grout, 'Geology and India', 53–58.

continued, 'this is a very different question from that of the fitness of the mountains for sanatory settlements occupied by those in the service of Government ... that, indeed, is no longer a question: a hundred applications for every vacant appointment in the mountains attest the "deep damnation" of a life in Hindoostan'.[49] While the foothills of the Himalaya – and the hill stations of Shimla, Mussoorie and Darjeeling – were to become central to colonial governance in India, the apparent impossibility of an 'Upper Empire' and the colonisation of the higher reaches of the mountains remained a source of insecurity.

Vertical Zoogeography

The limits of plants and people were rarely treated in isolation, and naturalists likewise traced zoogeographical distribution on a vertical globe. Brian Hodgson was the most prolific writer on Himalayan zoology in this period, and in his zoogeographical imagining of the mountains, he divided them into three vertical zones, based on those for plants.[50] For John Forbes Royle, it was also about plants first, and animals second, with the latter situated according to the geographical picture he had developed for vegetation: 'the animal kingdom affords many of the same indications of the Alpine nature of the country, as we have seen presented by the vegetable kingdom'.[51] In terms of absolute limits, Henry Strachey's observations also followed this hierarchy: 'the Zoology of all countries is probably proportionate to their Botany. Such any way is the case in Ladak; the stock of animal life being as scanty as the vegetation, and to the best of my knowledge, confined to the same limit of elevation vizt. 19,000 feet'.[52] These priorities reflect that colonial zoology was usually considered less important than colonial botany, in part because of its lack of obvious utility.[53]

It is thus unsurprising that the most sustained interest in high-altitude fauna in this period was directed towards a handful of commercial and

[49] Madden, 'The Turaee and Outer Mountains of Kumaoon', Vol 17, 423.

[50] Hodgson, 'On the Mammalia of Nepal', 338. David Arnold nevertheless notes that Hodgson initially avoided the terms alpine, temperate and tropical, likely because they seemed to be less easily applied to divisions of zoology than they did to botanical specimens. Arnold, *The Tropics and the Traveling Gaze*, 211.

[51] Royle, *Illustrations*, Vol 1, 24.

[52] Henry Strachey, 'Tibetan Commission Reports', British Library, IOR/F/4/2461/136807, f579.

[53] Arnold, *Science, Technology and Medicine*, 44.

domestic animals. One species with widely discussed (though ultimately unfulfilled) economic potential was the so-called shawl wool goat. As Hugh Falconer noted: 'the fine silky fleece, from which the Cashmeer shawls are wove, is abundantly developed at the roots of the long hairs of the domestic goat in the plains of Tibet, at, and upwards of, 16,000 feet above the level of the sea, where a highly rarefied atmosphere is combined with severe winter cold'.[54] Of Himalayan fauna, however, no species fascinated European travellers, nor was more important to high-altitude populations, than the yak (or its many, and indeed more common, hybrids). The most widely circulated depiction of a yak in the early nineteenth century comes from Samuel Turner's journey to Bhutan and Tibet in 1783, where it was depicted in front of an idealised Tibetan landscape (see Figure 6.1). Indeed, the yak would go on to become one of the key orientalised, exotic symbols of the Himalaya.[55] In the early decades of the nineteenth century, however, much was also made of the yak as a species seemingly exclusive to high mountains, restricted to the upper bands of the vertical globe. Turner did famously manage to transport a yak to England (another died in transit), where it survived for several years, and sired several hybrid offspring with domestic cows.[56] This would nevertheless become the exception that proved the rule, and other travellers, including John Wood and Joseph Hooker, reported that yaks quickly deteriorated and perished in the lowlands.[57] In his extended zoogeographical picture, Brian Hodgson noted that 'the *Bos grunniens* or Yak of Tibet likewise flourishes in the Kachár [alpine zone]: but not south of it'.[58] These elevational sensitivities could, however, be partially altered by cross-breeding, and as Hugh Falconer remarked, many if not most of these animals, especially at lower elevations, were hybrids rather than true yaks.[59]

Some naturalists also made more eclectic investigations into the vertical range of insects, birds and fish. They often expressed or affected surprise at the heights at which this fauna not only existed but also thrived. Henry Strachey remarked that 'the absolute height above the sea to which Birds are found ascending in these regions is remarkable. . . . [E]levated 18,400 feet my brother Richard and I saw large kites soaring high above us with the same slow and easy movement as in the plains of

[54] Falconer, *Palaeontological Memoirs*, Vol 2, 290.
[55] Bishop, *The Myth of Shangri-La*, 151–53; Teltscher, *The High Road to China*, 241–42.
[56] Turner, *Account of an Embassy to the Court of the Teshoo Lama*, 186–89.
[57] Wood, *Narrative of a Journey*, 322–23; Hooker, *Himalayan Journals*, Vol 1, 212–5.
[58] Hodgson, 'On the Mammalia of Nepal', 348.
[59] Falconer, *Palaeontological Memoirs*, Vol 1, 581–2.

Figure 6.1 'The Yak of Tartary'. Engraving based on a painting by George Stubbs, and figured in Samuel Turner's *Account of an Embassy to the Court of Teshoo Lama in Tibet* (1800). Joseph Hooker suggested that 'the artist is probably a little indebted to description for the appearance of its hair in a native state, for it is represented much too even in length'.[60] Image: © The Trustees of the British Museum.

India'.[61] Strachey went on to discuss insects, writing that even at high altitude, 'an entomologist would find something by searching; and I have been reminded of Humboldt's question *"Les insects sont ils moins haut que les plantes?"* by a beetle and a moth buzzing about my candle in camp above 17,000 feet, which is the highest I have ever seen insects'. He continued to note, however, that 'I do not think such a negative and isolated observation can warrant an answer to that question in the affirmative'.[62] As with botany, Strachey acknowledged his limitations as

[60] Hooker, *Himalayan Journals*, Vol 1, 214.

[61] Henry Strachey, 'Tibetan Commission Reports', British Library, IOR/F/4/2461/136807, f585-6.

[62] Henry Strachey, 'Tibetan Commission Reports', British Library, IOR/F/4/2461/136807, f589-90. ['Are insects less high than plants?'] John Forbes Royle also considered insects in some detail, see Royle, *Illustrations*, Vol 1, li.

a polymath, instead relying on his informants: 'I am incompetent to give a technical account of the zoology, but subjoin some notice of the principal animals known and named by the Tibetans'.[63] Thomas Thomson meanwhile observed the altitude limits of fish, explicitly comparing them to the range of humans in a three-dimensional world. As he recorded, 'the occurrence of fish in streams at 15,000 feet I considered at the time an exceedingly interesting fact [and] I don't think it likely that they can exist much higher – the same point seems to be about the highest level of human habitation & of cultivation'.[64] He also indicated that expectations based on received theories broke down, and '*à priori*, it would scarcely have been expected that they would have existed', especially 'as it would certainly not have been very surprising that air at that elevation should, from its rarity, be insufficient for the support of life in animals breathing by gills'.[65]

And what of that most notorious of Himalayan fauna, the yeti? In the period before 1850, the only reference in European accounts is a fleeting mention by Brian Hodgson in a footnote to 'On the Mammalia of Nepal', published in the first volume of the *Journal of the Asiatic Society of Bengal* in 1832. In this, he wrote: 'my shooters were once alarmed in the Kachár [alpine region] by the apparition of a "wild man," possibly an ourang, but I doubt their accuracy. They mistook the creature for a câcodemon or rakshas, and fled from it instead of shooting it. It moved, they said, erectly: was covered with long dark hair, and had no tail'.[66] Here Hodgson records the anecdote as information obtained from his assistants, who were in turn likely basing their depiction on existing Nepalese stories. Hodgson was immediately dismissive of the claims, while John Forbes Royle offered his own interpretation: 'the improbability of finding a real Ape in such a situation led him [Hodgson] to question the truth of the report' though 'it is well known that the woods of the lower ranges to the east of Nepal contain at least one species of Gibbon, *Hylobates Scyritus*, called *Hooloo* or *Hooloc* by the Assamese', and moreover it is 'not improbable that individuals may occasionally wander to the higher and more remote forests of the Central Hills'.[67] As Peter Bishop has shown, nothing more on the yeti appears to have been written or recorded by European travellers until the 1880s, which was in itself strange as 'stories about wild hairy men had long been integral to the folklore of

[63] Henry Strachey, 'Tibetan Commission Reports', British Library, IOR/F/4/2461/136807, f579.

[64] Thomas Thomson to William Jackson Hooker, 26 April 1848, Kew Archives, DC/54/502.

[65] Thomson, *Western Himalaya and Tibet*, 153, 165.

[66] Hodgson, 'On the Mammalia of Nepal', 339.

[67] Royle, *Illustrations*, Vol 1, lx.

the Tibetans and other Himalayan peoples'.[68] Indeed, given the prevalence of records of other myths such as those around Nanda Devi or the *bijli ki har*, this singular reference is perhaps surprising. Whatever the explanation, even if the yeti would go on to become one of the most enduring orientalised symbols of the Himalaya, in the period before 1850 it did not yet roam the imperial imagination.[69]

Time, Tropicality and the Line of Perpetual Snow

Investigating the distribution of plants and animals across the vertical globe meant paying attention to not just space but also time. Understandings of time were themselves in flux in this period, which was especially important for geology as discussed previously, but also for plant geography. As Marie-Noëlle Bourguet has shown, 'as much as a natural science, the study of plant distribution was to become a historical discipline'.[70] Elsewhere, scholars have argued that for Humboldt 'a study of the spatial relations of plants could yield a picture of the earth's geological, botanical, zoological, even human history'.[71] Observers were thus increasingly aware that the natural world they were attempting to map was far from static. Given a long-enough time frame, mountains moved, and vegetation moved with them. The study of plants that coexisted in a place not just in the present, but also in the past was thus a growing concern. As William Griffith wrote, 'the investigation of the real nature of our Indian fossil flora has now become a matter of paramount interest' (not least for tracing possible coal deposits).[72] He went on to speculate that 'further discoveries may prove the flora of the globe at a certain remote period to have been entirely tropical. At any rate it is quite certain that such floras of the now tropical countries were never boreal, or even temperate'.[73]

Bound up in Griffith's speculations was imaginative disconnect around the idea of 'tropicality', and how this related to the Himalaya. Historian David Arnold has convincingly laid out the way that India was rendered

[68] Bishop, *The Myth of Shangri-La*, 157. [69] Bishop, *The Myth of Shangri-La*, 156–58.

[70] Marie-Noëlle Bourguet, 'Measurable Difference: Botany, Climate, and the Gardener's Thermometer in Eighteenth-Century France', in *Colonial Botany: Science, Commerce, and Politics in the Early Modern World*, ed. Londa Schiebinger and Claudia Swan (Philadelphia: University of Pennsylvania Press, 2005), 286. See also Deborah R. Coen, *Climate in Motion: Science, Empire, and the Problem of Scale* (Chicago: University of Chicago Press, 2018), 274–311.

[71] Dettelbach, 'Global Physics and Aesthetic Empire', 267.

[72] Griffith, *Itinerary Notes*, 398.

[73] Griffith, *Itinerary Notes,* 398. For the broader context, including debates about the discoveries of possibly tropical vegetation in the Arctic, see Rudwick, *Worlds Before Adam*, 55–57.

'tropical' in the first decades of the nineteenth century, with the 'temperate' nature of the Himalayan foothills exaggerating this. Here 'tropicality' was imagined rather than real, 'a collection of ideas' that amounted to 'an especially potent and prevalent form of othering' or orientalising in contrast to the temperate norms of Europe.[74] Where the Himalaya fitted into this tropical picture was nevertheless something of a vexed question, not only in relation to the imperial imagination, but also in material terms of latitude and vegetation. Indeed, even leaving aside the cultural and moral connotations of 'tropicality', global comparisons based on the language of latitude pointed to an inconsistent picture. As Royle argued, in the Himalaya, altitude could be seen as 'counteracting the effects of latitude', or put differently, 'elevation, as in other tropical parts of the world, compensates for lowness of latitude, and allows the existence of plants of the temperate zone'.[75] William Griffith made a similar point: 'plants of very high latitudes require corresponding high elevations to cause their appearance at or near the tropics', and thus 'the genera found constituting the flora of Melville Island may be expected to be deficient on the highest land known to us near the equator, at least the extreme altitude required for their existence at the equator, would necessarily be very great, perhaps greater than we may really know to occur. Yet it is to great elevation in such low latitudes that we are to look for our deficient genera'.[76] The Himalaya thus presented seemingly insoluble contradictions, and as Brian Hodgson summarised: 'the suite of the seasons is tropical, as before; and, occasionally, the heat is extreme. But the season of heat is short; and, upon the whole, the climate of this region more nearly resembles that of high than that of low latitudes. It has nothing tropical about it but the course of the seasons'.[77]

Closely connected with these questions of latitude were investigations into the line of perpetual snow. What was meant by 'perpetual' remained a source of considerable uncertainty, exacerbated by the lack of awareness

[74] See Arnold, 'India's Place in the Tropical World, 1770–1930'; Arnold, *The Tropics and the Traveling Gaze*, 193–201. In the case of Royle, Arnold argues that his focus on identifying analogous regions led him to overlook the actual links between places which were ultimately essential for understanding plant distribution. For a different discussion of the tropicality of the Himalaya, see Mohammed Sohrabuddin, 'Construction of the "Himalayas": European Naturalists and the Oriental Mountains', in *Force of Nature: Essays on History and Politics of Environment*, ed. Sajal Nag (Abingdon: Routledge, 2018), 87–108. See also Felix Driver and Luciana Martins, *Tropical Visions in an Age of Empire* (Chicago: University of Chicago Press, 2005); Bernhard C. Schär, 'On the Tropical Origins of the Alps: Science and the Colonial Imagination of Switzerland, 1700–1900', in *Colonial Switzerland: Rethinking Colonialism from the Margins*, ed. Patricia Purtschert and Harald Fischer-Tiné (London: Palgrave Macmillan, 2015), 29–49.

[75] Royle, *Illustrations*, Vol 1, 62, 78. [76] Griffith, *Itinerary Notes*, 394.

[77] Hodgson, 'On the Mammalia of Nepal', 338.

of some earlier travellers about the nature of glaciers. Likewise, significant differences when compared to the Andes and the Alps had for some time been a source of controversy in both India and globally. Alexander Gerard mused, for example, that 'the inferior limit of perpetual snows does not appear to follow the same regularity on the Himalaya mountains as on the Andes, and mountains in Europe, there being a very great difference between the outer and inner ranges'.[78] His brother James elsewhere recorded that 'the marginal limit of the snow, which upon the sides of *Chimborazo* occurs at fifteen thousand seven hundred feet, is scarcely permanent in *Thibet* at nineteen thousand'.[79] As discussed earlier, these incorrect assumptions played a significant role in initial scepticism that the Himalaya might be the highest mountains in the world. Ongoing confusion, partly arising from imprecise use of terminology, also led to a rather bombastic disagreement in the pages of the India-based *Calcutta Journal of Natural History* and *Journal of the Asiatic Society of Bengal* in the 1840s. This played out between Thomas Hutton, John Hallett Batten and Richard Strachey over apparent differences in the height of the snow-line on northern and southern slopes, and indicated that uncertainties were far from resolved at the mid-century.[80]

Indeed, from Humboldt to Buckland and beyond, various theories that attempted to predict the elevation of the line of perpetual snow based on latitude came unstuck in the Himalaya. Here we again find James Gerard gleefully taking on metropolitan savants, writing for example:

> The barometer gave for the highest field 14,900 feet of elevation. . . . The *yaks* and shawl goats at this village seemed finer than at any other spot within my observation. In fact, both men and animals appear to live on and thrive luxuriantly, in

[78] Gerard, *Account of Koonawur*, 158.

[79] Gerard, 'Observations on the Spiti Valley', 254. Later observations, backed up by precise barometrical measurements, suggested that in the southern parts of the range, despite their near-tropical latitude the snowline was not less than 16,000 feet, while in the northern Himalaya it could be as high as 20,000 feet in places. See Cunningham, *Ladak, Physical, Statistical, and Historical*, 77; Hooker, *Himalayan Journals*, Vol 2, 394–96.

[80] Thomas Hutton, 'Correction of the Erroneous Doctrine That the Snow Lies Longer and Deeper on the Southern than on the Northern Aspect of the Himalaya', *Calcutta Journal of Natural History* 4 (1844): 275–82; John Hallett Batten, 'Extract of a Letter from J. H. Batten, Esq. Bengal Civil Service, Dated Camp Semulka on the Cosillah River, Kumaon, December 28th, 1843', *Calcutta Journal of Natural History* 4 (1844): 537–39; Thomas Hutton, 'Note on the Snow Line on the Himalaya', *Calcutta Journal of Natural History* 5 (1845): 379–83; John Hallett Batten, 'Line of Perpetual Snow', *Calcutta Journal of Natural History* 5 (1845): 383–88; Thomas Hutton, 'To the Editor of the Calcutta Journal of Natural History', *Calcutta Journal of Natural History* 6 (1846): 56–59; Strachey, 'On the Snow-Line in the Himalaya'; Hutton, 'Remarks on the Snow Line in the Himalaya'; Joseph Davey Cunningham, 'Note on the Limits of Perpetual Snow in the Himalayas', *Journal of the Asiatic Society of Bengal* 18, part 2 (1849): 694–97.

> spite of Quarterly Reviewers, and Professor Buckland, who had calmly consigned those lofty regions, and those myriads of living beings to perpetual ice and oblivion. What would have become of the beautiful shawl goats which furnish those superb tissues, that adorn the ivory shoulders of our fair countrywomen, had the Professor and the Quarterly the management of these matters their own way![81]

Elsewhere, Gerard remarked similarly that 'the Himalaya peer over the Andes, laugh at philosophers and closet speculators, and dwindle Dr. Buckland and his fossil bones into utter insignificance. The phenomena which are presented in obscure caves in Europe are appealed to in the mountains of Asia, but they answer by exhibiting a superb contrast'.[82] Gerard thus points to the absurdity of theories received from Europe, when compared with his own direct experience. Such moments, when the theories of savants broke down, were seized upon (and sometimes exaggerated and caricatured) to argue for the value of scientific work being done in displaced locations in Asia. Indeed, Gerard expounded on the existence of luxuriant forests above 10,000 feet, 'a limit beneath which on the equator (according to Baron Humboldt) the larger trees of every kind shrink; a limit which Mr. Colebrooke and clever reviewers placed close to the marginal snow in the region of the torpid lichen'. Here it is not only those in Europe, but also those in Calcutta, who are impugned as unreliable purveyors of knowledge. These comparisons indicated that Humboldt's theory (drawing on Henry Thomas Colebrooke) that predicted the relationship of the height of the perpetual snowline to latitude failed spectacularly when transferred to the Himalaya. As Alexander Cunningham put it at the mid-century: 'the long-unsettled question of the snow-line, which, on the joint authority of the great Humboldt and the learned Colebrook, had been fixed at 13,000 feet, between 30½ and 32° of latitude', was patently too low.[83] Thomas Thomson went out of his way to excuse Humboldt, noting that 'whether the error originated in India or in England I have no means of ascertaining. There can at all events be no doubt that on the part of Baron Humboldt, it arose from an over confidence in the accuracy of Mr Colebrooke'.[84] Regardless of where

81 Gerard, 'Letter from a Correspondent in the Himalaya', 109. Here he was referring to the 'Himalaya Mountains and Lake Manasawara', 437–39; Buckland, *Reliquiae Diluvianae*, 222–23.

82 [James Gilbert Gerard], 'The Himalaya Country', 110.

83 Cunningham, *Ladak, Physical, Statistical, and Historical*, 77. For his part, Colebrooke was wary of using theory in explaining these kinds of phenomena. See Colebrooke, 'On the Limit of Constant Congelation in the Himalaya Mountains', 43.

84 Thomas Thomson, 'Report on Western Tibet', 10 October 1849, British Library, IOR/F/4/2461/136806, f163. Richard Strachey also excused Humboldt for reliance on

the fault lay, in this case global comparisons – however necessary – initially led to more rather than less confusion.

While the distribution of plants allowed for reflection on a range of timescales, the limits of the perpetual snowline particularly saw discussions of the more recent past. As with the related question of glaciers, this sometimes relied on knowledge from the inhabitants of the mountains, and intergenerational oral histories. For example, the deputy commissioner of Kumaon, George Traill, recorded in 1832 that 'the interior of the *Himalaya*, except at the passes and paths in question, is inaccessible, and appears to be daily becoming more so from the gradual extension of the zone of perpetual snow'. On this he continued that 'the *Bhotias* bear universal testimony to the fact of such extension, and point out ridges now never free from snow, which, within the memory of man, were clothed with forest, and afforded periodical pastures for sheep', and 'they even state that the avalanches, detached from the lofty peaks, occasionally present pieces of wood frozen in their centre'.[85] Here both anecdotal reporting and material evidence were called on to demonstrate changes in climate and the snowline, which were in turn explicitly linked to consequences for human habitation.

Hugh Falconer later picked up on this account and offered a word of caution, especially when 'the circumstance of most weight is the assertion that pieces of wood are found frozen in the centre of the avalanches detached from the lofty peaks' (with the assumption this could only have been brought about by a descending snowline). But as he continued, this could potentially occur in two ways: 'by the line of perpetual snow being actually lowered to the level of the sea, or, supposing it to maintain a constant mean height, by an elevation of the mountain belt into the snow zone; either of which would produce, in appearance, the same effects'.[86] Falconer thus points to the possibility of an alternate explanation, arising from his discussion of the geological processes that had upheaved the Himalaya. Falconer expanded on this reasoning in noting that the snow line should not oscillate 'more than the mean temperature of a place does' and relating it back to Humboldt's failed theory linking the line of perpetual snow to latitude: 'in the Himalayah Mountains the present elevation of the line of perpetual snow is a huge anomaly, the plane being upwards of an English mile in excess of the amount yielded by calculation, with a formula for the latitude and height above the sea'. As he continued, 'if, therefore, we suppose that the pieces of timber

imperfect sources and noted that he had revised his views by the 1840s. See Strachey, 'On the Snow-Line in the Himalaya', 287–88.

[85] Traill, 'Statistical Report on the Bhotia Mehals of Kemaon', 3.

[86] Falconer, *Palaeontological Memoirs*, Vol 1, 183–84.

mentioned by Mr. Traill got enveloped in an avalanche by a lowering of the zone of perpetual snow, it would necessarily be implied that the plane of congelation was formerly more elevated', and this 'would involve a still greater irregularity than the enormous extent at present ascertained, a position which it would be unphilosophical to admit, except on the strongest grounds'.[87] Falconer is thus cautious, noting the intense comparative contradiction the snowline already presented, and goes on to prefer the explanation that saw forests upheaved to the elevation of the perpetual snow.

While looking to longer timescales to try and settle this point, Falconer did not necessarily think these geological explanations ruled out the possibility of a more modest descent of the snowline within the cultural memory of Himalayan people. In a footnote to his essay, other evidence was compiled in support on this idea, for example that 'there is an artificial mound, at a place called Kutlean Kotee, which the Puliarees say is the remains of a large hill city that became deserted in consequence of the increased cold or descent of the snow zone' and because 'charcoal and remains of pottery are found in it. ... [Michael] Edgeworth says the mound is, beyond all doubt, artificial'.[88] Falconer thus may not have had all the answers, but he concluded his essay by restating Traill's claim and the value of local knowledge in demonstrating 'that the zone of perpetual snow is gradually extending; and that ridges which, within the memory of man, were clothed with forest and afforded periodical pasture for sheep, have an obvious and important bearing on the question'.[89]

Cultivation, Habitability and Movement in the Mountains

The centrality of Bhotiya testimony to Traill's and Falconer's speculations about the perpetual snowline was tacit acknowledgement that the high spaces of the Himalaya had long been lived (and managed) landscapes. It was clear to European travellers, who were always dependent on pre-existing routes and networks of labour, that human beings had inhabited the high mountains for millennia; just how high nevertheless remained an open question. James Gerard, placing his observations in a comparative context, thought that given the limits of knowledge in the 1820s, 'it is no vague conjecture to entertain that tracts of land will one day be discovered, where the abodes of mankind and cultivation

[87] Falconer, *Palaeontological Memoirs*, Vol 1, 184.
[88] Falconer, *Palaeontological Memoirs*, Vol 1, 185.
[89] Falconer, *Palaeontological Memoirs*, Vol 1, 185. Here Falconer was explicitly paraphrasing Traill's earlier reports.

surpass in height the summits of the Andes'.[90] While travellers paid attention to the altitudinal ceilings of all types of vegetation, they were especially drawn to the limits of crops and cultivation. They also had a vested interest in any vegetation that could be used as firewood, a staple of existence in the high mountains (not to mention a necessity for boiling thermometers to measure altitude). The material possibilities for obtaining food and fuel (as well as guides and local knowledge of routes and conditions) thus circumscribed movement through the sparsely populated highlands. The altitude limits of these resources were bound up with questions about the potential for colonisation in the mountains, and intermittent insecurities around the frontier. These political and practical realities are thus inseparable from the broader scientific imaginings of the Himalaya that emerged in the first half of the nineteenth century.

Surveyors were keenly interested in the limits of cultivation and habitation, especially when operating near the high frontiers. As Alexander Gerard recorded in 1821, the village of Nako is 'the highest village that occurs to the traveller who traces round the frontier of Busahir. Separate measurements ... indicate a little above 12,000 feet from the level of the sea' and 'yet there are produced the most luxuriant crops of barley and wheat: rising by steps to nearly 700 feet higher, where there is a Lama's residence occupied throughout the year'.[91] Elsewhere, he wrote that the fields he had seen 'at 13,600 feet were very poor, and the people said they would never be properly ripe, although in Chinese Tartary grain comes to maturity in the vicinity of Koongloong, which must be almost 16,000 feet above the level of the sea, and within the circle of congelation'.[92] Gerard concluded that 'nature has adapted the vegetation to this extraordinary country, for did it extend no higher than on the southern face of the Himalaya, Tartary would be uninhabitable either by man or beast'.[93] Later naturalists like Thomas Thomson built on these observations and added their own, to determine the maximum heights for successful cultivation and subsistence farming: 'in favourable exposures, and sheltered spots, villages may even be seen as high as 14000 feet, and a few fields as high as 14500 feet', while 'in the neighbourhood of monasteries which are at times built at higher levels than cultivators ever venture to ascend to attempts at cultivation are now and then made as high as 15000 feet,

[90] Gerard, 'Observations on the Spiti Valley', 256.
[91] Lloyd and Gerard, *Narrative of a Journey*, Vol 2, 166–67.
[92] Gerard, *Account of Koonawur*, 64.
[93] Alexander Gerard, 'Remarks Regarding the Geological Specimens, Collected in 1821 by J.G. Gerard and A. Gerard', British Library, Mss. Eur. D137, f77.

with what success I do not know'.[94] When recording the grains that could be successfully grown at these elevations, Thomson also drew explicitly on the agricultural practices of his Bhotiya guides: 'a variety of common barley, with the seeds lower in the ear called in Ladakh Shirokh, is that which thrives best at high elevations according to the inhabitants'.[95]

The limits of the habitable world, for both permanent residents and itinerant visitors, were also circumscribed by a mundane yet critical relationship with firewood. As Alexander Gerard wrote of Lake Manasarovar, which was just beyond the assumed frontier with Tibet: 'the only firewood near Mansurowur is the prickly bush before mentioned'; however, 'notwithstanding the extraordinary altitude of this spot, lamas and nuns, who subsist chiefly by the offerings of pilgrims, reside in houses on the bank of the lake throughout the whole year; and this is most likely the highest inhabited land on the face of the whole globe'.[96] At altitudes where vegetation was scarce or non-existent, travel could become impossible or dangerous, deepening the dependency of travellers on local knowledge of route conditions. For example, Alexander Gerard wrote that 'the distance of today's march is only 3 miles, & we might have gone much farther but the guides objected, as there was no firewood nearer the Pass'.[97] Certain passes were rendered notorious by their lack of firewood, such as the Shatul, where James Gerard's notebooks (not to mention the lives of two of his porters) had been lost. Indeed, although relatively low in elevation, the Shatul Pass was considered highly dangerous because there was no firewood for thirteen miles on the approach.[98] Withholding firewood was also one of the key strategies employed by Tartars (seconded as border guards by the Qing empire) to prevent East India Company surveyors entering Tibet. As Alexander Gerard described his unsuccessful attempt to penetrate the Tibetan frontier in 1821: 'the chief person of the place paid me a visit, and informed me, that orders had been received from Lahassa, some months ago, to make no friends of Europeans, and to furnish them neither with food nor firewood'.[99]

[94] Thomas Thomson, 'Report on Western Tibet', 10 October 1849, British Library, IOR/F/4/2461/136806, f93-4.
[95] Thomas Thomson, 'Report on Western Tibet', 10 October 1849, British Library, IOR/F/4/2461/136806, f96.
[96] Gerard, *Account of Koonawur*, 133.
[97] Alexander Gerard, 'Remarks Regarding the Geological Specimens, Collected in 1821 by J.G. Gerard and A. Gerard', British Library, Mss. Eur. D137, f16.
[98] Gerard, *Account of Koonawur*, 41–42.
[99] Lloyd and Gerard, *Narrative of a Journey*, Vol 2, 155.

That firewood was a long-standing concern among Himalayan peoples is evident in naming practices. Alexander Gerard recorded, for example, a traveller's resting place called 'Nama Cheen', which he remarked 'is named after the species of Juniper called Nama, which is the only Wood for fuel found in the vicinity'.[100] Thomas Thomson meanwhile remarked in a footnote that 'there is a village marked Shing in [Godfrey] Vigne's map, at the bend of the Indus, but as Shing is simply the Tibetan for wood, it may be inferred that his informants meant that fuel was procurable at the place in question, and that it was in consequence a habitual halting place'.[101] One of the most important species in the pantheon of Himalayan firewood was the so-called 'Tartaric furze' (*genista versicolor*), which Alexander Gerard recorded was called *Tama* or *Tamak* by the Tartars. Gerard continued that he 'could not get the upper limit of furze on this (the Tartaric) side, but I reckon it fully 17,000 feet'.[102] This was important because 'it is the only kind of fire-wood, and ... it blazes like turpentine. How fortunate for the travellers who cross these bleak and frozen mountains to be so well accommodated!'[103] As he continued, 'were it not for this provision of nature, these lofty Passes would only be encountered by the intrepidity of a few'.[104] Joseph Dalton Hooker, who would go on direct the Royal Botanic Gardens at Kew as one of the most important botanists of the nineteenth century, also reveals himself as a connoisseur of firewood. Writing at the mid-century he suggested that 'as the subject of fire-wood is of every-day interest to the traveller in these regions, I may here mention that the rhododendron woods afford poor fires; juniper burns the brightest, and with least smoke. *Abies Webbiana*, though emitting much smoke, gives a cheerful fire, far superior to larch, spruce, or *Abies Brunoniana*'.[105] For those living and operating in the high mountains, the limits of vegetation was thus far from only an abstract scientific interest. As Richard Strachey put it, knowledge of vegetation 'is of no less importance to the mountain shepherd than of interest to the naturalist'.[106] He might have added that it was of no less importance

[100] Alexander Gerard, 'Remarks Regarding the Geological Specimens, Collected in 1821 by J.G. Gerard and A. Gerard', British Library, Mss. Eur. D137, f138.
[101] Thomas Thomson, 'Report on Western Tibet', 10 October 1849, British Library, IOR/F/4/2461/136806, f108.
[102] Lloyd and Gerard, *Narrative of a Journey*, Vol 2, 105, 118; Gerard, *Account of Koonawur*, 99.
[103] Lloyd and Gerard, *Narrative of a Journey*, Vol 2, 118.
[104] Lloyd and Gerard, *Narrative of a Journey*, Vol 2, 118.
[105] Hooker, *Himalayan Journals*, Vol 2, 150.
[106] Strachey, 'On the Physical Geography', 77.

to the prospects of empire, given the way vegetation circumscribed the ability to move safely over the high passes, secure the frontier and entertain thoughts of colonisation in the higher reaches of the mountains.

Visual Representations of Verticality

While continuing to anecdotally trace the altitude limits and zones of the vertical globe, Himalayan naturalists increasingly also turned to attempts to visually represent their data. Graphs and charts not only represented relationships, they also became tools to investigate and test ideas about plant geography and various other scientific phenomena. Alexander von Humboldt explained this with reference to his highly influential profile of Chimborazo, which was 'useful not only for developing new ideas regarding the geography of plants; I believe that it could also help us understand the totality of our knowledge about everything that varies with the altitudes rising above sea level'.[107] In mapping and graphing their data, Himalayan naturalists were certainly inspired by designs popularised (if not necessarily invented) by Humboldt.[108] As William Webb remarked in an account of the survey of Kumaon: 'since M. HUMBOLDT's account of *New Spain* has been published, and from other considerations, it is probable that the work will be thought incomplete, if not accompanied by vertical sections'.[109] Representations of the Himalaya produced in this period were nevertheless eclectic in style and content. Some depicted only the limits of a particular species, while others featured an encompassing range of characteristics, such as the line of perpetual snow, geological features, places of habitation, the range of fauna and even important ascents and feats of exploration. At the same time, these various attempts at visual representation often reflect the limits of mastery and the

[107] Alexander von Humboldt and Aimé Bonpland, *Essay on the Geography of Plants*, ed. Stephen T. Jackson, trans. Sylvie Romanowski (Chicago: University of Chicago Press, 2013), 99. For more, see Anne Marie Claire Godlewska, 'From Enlightenment Vision to Modern Science? Humboldt's Visual Thinking', *Geography and Enlightenment*, ed. David N. Livingstone and Charles W. J. Withers (Chicago: University of Chicago Press, 1999), 236–76; Sylvie Romanowski, 'Humboldt's Pictorial Science: An Analysis of the Tableau physique des Andes et pays voisins', in *Essay on the Geography of Plants*, ed. Stephen T. Jackson (Chicago: University of Chicago Press, 2013), 157–98.

[108] Dettelbach, 'The Face of Nature', 481–82; Nicholaas Rupke, 'Humboldtian Distribution Maps: The Spatial Ordering of Scientific Knowledge', in *The Structure of Knowledge: Classifications of Science and Learning Since the Renaissance*, ed. Tore Frängsmyr (Berkeley: Office for History of Science and Technology, University of California, 2001), 93–116.

[109] Webb, 'Memoir Relative to a Survey of Kumaon', 294.

challenges representing the unprecedented verticality and scale of the Himalaya on a flat page. Indeed, these visualisations of altitude often come across less as confident expressions of imperial knowledge and more as evidence of the ongoing scientific uncertainties and imaginative reconfigurations that needed to be navigated in adapting imperial geography to a three-dimensional globe.

William Griffith was one of the most enthusiastic early experimenters with representing verticality visually in the Himalaya. His notebooks contain a variety of styles of elevational charts and graphs, many of which were later included in his *Journal of Travels* (1847) and *Itinerary Notes* (1848). (These were compiled posthumously by EIC surgeon and naturalist John McClelland after Griffith's untimely death in Malacca in 1845.) An example of the simple style of chart Griffith employed most frequently can be seen in Figure 6.2, covering the vertical movements associated with one day's travel in the Himalaya. Though never offering an extensive explanation of his methodology, Griffith did comment on his intentions with this type of graph: 'the annexed table of the distributions of plants in relation to altitudes ... may render the subject of the preceding observations more clear and distinct. The dotted line along the left-hand margin represents the elevation of the mountains'.[110] John McClelland later worked these up into more comprehensive diagrams, presented as 'constructed from Information contained in Private Journals and Itinerary Notes of William Griffith'.[111] Produced in India and consumed by scientific audiences both there and in Europe, these represented entire sections of the mountains (as can be seen in the example for Bhutan in Figure 6.3). Though inevitably highly simplified, these nevertheless represented sustained attempts – and a desire – to think about altitude holistically, and to understand it as a critical variable in natural history and plant geography.

For a different attempt at visualising the verticality of scientific relationships in the Himalaya, it is also worth considering the mid-century efforts of Richard Strachey. While employed by the EIC, he undertook a major expedition with botanist James Edward Winterbottom, sneaking across the frontier and surveying the lakes of Rakas Tal and Manasarovar in 1848. Strachey and Winterbottom collected some 2,000 specimens of plants at various elevations along the way, many of which are now preserved at Kew Gardens in

[110] Griffith, *Journals of Travels*, 162. [111] Griffith, *Itinerary Notes*, 403.

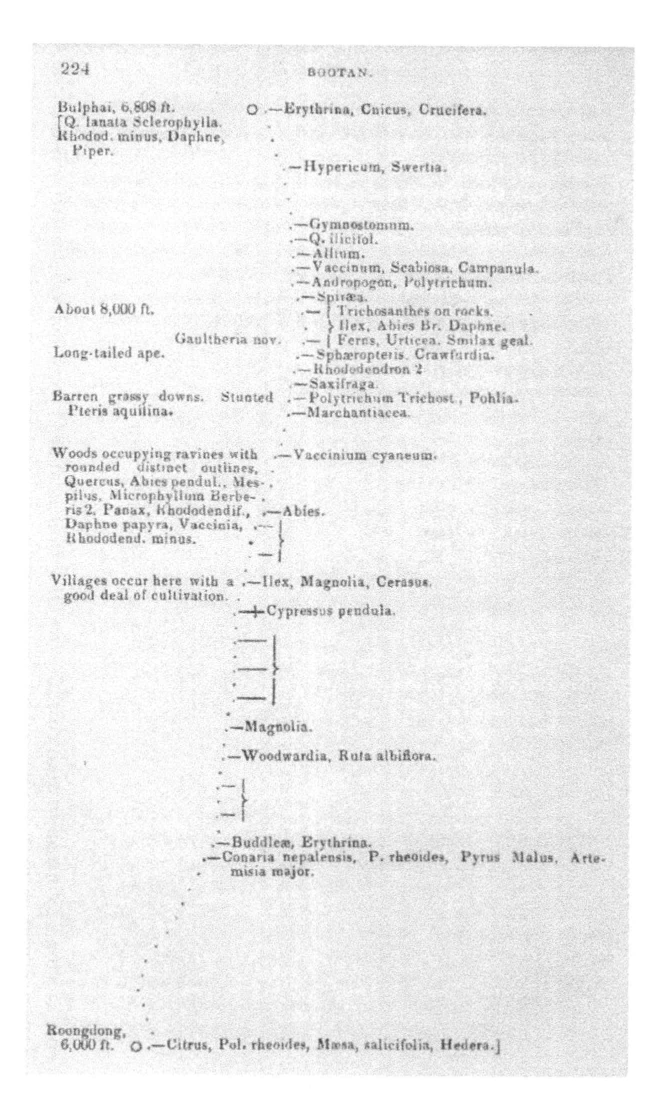

224 BOOTAN.

Bulphai, 6,808 ft. ○ .—Erythrina, Cnicus, Crucifera.
[Q. lanata Sclerophylla.
Rhodod. minus, Daphne,
Piper.
.—Hypericum, Swertia.

.—Gymnostomum.
.—Q. ilicifol.
.—Allium.
.—Vaccinum, Scabiosa, Campanula.
.—Andropogon, Polytrichum.
.—Spiræa.
About 8,000 ft. .— { Trichosanthes on rocks.
} Ilex, Abies Br. Daphne.
Gaultheria nov. .— { Ferns, Urticea. Smilax geal.
Long-tailed ape. .—Sphæropteris. Crawfurdia.
.—Rhododendron 2.
.—Saxifraga.
Barren grassy downs. Stunted .—Polytrichum Trichost., Pohlia.
Pteris aquilina. .—Marchantiacea.

Woods occupying ravines with .—Vaccinium cyaneum.
rounded distinct outlines,
Quercus, Abies pendul., Mes-
pilus, Microphyllum Berbe-
ris 2, Panax, Rhododendif., .—Abies.
Daphne papyra, Vaccinia,
Rhododend. minus.

Villages occur here with a .—Ilex, Magnolia, Cerasus.
good deal of cultivation.
.—+Cypressus pendula.

.—Magnolia.

.—Woodwardia, Ruta albiflora.

.—Buddleæ, Erythrina.
.—Conaria nepalensis, P. rheoides, Pyrus Malus, Arte-
misia major.

Roongdong,
6,000 ft. ○ .—Citrus, Pol. rheoides, Mæsa, salicifolia, Hedera.]

Figure 6.2 Chart from William Griffith's *Journal of Travels*.[112] The dotted line indicates relative elevations in Bhutan. This chart represents the ascents and descents during a day's journey, from 'Bulphai' to 'Roongdong', and serves to illustrate and quantify the accompanying journal. The majority of observations refer to plant species, but notes on fauna, habitation and cultivation also feature.

[112] Griffith, *Journals of Travels*, 224.

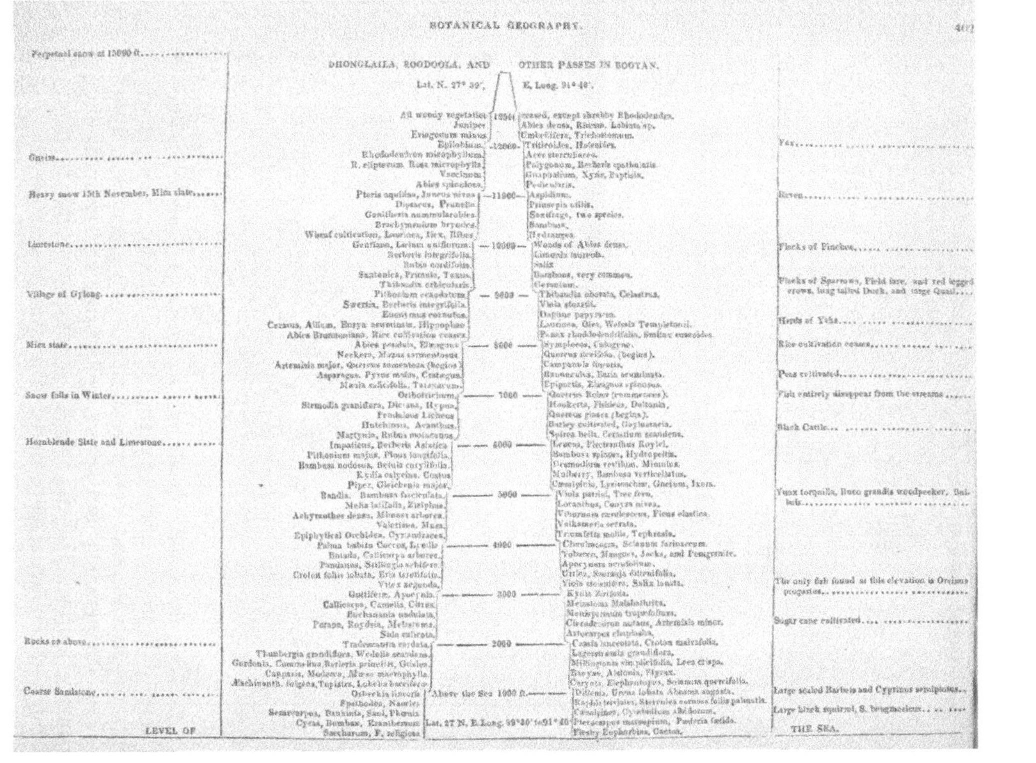

Figure 6.3 Composite of elevational information from Griffith's travels to Bhutan in 1837 compiled by John McClelland.[113] Though perhaps less aesthetically pleasing than Humboldt's famous profile of Chimborazo, this nevertheless contains an enormous amount of information about elevational relationships, and organises the natural world in terms of verticality. The chart incorporates the line of perpetual snow, the geology of the mountains, the limits of fish, yaks and various types of cultivation, culminating at the point where 'all woody vegetation ceased, except shrubby Rhododendron'.[114]

[113] Griffith, *Itinerary Notes*, 402. [114] Griffith, *Itinerary Notes*, 402.

London.[115] Based on this journey and the specimens gathered, as well as those sourced from others, Strachey attempted to create a chart that would represent a single continuous line section of the Himalaya, from the plains through to the summit of Nanda Devi (25,643 feet, though the peak itself was well beyond the altitude range of naturalists in this period). An unpublished EIC report indicated that 'the principal object in view was the completion of a sectional drawing of the Himalaya illustrative of the Botanical Geography of the mountains exhibiting in a graphical way the distribution of the plants at different altitudes'.[116] With a largesse that contrasted with that shown to many of his predecessors, the socially well-connected Strachey's endeavours were generously supported by the EIC. His involvement was couched in apologetic terms with the suggestion that the sections 'will throw much light on the natural history and geology of a part of the British territories which is the subject of curiosity on the continent of Europe & towards the scientific examination of which the British Government has up to the present time contributed little or nothing'.[117] The intention was that once the botanical section was complete, Strachey would prepare equivalent sections for the zoology and geology of the mountains, while sections for meteorology and magnetism were also mooted.[118] At the mid-century then, these sections thus sought to imagine the vertical world of the Himalaya graphically, and to consolidate the vast amounts of scientific knowledge about the mountains (and the EIC's northern frontier) that had been acquired in the preceding decades.

While representing a notably ambitious attempt to address the ongoing imaginative incoherence of the Himalaya, the project was beset by difficulties and delays. A draft was eventually completed, of which Strachey explained: 'I propose to divide this into 4 parts each of a size that can be lithographed easily' (the first of these can be seen in Figure 6.4). However, as Strachey continued, while 'the drawings can only be looked upon as provisional they will I think be quite well enough done in this way, and in my opinion a more careful execution of them would not only give rise to expense that would misdirected, but tend to produce a false idea of

[115] See Richard Strachey, *Catalogue of Plants Found in Kumaon and Garhwal and the Adjoining Parts of Tibet*, ed. John Firminger Duthie (London: Lovell, Reeve, 1906).
[116] 'Drafts of reports and correspondence on Richard Strachey's geological and botanical explorations in Kumaon', c. 1849–50, British Library, Mss. Eur. F127/183, f2.
[117] J. Thornton to G.A. Bushby, 29 December 1847, British Library, IOR F/4/2356/124635, f10.
[118] 'Extract Agra Public Narrative', 10 April 1848, British Library, IOR F/4/2356/124635, f1-2. These appear not to have been completed.

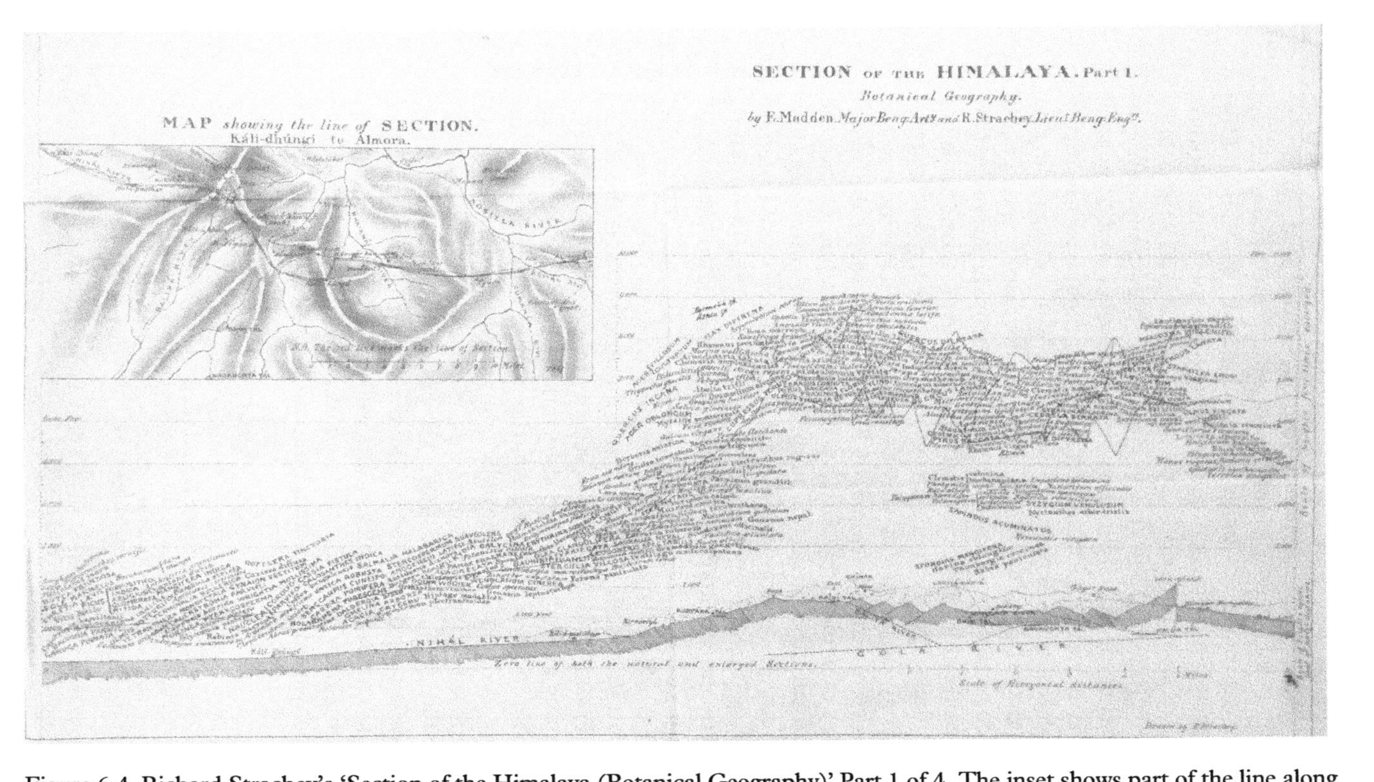

Figure 6.4 Richard Strachey's 'Section of the Himalaya (Botanical Geography)' Part 1 of 4. The inset shows part of the line along which Strachey collected plant specimens, overlaid onto his map of the mountains. While broadly perpendicular to the range, the line nevertheless meandered to follow weaknesses in both the terrain and frontier politics, much as exploration of the mountains did more generally. Strachey intended to produce similar sections for the geology and zoology of the mountains along the same line of section, but never completed them. British Library, Mss. Eur. F127/202. Courtesy of the Society of Authors as agents of the Strachey Trust.

the degree of accuracy at which they aim'.[119] In taking on these sections, Strachey freely acknowledged his limitations, noting that the government 'was aware that I made no pretension to any knowledge of Botany' (the same was true of his brother Henry who, as discussed in Chapter 5, resorted to requesting a collector from Saharanpur).[120] The official correspondence thus made it clear that it was to be a collaborative effort: '[Strachey] will have on the spot the assistance of Doctor Jameson, of Major Madden ... and of Mr J.H. Batten, and he will also be able to consult with Doctor Falconer, Mr Edgeworth & Major Cautley'.[121]

Completing the sections in India nevertheless proved challenging, much as had the production of botanical texts at the 'mountain' gardens. Indeed, Strachey quickly found himself coming up against the limits of working in a displaced location: 'assistance has on all occasions been most freely given but to examine and name so large a collection of specimens as I have is not only an operation that requires much time & labor but is impossible without books or a properly named herbarium for comparison which do not exist at Almora'.[122] While the sections were enormously detailed and represented a consolidated understanding of Himalayan plant geography that corrected some of the uncertainties, speculations and failed theories of earlier decades, Strachey was nevertheless very conscious of the ongoing limits of knowledge: 'the more northern part of the line of section it is still impossible for me even to attempt as it had never before been visited by Botanists and a very great proportion of the plants collected on it are still not named'.[123] These tensions also reflect the ongoing challenge of designing diagrams to represent the scale and verticality of the mountains, which sometimes overwhelmed not only the senses, but also the available techniques of drawing and visual representation. Richard Strachey's sections, as much as they were intended to demonstrate scientific and imaginative mastery over the previously 'blank spaces' of the mountains, thus ultimately reveal the scope of ongoing limits – not least because they remained incomplete and unpublished.

[119] 'Drafts of reports and correspondence on Richard Strachey's geological and botanical explorations in Kumaon', c. 1849–50, British Library, Mss. Eur. F127/183, f6.
[120] 'Drafts of reports and correspondence on Richard Strachey's geological and botanical explorations in Kumaon', c. 1849–50, British Library, Mss. Eur. F127/183, f5.
[121] J. Thornton to G.A. Bushby, 29 December 1847, British Library, IOR F/4/2356/124635, f10.
[122] 'Drafts of reports and correspondence on Richard Strachey's geological and botanical explorations in Kumaon', c. 1849–50, British Library, Mss. Eur. F127/183, f5.
[123] 'Drafts of reports and correspondence on Richard Strachey's geological and botanical explorations in Kumaon', c. 1849–50, British Library, Mss. Eur. F127/183, f6.

Conclusion

Whatever their limitations, these developments in understanding the plant geography of the Himalaya, and the global comparisons that they allowed, had broader implications for science. These are apparent in correspondence between two of the most influential naturalists of the nineteenth century, Joseph Dalton Hooker and Charles Darwin. In 1848, early in his visit to the Himalaya, Hooker wrote Darwin the following: 'I will give you but one Botanical fact, & that is regarding the vegetation of heights; You have often asked if Mts, especially isolated ones, in the Tropical & S. Latitudes had closely allied representatives of Arctic or N. Temperate forms'.[124] Here Hooker referred to an important problem in explaining the migration of species (and ultimately in explaining evolution): how did 'arctic' plants move from one high mountain to another high mountain (especially isolated peaks) when they were unable to grow in the warm intervening valleys?[125] Hooker later discussed this in terms of changes in climate brought about by the upheavement or subsidence of the mountains, but as he wrote to Darwin in 1850: 'I have been somewhat disappointed in my expectations of finding that Sikkim would tend to clear up your doctrines to my mind' because 'I thought that the transitions from one form to another would be more apparent in a country where under a perfectly equable climate the floras of the tropical temperate & arctic zones blend in the same Longitude & Latitude' but 'such has not been the case I think'.[126] Mountains mattered to Darwin's not yet fully formed 'doctrine', Hooker knew, because 'a country combining the botanical characters of several others, affords materials for tracing the direction in which genera and species have migrated [and] the causes that favour their migrations'.[127] Even if Hooker was not yet quite satisfied with how this evidence fit with Darwin's theory, the global comparison of plant geography in mountains thus underpinned a key part of the framework by which he ultimately would be.

But verticality mattered not only because it provided the tools to advance scientific understandings of the globe. In the vertically ordered world that was laid out in the first half of the nineteenth century, almost nothing was off limits to elevational speculation. Indeed, travellers had

[124] Hooker to Charles Darwin, 20 February 1848, Kew Archives, JDH/1/10, f52-54.

[125] For Darwin's later formulation of this problem, see Charles Darwin, *On the Origin of Species by Means of Natural Selection, or the Preservation of Favoured Races in the Struggle for Life* (London: John Murray, 1859), 375. See also Browne, *The Secular Ark*, 124–28; Arnold, *The Tropics and the Traveling Gaze*, 199–200; Reidy, 'From the Oceans to the Mountains', 31–32.

[126] Hooker to Charles Darwin, 6 April 1850, Kew Archives, JDH/1/10, f274-276.

[127] Hooker, *Himalayan Journals*, Vol 2, 37–8.

few qualms extending observations of verticality in the natural world to various forms of often geographically determinist cultural comparison. Thomas Thomson, for example, tied altitude to ethnography: 'the gradual transition, in ascending the Sutlej, from Hinduism to Buddhism, is very remarkable, and not the less so because it is accompanied by an equally gradual change in the physical aspect of the inhabitants' with 'the Hindus of the lower Sutlej appearing to pass by insensible gradations as we advance from village to village, till at last we arrive at a pure Tartar population'.[128] Religion and physiology are here explicitly linked to elevational change, ideas that would coalesce into myths around the supposed purity and morality of mountain peoples in the later nineteenth and twentieth centuries. Representing the world on a vertical scale, with precise elevations secured by new instrumental practices, thus had wide-ranging consequences for the long-term understanding of mountains and those who lived in them. Scholars of 'Zomia' have demonstrated, for example, that mountains and mountain peoples are often pushed to the peripheries of states and empires.[129] A central concern of this chapter has been that the new language – or vocabulary – of verticality borrowed heavily from the horizontal, and that this reinforced lowlands as the point of reference. As we will see in a series of European atlases that bring this story full circle in the conclusion, this process was ultimately insidious, and meant that the mountains were subsumed into a framework of empire as aberrant and marginal places.

[128] Thomson, *Western Himalaya and Tibet*, 109.
[129] For an overview, see Michaud, 'Editorial – Zomia and Beyond'.

Conclusion: A Vertical Globe

This book has been about the Himalaya, and how they were examined and drawn into a global framework of mountains in the first half of the nineteenth century. At the same time, it has demonstrated that this imagined framework – or 'vertical globe' – was itself in the making, and co-constituted with new scientific understandings of the Himalaya. Focusing on the comparison of mountain phenomena in a context of empire has thus allowed for a window into the making and remaking of what were – and what could only be – global sciences. Historicising these has required following a series of interrelated scientific practices across six chapters, rather than focusing comprehensively on just one. Only by looking at the measurements that made global comparisons possible, the uncertain physiology of high places, and by considering the geological and botanical questions that underpinned three-dimensional plant geography, has it been possible to see the emergence of verticality as a central organising principle for both human and non-human worlds. Much as for knowledge-making and science in this period more widely – from China to Egypt, and from the Galapagos to Tahiti – these practices were intimately and inextricably linked with imperial expansion and control. Historicising global sciences thus matters, as I have argued in this book, because only by doing so can we understand the imperial utility of remaking categories as large and as allegedly universal as mountains. In this way, studying verticality adds a new dimension (quite literally) to the 'spatial turn' that continues to animate our histories of science and geography, but it also gives greater urgency to integrating these with innovative and activist new approaches to global history.

In taking the view from the Himalaya, it has been evident that understanding mountains was an inherently global and comparative process. Whether in the distribution of plants, the behaviour of glaciers or the difficulty of respiration, scientific practice in the Himalaya always involved figuring out where Asia's mountains fit on an imagined globe that already prominently featured other peaks, most canonically Mont Blanc and Chimborazo. Throughout, I have nevertheless emphasised that making the Himalaya globally commensurable was far from a story of the seamless accumulation of knowledge, and that theories developed based on norms derived from the Alps and the Andes often initially failed to account for the Himalaya. We have seen travellers finding crops in regions purportedly locked in ice, volcanoes assumed into being where none existed and instrumental scales revealed as laughably inadequate. These moments of failure were often compounded by the East India Company's unsystematic approach to science, and the equivocal social status of its employees, especially those grafting eclectic scientific interests onto their official duties, and those operating away from 'centres in the periphery' with limited resources and a want of up-to-date libraries, instruments and herbariums. Together, these overlapping limits of imagination and practice have demonstrated the necessity of tracing global histories of science through unsuccessful as much as successful moments of comparison. Indeed, these failures and dead ends, and reorientations and reconfigurations, ultimately serve to reveal the uneven, incomplete and contested processes that underlay the making of supposedly universal categories in an age of empire.

While necessarily expansive, the range of sciences treated in this book has by no means been exhaustive either in the individual cases or collectively. Aspects of scientific practice which might have received expanded treatments in a larger volume include cartography, medicine and zoology. Other sciences entirely – for example, meteorology or hydrology or magnetism – were also all of growing importance in consolidating understandings of the vertical globe by the mid-century. Similarly, adjacent and often entangled strands of imperial knowledge-making, from ethnography to linguistics to antiquarianism, were also sometimes explicitly related to altitude and placed on the vertical globe. This is evident for example in Thomas Thomson's characterisation of the change in religion with altitude as one climbed towards Tibet, or in the supposed similarities surveyors like the Gerard brothers imputed between Bhotiyas and Scottish highlanders, itself and telling and consequential form of cultural equivalency.[1] Here the works of scholars like Emma Martin and Jayeeta Sharma on Darjeeling as a transcultural borderland is instructive, and there is also scope for more work on the higher mountains, such as in tracing the linguistic contributions of Hungarian scholar Csoma de Körös.[2] These are rich and complex stories, which might add additional facets to the story of the reordering of global space in three dimensions. It is nevertheless not my intention to

[1] See Gerard, *Account of Koonawur*, 85, 113.

[2] Emma Martin, 'Translating Tibet in the Borderlands: Networks, Dictionaries, and Knowledge Production in Himalayan Hill Stations', *Transcultural Studies* 7, no. 1 (2016): 86–120; Sharma, 'A Space That Has Been Laboured On'.

examine all of the myriad links between scientific and scholarly practice, or global cultural comparisons around mountains and mountain peoples here. Rather, in this conclusion I want to reflect on some of the ways that studying the emergence and consolidation of the vertical globe through the practice of natural history and measurement can help us to see empire and its inequities anew.

In particular, tracing the processes by which the Himalaya were made compatible with an imagined vertical globe in the first half of the nineteenth century requires us to re-examine shifts towards universalisation in imperial science and geography. As Michael Reidy argues: 'ordering the natural environment enable[d] imperial regimes to project power more efficiently across space ... for this reason, Western imperial powers attempt[ed] to standardize quantities of all types, both physical and imaginary' such that in this period 'vertical and horizontal spaces were created anew, a conceptualization that legitimized both the spatial turn in science and the expansionist programs of the enabling imperial power'.[3] This book has argued similarly, but it also goes a step further by considering how these standardising imperatives shaped imperial visions of the globe itself. In the first half of the nineteenth century, a commensurable 'globe' was a growing necessity for imperial governance as much as it was for science (regardless of how uneven this remained in practice, and the way it sometimes unhelpfully mirrored the assumed cultural and epistemological superiority of its makers). Here Simon Naylor and Simon Schaffer explain that this increasingly global orientation could take the form of a self-fulfilling prophecy, where surveyors 'defined phenomena and systems as worldwide in principle, then in an intriguingly circular tactic of self-validation drew their legitimacy and their resources from this very definition of global extension'.[4] However, if scholars are increasingly concerned by a now-widespread tendency to equate 'the imperial' and 'the global' – sometimes deliberately, sometimes carelessly, and with all the Eurocentrism this entails – then the story of altitude sciences reminds us of the utility of this symbiosis for nineteenth-century empires. Collapsing the global and the imperial through the lens of science was thus in many ways the ultimate expression of centuries of attempts at legitimation and appropriation. A sustained examination of the practices and processes that enabled this – as I have sought to provide in this book – thus forcefully reminds us why nineteenth-century global and globalising visions need to be

[3] Reidy, 'From the Oceans to the Mountains', 17, 34. See also Cosgrove, *Apollo's Eye*; Coen, *Climate in Motion*.

[4] Naylor and Schaffer, 'Nineteenth-Century Survey Sciences', 141.

historicised and denaturalized. Doing so ultimately demonstrates that in our histories of imperial science and geography, and their legacies, 'the globe' needs to be understood not only as method and a means for understanding ostensibly universal phenomena and our planet holistically, but also as a powerful tool of empire.

Imposing the Global

Having spent the previous six chapters firmly within the mountains and valleys of the Himalaya, and focused on the making of knowledge within these particular spaces, it is time to expand the lens. In particular, it is worth taking a moment to invert the story by considering global comparisons made not by surveyors or naturalists from the altitude-sickness-inducing high passes of the Himalaya themselves, but by atlas makers and publishers a world away in Europe. In so doing, it becomes possible to examine how the observations and comparisons made by actors in the Himalaya were absorbed into a broader story of global verticality, one which inevitably – and in many ways insidiously – flattened the nuances of local observations and the laboriousness of the processes that the preceding chapters have detailed.

Across the period of this study, European publishers drew extensively on material acquired via the imperial networks this book has traced to create new types of images of mountains (as seen for example in Figure 7.1). In this image, the three mountains selected to represent the Himalaya are Dhaulagiri (believed at the time to be the highest mountain in the world), Nanda Devi (not, as discussed in Chapter 4, a volcano) and Reo Purgyil (on which the Gerard brothers' high point, as featured in Chapter 2, is clearly marked). Ambiguously straddling the highlands and the lowlands (see Chapter 5), Saharanpur is marked on the right-hand side, as are Landour (a village on the same ridge as Mussoorie) and Nahan (the site of George Govan's temporary 'hill nursery'). Various limits of cultivation and crops are depicted, as well as zones of vegetation more broadly. The latitude of the Himalaya, which vexed theories about the line of perpetual snow (see Chapters 1 and 6), is prominently displayed below the section. At the bottom of the frame, some of the sources of the information used to compile the image are acknowledged, including several key figures in this book, namely: James Herbert, Alexander Gerard, George Govan and John Forbes Royle. In combining these diverse phenomena and sources of information, the image thus presents a holistic vision of the mountains. In the preceding chapters we have seen such all-encompassing approaches in practice, and here we see their culmination in an ecological picture of the world *avant la lettre*.

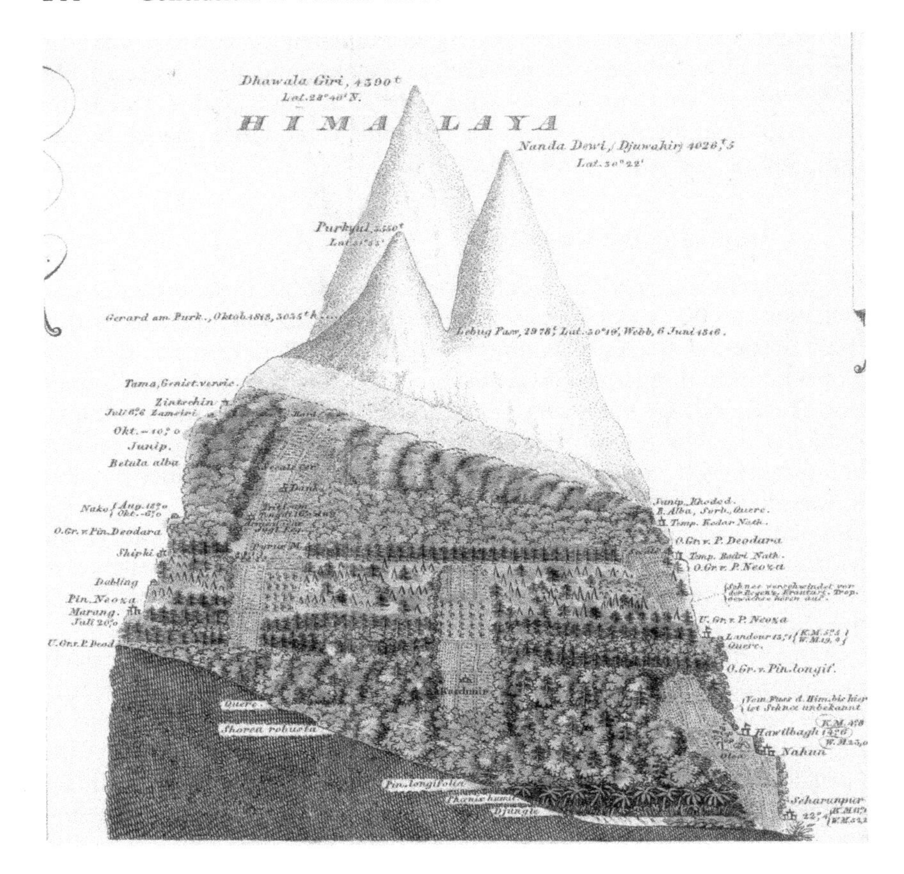

Figure 7.1 'Himalaya' section from the 'Umrisse der Pflanzengeographie [Outline of Plant Geography]' produced in 1838 and published in Heinrich Berghaus, *Dr. Heinrich Berghaus' Physikalischer Atlas* (Gotha: Justus Perthes, 1845). Reproduced by kind permission of the David Rumsey Map Collection, David Rumsey Map Center, Stanford Libraries.

In both layout and aspect, Figure 7.1 is a composite and highly stylised image, and one which imagines relationships of all kinds through the lens of verticality. In depicting the Himalayan mountains in this manner, they are laid out in a form which can then be compared globally. The result of this is evident in the full tableaux (see Figure 7.2). This includes, as well as the Himalaya, profiles of the Alps, the Andes, Lapland and Tenerife, all of which became exemplar ranges for constructing a picture of global verticality.

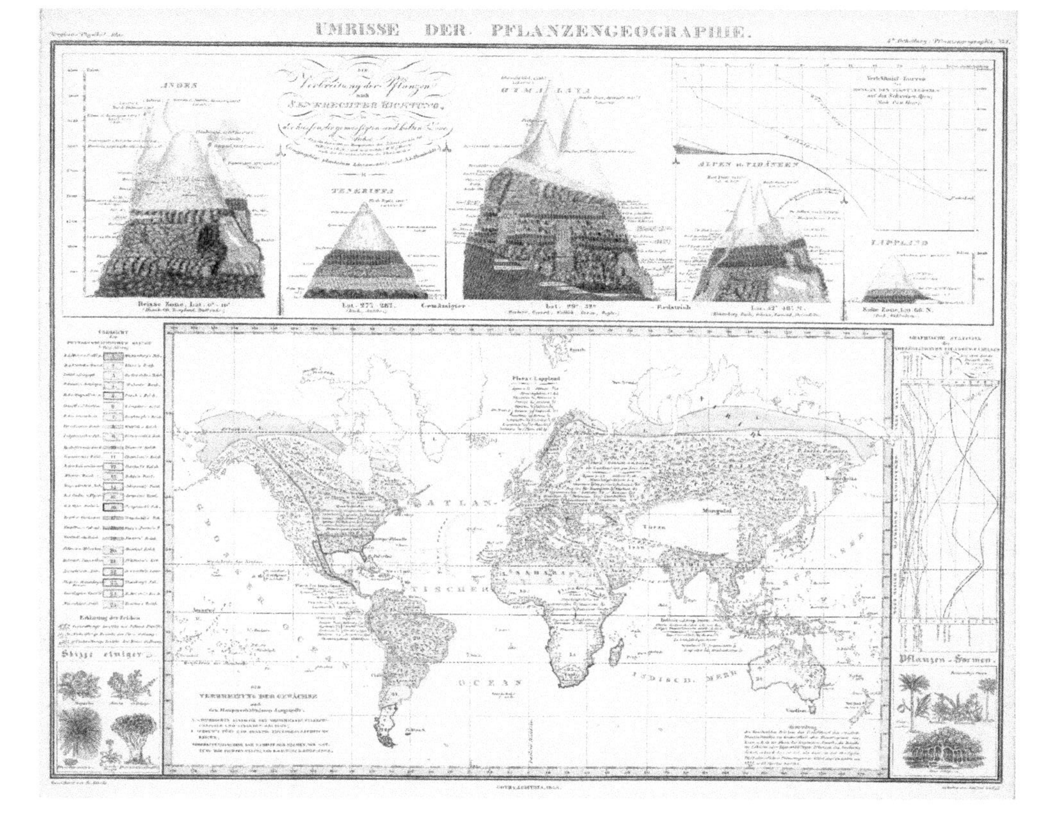

Figure 7.2 The full version of the ‘Umrisse der Pflanzengeographie’ (1838), which maps botany in terms of both altitude and latitude, that is, on globes both vertical and round.
Source: Heinrich Berghaus, *Dr. Heinrich Berghaus’ Physikalischer Atlas* (Gotha: Justus Perthes, 1845). Reproduced by kind permission of the David Rumsey Map Collection, David Rumsey Map Center, Stanford Libraries.

In these profiles, it is noteworthy that all information is presented as equally reliable in order to achieve a totality of expression. This undifferentiated approach results in contradictions, and to take one example, glaciers cascade down the Alps but in the Himalaya there are as yet none. At the same time, it is notable that in the full world map below the mountain profiles, the Tibetan plateau and the uplands beyond the Himalaya remain conspicuously blank, serving as a reminder of the way the frontier continued to circumscribe the limits of knowledge. This 'blank space' stands out as a stark affront to imperial mastery, and a source of frequent, if intermittent, frontier insecurity, which was a key driver of exploration and the scientific imagining of the mountains across this period.

The production of these sorts of global and comparative images was given considerable impetus, if not actually invented, by Alexander von Humboldt with his widely circulated cross section of Chimborazo.[5] This influence continued across the first half of the nineteenth century, and indeed Henrich Berghaus worked closely with Humboldt, and his atlas was originally intended as a visual counterpart to the Prussian's major late career work *Cosmos*.[6] Jon Mathieu is thus right to suggest that with Humboldt's 'generation there began a new phase in the globalisation history of natural observation and especially mountain perception', even if, as this book has demonstrated, accounting for Humboldt's influence and place in this story is a complex issue.[7] Indeed, as I have argued, the frequently and sometimes uncritically adopted trope of 'Humboldtian science' can obscure as much as it illuminates when it comes to explaining heterogeneous scientific practices on the edges of empires. As this book has argued, Humboldt certainly matters to the story of altitude – not least because so many contemporary surveyors and naturalists agreed that he did – but it is only by resisting the pull of biographical studies of 'great men' of science (whether hagiographic, critical or contextualised) that we can understand something like the uneven process of remaking global space in three dimensions in the nineteenth century. As Mathieu continues, envisioning mountains as a global category was not inevitable and 'it takes some imagination to bring them together and to see them as one distinct region on a global scale', and ultimately, 'the idea of viewing these regions as a universal whole does not arise through

[5] More properly the 'Tableau physique des Andes et pays voisins', this was published in Humboldt and Bonpland, *Essai Sur La Géographie Des Plantes*.

[6] See Jane R. Camerini, 'The Physical Atlas of Heinrich Berghaus: Distribution Maps as Scientific Knowledge', in *Non-Verbal Communication in Science Prior to 1900*, ed. R. Mazzolini (Florence: L. S. Olschki, 1993), 479–512; Rupke, 'Humboldtian Distribution Maps', 95–96. See also Godlewska, 'From Enlightenment Vision to Modern Science?'; Romanowski, 'Humboldt's Pictorial Science'.

[7] Mathieu, *The Third Dimension*, 22.

simple observation'.[8] While I would agree, this was far from only the work of one individual or even several. As the preceding chapters have shown, whether in hauling instruments to high points, drawing vertical sections or hunting fossils in bazaars, making the mountains globally commensurable required enormous amounts of labour, both intellectual and physical. Moreover, whatever these impetuses towards a global picture, it is also worth remembering, as Thomas Simpson reminds us, that because mountains 'have rarely if ever seemed lifeless or inert . . . no variant of modernity has flattened them through entirely subsuming them into universal schemes'.[9]

This is not to say that atlas makers did not try, squeezing the world's mountains onto a single page and asserting an encyclopaedic sense of the known. Especially from the 1810s onwards, all-encompassing comparative tableaux of mountains (and often also rivers) enjoyed considerable popularity in European atlases, and were circulated widely by the middle of the nineteenth century.[10] Perhaps the most fully realised example of this style of visual representation can be seen in Figure 7.3. This evocative and stylised image depicts elevational relationships of all kinds, from the line of perpetual snow through to Joseph Louis Gay-Lussac's famous hot air balloon ascent to 23,018 feet over Paris in 1804. Alexander von Humboldt's high point on Chimborazo is clearly marked, although unlike in Figure 7.1, the Gerards' competing elevational record is not. Just as in Berghaus's *Atlas*, Dhaulagiri is here depicted as the highest mountain in the world (it is now known to be only the seventh highest). Produced in 1836, only around twenty years since Chimborazo had been displaced as the world's 'highest' mountain, this tableau indicates just how rapidly the vertical globe was evolving in the first half of the nineteenth century.[11] However, that this is a snapshot of a mountain world in the making is unacknowledged, because to do so would be counter to purpose.

[8] Jon Mathieu, 'Globalisation of Mountain Perception: How Much of a Western Imposition?', *Summerhill: IIAS Review* 20, no. 1 (2014): 8. See also Mathieu, *The Third Dimension*, 30.

[9] Simpson, 'Modern Mountains from the Enlightenment to the Anthropocene', 28.

[10] These then declined and disappeared in the later nineteenth century, as topographic approaches centred around watersheds became preferred. For more on the production of these tableaux, see John A. Wolter, 'The Heights of Mountains and the Lengths of Rivers', *The Quarterly Journal of the Library of Congress* 29, no. 3 (1972): 186–205; Mathieu, *The Third Dimension*; Jean-Christophe Bailly, Jean-Marc Besse, and Gilles Palsky, *Le Monde Sur Une Feuille: Les Tableaux Comparatifs de Montagnes et de Fleuves Dans Les Atlas Du XIXe Siècle* (Lyon: Fage Éditions, 2014); Baptiste Hautdidier, 'The Comparative Tableau of Mountains and Rivers: Emulation and Reappraisal of a Popular 19th-Century Visualization Design', *Environment and Planning A* 47, no. 6 (2015): 1265–82.

[11] This version was based on an image originally made in 1829. See also Mathieu, 'Globalisation of Mountain Perception', 12.

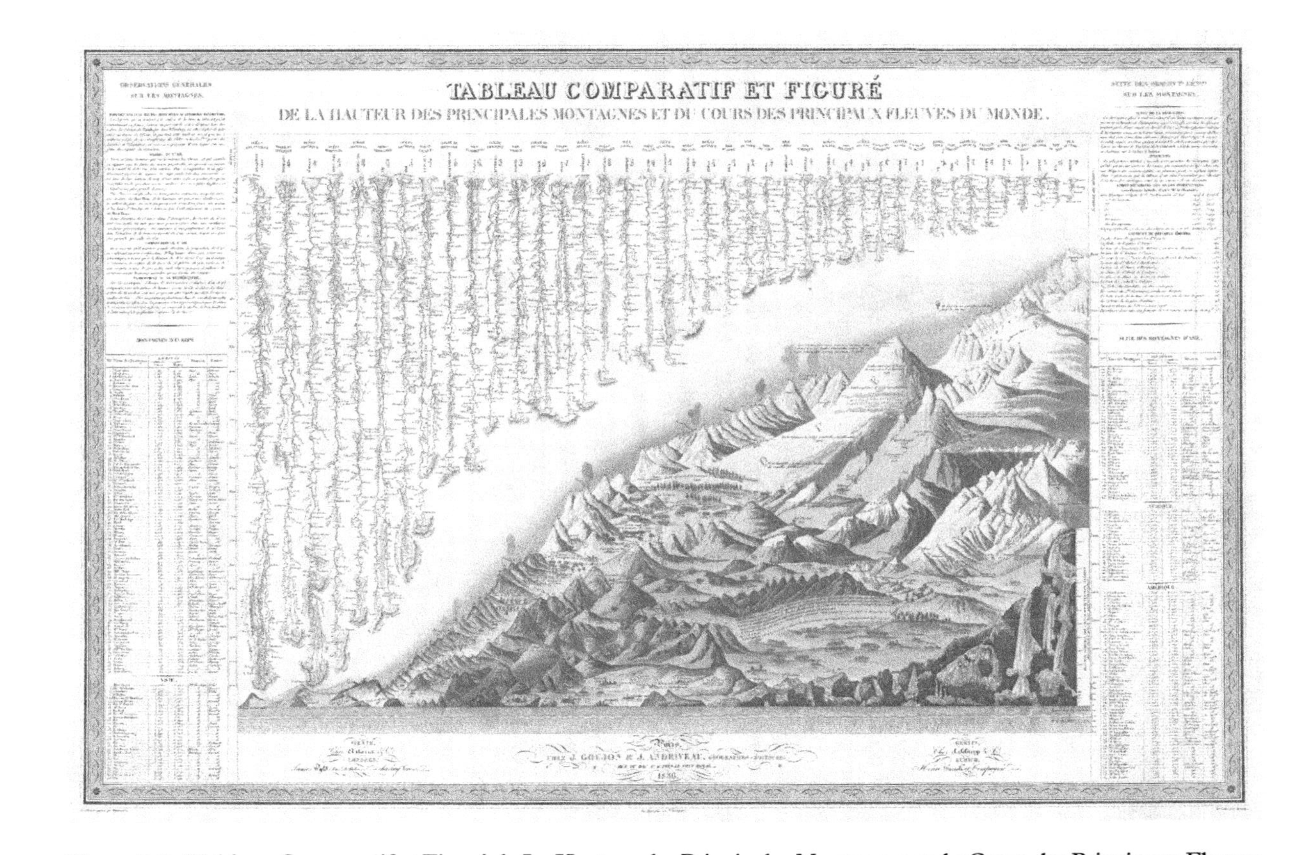

Figure 7.3 'Tableau Comparatif et Figuré de La Hauteur des Principales Montagnes et du Cours des Principaux Fleuves du Monde'. Paris: Chez J. Goujon & J. Andriveau (1836). Reproduced by kind permission of the David Rumsey Map Collection, David Rumsey Map Center, Stanford Libraries.

Indeed, while these comparative tableaux present an ordered and orderly world, they belie the way understandings of the vertical globe remained in flux, not least because of ongoing efforts to fit the unprecedented heights of the Himalaya into the picture. This is graphically illustrated by an updated version of the 'Umrisse der Pflanzengeographie' produced in 1851 (see Figure 7.4). In this version of the image, Kanchenjunga – by this time

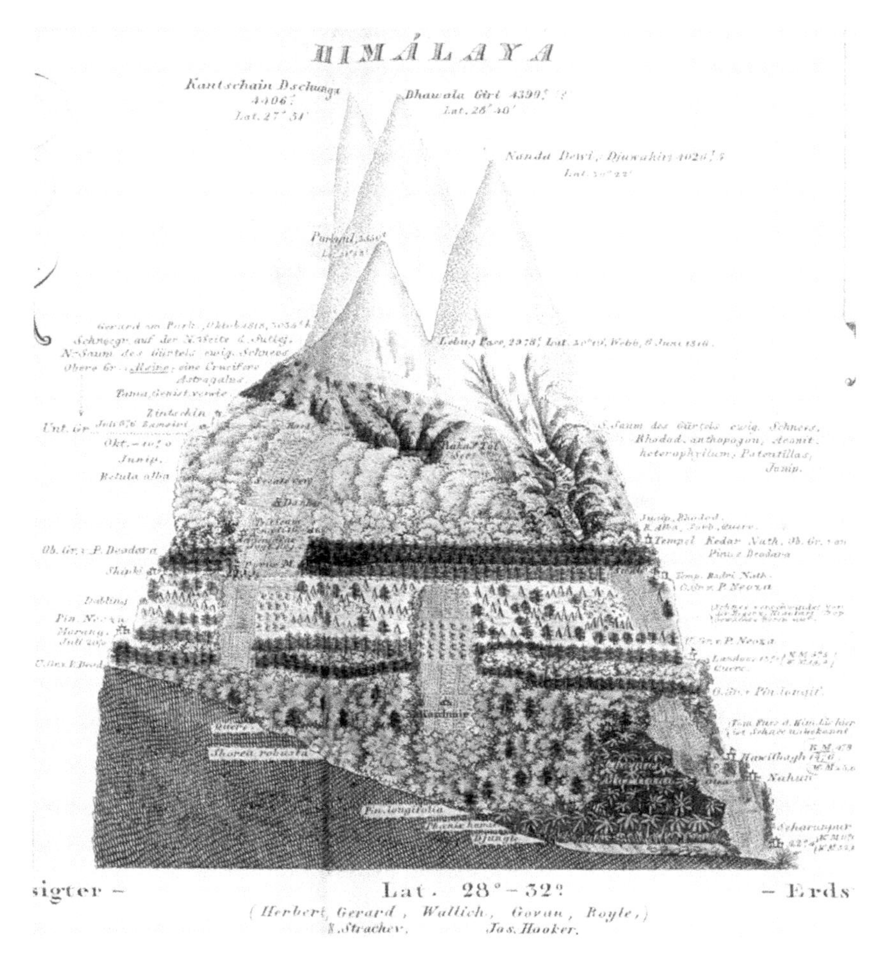

Figure 7.4 Revised 'Himalaya' section of the 'Umrisse der Pflanzengeographie', produced in 1851 and published in the 1852 edition Berghaus's *Atlas*. Heinrich Berghaus, *Dr. Heinrich Berghaus' Physikalischer Atlas,* 2nd ed. (Gotha: Justus Perthes, 1852). Courtesy of the Royal Danish Library, Photo Collection.

measured and confirmed as higher than Dhaulagiri – is awkwardly tacked onto the original profile.[12] Inserted on the side and behind, it appears as a clumsy addition to the original artistic rendering.[13] Little did the artist know, however, that in less than five years the Himalaya would have to be redrawn again, and space finally made for Mount Everest. Kanchenjunga is not, however, the only addition, and the by this time celebrated Pindari Glacier (see Chapter 4) has been inserted, flowing down into the vegetated zone, as has the lake of Rakas Tal, now surveyed by the Strachey brothers (see Chapters 1 and 6). As these sometimes-awkward additions indicate, the Himalaya were constantly being made and remade, as new measurements and specimens were acquired, and new scientific theories rose and fell. In the updated image, the names of Richard Strachey and Joseph Hooker, whose contributions to Himalayan plant geography and geology occurred after the original profile was produced, have also been added onto the bottom of the frame. (Notable names remain absent, especially Victor Jacquemont and William Griffith, perhaps reflecting the state of their collections following their untimely deaths in Asia.) With these additions, different generations of travellers are nevertheless linked, and presented as contributors to a supposedly coherent project. Indeed, beyond simply failing to account for a vertical globe in flux, these images suggest a completeness to imperial and global visions and actions that mask material differences between the actors – from French savants to Scottish surveyors – that produced them. These images present a homogenous picture of a mountain world, glossing over the way it emerged from and for disparate purposes: imperial and scientific, aesthetic and cultural, religious and economic.

Perhaps even more significant than what is added, however, is the way that imposing global commensurability brought about multiple erasures, both intentional and unintentional. As the preceding chapters have demonstrated, the appropriation of botanical, geological and physiological knowledge about the mountains was beset by uncertainties, especially around scales and frontier insecurities. Scholars have rightly argued that in this period 'both geographical distance and cultural diversity came to be regarded as obstacles to scientific practices when they would not allow for meaningful comparisons. Instruments, measures and data were meant to travel and provide templates for standardisation'.[14] However, this book

[12] For this problem in comparative tableaux more broadly, see Mathieu, *The Third Dimension*, 32.

[13] There is also an 1854 version of the profile, this time substantially redrawn with Kanchenjunga in the middle, and Herbert and the Gerards deleted for 'Hooker, Thompson, Jacquemont, Royle &c'. This was titled 'Geographical distribution of indigenous vegetation' and published in Alexander Keith Johnston, *The Physical Atlas of Natural Phenomena*, 2nd ed. (Edinburgh: William Blackwood and Sons, 1856).

[14] Bourguet, Licoppe, and Sibum, *Instruments, Travel and Science*, 3.

has spent much of its length demonstrating and explaining the many ways that constituting the global sciences of mountains in this period was laborious and prone to breakdowns. Travellers' bodies were pushed to and beyond their physiological limits, and operating in the mountains required mobilising large expedition parties and negotiating with stubborn border guards. In this context, expedition sociability and frontier politics often reveal the limits of explorers' mastery, whether in the discarding or theft of geological samples, or in the inability to take instrumental measurements for fear of detection (not to mention the way these interactions sometimes threatened to destabilise expected hierarchies around race, class and gender). In focusing on displaced locations in the high mountains and on decentred institutions, this book has shown that breakage and repair, limited resources and theorisation without access to the most up-to-date journals was endemic to scientific practice in the Himalaya. These atlases erase such moments of disconnection and failure in preference for an orderly and aesthetic sense of completeness. As is evident in Figures 7.1–7.4, making the Himalaya into a coherent place capable of being compared globally necessarily meant erasing both ongoing uncertainties and the laboriousness of knowledge-making in the mountains themselves. The messy and contested way science was practised in response to the human and non-human worlds of the Himalaya is here suppressed, if not entirely deleted. These self-consciously global visions thus allow for the erasure of locality and the imposition of commensurability that made global comparisons possible, but they also occlude half of the story.

Most insidiously, these erasures are not equally applied. One of the key threads running throughout *Science on the Roof of the World*, in both argument and design, has been the centrality of Himalayan peoples' expertise and labour to the remaking of the mountains. Making the Himalaya globally commensurable involved locating, identifying and moving literally tons of material – dried and live plants, stuffed and pickled animals, rock and fossils, fieldbooks and journals – both into and out of the mountains. Similarly, knowledge of fossil locations, the availability of firewood for safe travel and changes in the perpetual snowline over generations, all depended on the expertise of Himalayan peoples. In this period, scientific practice was thus overwhelmingly reliant on eclectic networks of, among others, Bhotiyas, Lepchas and Tartars, all of whom are rendered invisible in these comparative tableaux. Some of the European explorers like the Gerards, Royle and Hooker survive the transition to these global atlases (even if the laboriousness and inconsistency of their enterprises do not). The many South Asian actors featured in this book – from Hari Singh to Murdan Ali, and Pati Ram to Rechu – are, however, entirely effaced. The practical, everyday interactions by which expeditions and institutions functioned have thus disappeared from the picture. Similarly, the ongoing cosmological significances of the Himalaya in South Asian traditions – such as Nanda Devi as site of pilgrimage or Rakas Tal as a place of scripture – are here overwritten and subsumed within a language and framework of scientific plant geography.

Paraphrasing Haitian academic Michel-Rolph Trouillot, there are multiple places at which silences enter our histories.[15] In this context, the production of knowledge in the mountains is one, these atlases represent another and the creation of the archives is a third (my choices in this book undoubtedly represent a fourth). If we have seen many small and large moments of silencing in the journals and fieldbooks of travellers and surveyors, in these atlases we see silencing on an even greater scale. However, as Trouillot continues, if there are multiple places at which silences originate, then this also means that there are multiple places at which silences might be recovered (however unequally and imperfectly). *Science on the Roof of the World* has sought to historicise the global sciences of mountains, and in turn historicise particular forms of 'the global' and their imperial utility. In so doing, it has sought to show how the global was made – sometimes painfully and laboriously – as much as it was found. These atlases indicate, however, that the idiosyncrasies of local practice and South Asian agency were ultimately anathema to imposing global visions. The constitution of global mountains in this period was thus both an intensive exercise in material terms, but also an intensely imperial form of globality predicated on erasure and silencing.

Marginalising Uplands

This book has focused on the early nineteenth century, when increasing imperial access and periodic insecurity around the limits of knowledge converged in the Himalaya, in tandem with the recognition that they were the highest mountains on the globe. By the time we reach the mid-century, when Everest was confirmed as the world's highest mountain and this book takes its leave, some of the imaginative incoherence around the mountains that characterised this early period had been alleviated (even if imperial insecurities prevailed and particular scientific questions, such as around latitude and tropicality, continued to vex the coherence of the Himalaya's place in a global mountain world). Medical topographies had been sketched, but long-term acclimatisation remained little understood. New ascents were contemplated, but the summits continued to

[15] Trouillot, *Silencing the Past*, 22–30.

elude both naturalists and climbers. Uniformitarian explanations were increasingly adopted, but plate tectonics was not yet in geologists' toolboxes. Looking forward, a degree of mid-century confidence in the coherence of the Himalaya would give way to later-century doubts about whether the mountains were truly knowable at all.[16] Intractable limits to operating in decentred spaces of science and ongoing dependency on labour and expertise meant mastery continued to prove elusive (indeed, scientific engagement with the Himalaya under the Raj was often as unsystematic as under the EIC). Likewise, topography and politics continued to circumscribe knowledge production, and to impose limits in the mountains (sometimes magnified by the machinations of the so-called Great Game). Most notoriously, Tibet remained unattainable. As Scottish traveller Andrew Wilson (1831–1881) noted in his *Abode of Snow* (1875), any traveller kicking at its 'doors' was 'likely to find himself suddenly going down the mountains considerably faster than he went up them'.[17]

From the mid-century onwards, global comparison remained a key lens for unlocking the mountains, even as it was reconfigured. Equivalency was increasingly assumed, even if material realities continued to make this sometimes hard to reconcile. The topography of the vertical globe nevertheless evolved less rapidly – brief late-nineteenth century rumblings of the possible superiority of K2 over Everest notwithstanding – as more and more elevations were secured and confirmed (even while associated political and imaginative framings remained malleable and contested). More travellers also had opportunities to visit different ranges in person, and this brought about new configurations to rhetorical claims for authority (even as they drew on the language of verticality consolidated earlier in the century). As Andrew Wilson continued: 'I have had the privilege of discoursing from and on many mountains – mountains in Switzerland and Beloochistan, China and Japan – and would now speak', a lofty pedestal not afforded to many of those whose travels through the Himalaya have featured in this book.[18] (Wilson regretted, however, that 'the loftiest mountain of all, is out of the reach of nearly all travellers, owing to our weakness in allowing Nepal to exclude Englishmen from its territory'.[19]) Meanwhile, tourism and mountaineering, in both discourse and practice, added to the available vocabulary for interactions with the Himalaya, as did new technologies of visual representation like photography. While in

[16] See Bishop, *The Myth of Shangri-La*; Simpson, 'Clean out of the Map'.

[17] Andrew Wilson, *The Abode of Snow: Observations on a Journey from Chinese Tibet to the Indian Caucasus, Through the Upper Valleys of the Himalaya* (Edinburgh: William Blackwood and Sons, 1875), 19.

[18] Wilson, *The Abode of Snow*, 3.

[19] Wilson, *The Abode of Snow*, 14.

some cases beginning in the period of this study – as evident in travels like those of James Baillie Fraser – these activities took on new significance in the second half of the nineteenth century. In particular, hill stations emerged as important centres of governance and medicine, and the higher mountains increasingly served as canvases for imperial leisure, and the performance of hunting and sporting prowess.

In summing up this increased access, Andrew Wilson wrote that 'the change in modern travel has brought the most interesting, and even the wildest, parts of India within easy reach … nowadays, old ladies of seventy, who had scarcely ever left Britain before, are to be met with on the spurs of the Himálaya'.[20] In this gendered lament, Wilson comes across as vaguely disappointed, his words anticipating turn-of-the-century melancholy around the filling in of the last 'blank spaces' and Joseph Conrad's gloomy reflections on an era of 'Geography Triumphant'.[21] The opening up of the mountains thus heaped even greater imaginative resonances on Tibet, and those high places that remained isolated. As Wilson noted, 'the valley of Spiti is secluded in such a very formidable manner from the civilised world that it has very few European visitors' and elsewhere remarked that it is 'tolerably well raised out of the world'.[22] In such formulations, it is significant that the lowlands are the 'civilised world' and the uplands of the Himalaya are a place beyond, and a land apart. This imaginative geography of Shangri-La-type isolation and 'sanctuary' increasingly appealed to the imperial imagination in the period after the mid-century.[23] Though not necessarily negative – being seen as spaces of purified air and insight, free from disease and political corruption – these were nevertheless heavily 'othered' places. Indeed, these tropes endure, and whether in the context of mountaineering or tourism or art, mountains continue to be viewed as exceptional (if not aberrant) spaces – places to visit perhaps, and to briefly escape from or contemplate our horizontal lives. But as this book has shown, this was not necessarily inevitable, nor the only way of interacting with mountains. After all, the Himalayan massif was thrust up out of the Tethys Sea only relatively recently on the scale of deep time, and the Himalaya only became the semiotic 'roof of the world' considerably more recently than that. If the late eighteenth and early nineteenth centuries provided the tools for privileging a vertically oriented view, and the language to describe human and non-human worlds in three dimensions,

[20] Wilson, *The Abode of Snow*, 5.

[21] For more on Conrad, and the *longue durée* history of this tension between 'Geography Militant' and 'Geography Triumphant', see Driver, *Geography Militant*, 3–8.

[22] Wilson, *The Abode of Snow*, 242.

[23] Bishop, *The Myth of Shangri-La*, 115–23.

we begin to see here some of the consequences of enshrining the lowlands as the point of reference.

It becomes necessary to return now, one last time, to the high spaces of the Himalaya, and to consider the implications of imposing not only global commensurability, but also distinctions between uplands and lowlands. While the mutual formulation of the Himalaya and the global sciences of mountains has been the key thread running throughout this book, the mutual formulation of the Himalaya and the subcontinent is no less important. Placing bodily performance, the possibilities for cultivation and the occurrence of glaciers into a globally commensurable framework was not a priority for those who had long traversed the high passes, farmed the valleys and navigated the masses of moving ice. This study has thus been about the making of the Himalaya into a globally comparable region, even while acknowledging the potential lack of coherence of this concept to many of the historical actors. Indeed, the imperial idea of the 'Himalaya' as it emerged in this period would not necessarily have been meaningful or useful to many of those who lived in the mountains.[24] To gaze down on an image of a contiguous mountain belt from high is, after all, the conceit of a modern observer (or perhaps, as we have seen, a mid-nineteenth-century European atlas maker). More broadly, taking this perspective might inadvertently perpetuate Eurocentric legitimisations of empire, and lowlanders' perceptions of their own cultural and civilizational superiority.

Indeed, homogenising the uplands – whether as spaces of shared experience or as commensurable within global frameworks – was central to the process of 'othering' that confirmed the mountains as the margins. The area that this study has encompassed is hugely diverse in terms of demography – linguistically, ethnically, culturally and religiously – as well as in terms of climate and environment. Any simplistic distinctions between upland and lowland geographies and populations thus elide great deals of difference. As Chetan Singh argues, 'the study of mountain societies has usually carried with it some implicit assumptions. To begin with is the commonly held view that mountainous physiography was itself reason enough to delineate highlands as distinct geographical regions' such that even 'unconnected and distant highland cultures' have often been seen as related and 'their difference from non-highland regions has come to be perceived as the basis of similarities'.[25] He goes on to note that 'even though they occupy a position of considerable significance in the

[24] See Michaud, 'Editorial – Zomia and Beyond'; Shneiderman, 'Are the Central Himalayas in Zomia?'
[25] Singh, *Recognizing Diversity*, 1.

popular imagination of most South Asian societies, the Himalaya have remained marginal in almost every other respect'.[26] While simultaneously the cosmological home of the gods, the geographical source of life-giving rivers and the barrier that could protect empires, the Himalaya and their peoples had long been considered peripheral. The global scientific visions and imperial insecurities that were applied in the early nineteenth century only amplified this, as new imaginative geographies of the 'roof of the world' enshrined the worldviews of their lowland makers.

In this book, a mountain-centric approach has thus helped to frame the perspective of the Himalaya when viewed from the subcontinent and vice versa, even while acknowledging significant diversity within these perspectives. It has pointed to the way that the mountains could, in a sense, only exist when constituted simultaneously and in contrast with the lowlands. This early phase of the remaking of the Himalaya as both the highest mountains in the world and an insecure frontier was thus inevitably bound up with the accelerating imperial appropriation of the subcontinent, which – by default – cast the mountains as peripheral. Whether in applying horizontal divisions of latitude to vertical changes in vegetation, delineating 'normal' bodily reactions to the atmosphere or determining the location for a 'northern' garden, the lowlands remained the point of reference. The mapping of these phenomena through the norms of the plains (and of temperate Europe) was as pervasive as it was ultimately insidious. Indeed, the imaginative, scientific and political engagements with the mountains that this book has traced ultimately only served to confirm them as marginal spaces and peripheral places.

[26] Singh, *Recognizing Diversity*, 3.

Bibliography

Archival Collections

British Library (London)

India Office Records (IOR)
- Bengal Proceedings
  - P/9/8
  - P/10/10
  - P/11/46
  - P/12/28
- Board's Collections
  - F/4/552/13384
  - F/4/587/14218
  - F/4/660/18324
  - F/4/750/20517
  - F/4/955/27123(2)
  - F/4/957/27123(24)
  - F/4/1017/27954
  - F/4/1068/29191
  - F/4/1191/30877
  - F/4/1732/69955
  - F/4/1828/75444
  - F/4/2356/124635
  - F/4/2461/136806
  - F/4/2461/136807
  - F/4/2498/141673
  - F/4/2528/145895
- Private Papers/European Manuscripts (Mss. Eur.)
  - B336
  - C951
  - D137
  - E96

F127
Selections from the Records of the Government of the North Western Provinces
V/23/121 Pt 37 Art 4
Additional Manuscripts (Add. Ms.)
Napier Papers
Western Manuscripts
Western Drawings (WD)

Natural History Museum (London)

Palaeontology Manuscripts
Watercolours

Royal Botanic Gardens, Kew (London)

Library, Art and Archives
Director's Correspondence
DC/53/68
DC/53/102
DC/54/231
DC/54/336
DC/54/337
DC/54/352
DC/54/485
DC/54/502
Joseph Dalton Hooker Collection
JDH/1/10
Madden Papers
Nathaniel Wallich Collection
William Griffith Papers
Herbarium Specimens

University of Nottingham Archives (Nottingham)

William Bentinck Papers

National Archives of India (New Delhi)

Survey of India Records (SOI)
Dehra Dun Volumes (DDn.)

128
130
135
145
150
152
220
231
Fieldbooks (Fdbk.)
87
91
97
113
Memoirs (Mem.)
Miscellaneous (MRIO)

Survey of India Museum (Dehradun)

Instruments and papers

Published Primary Sources

'Account of a Volcano in the Himalayah Mountains. Communicated to Dr Brewster by a Correspondent in India'. *Edinburgh Journal of Science* 4, no. 2 (1826): 209–11.

'Account of Hot Springs and Volcanic Appearances in the Himalaya Mountains'. *Edinburgh Journal of Science* 7, no. 2 (1827): 55–56.

[Amicus]. 'Remarks on the Himalaya Mountains'. *Asiatic Journal* 5 (1818): 319–25.

'Asiatic Society of Calcutta – Physical Class, 8 June 1832'. *Asiatic Journal (New Series)* 7 (1832): 147–50.

Batten, John Hallett. 'Extract of a Letter from J. H. Batten, Esq. Bengal Civil Service, Dated Camp Semulka on the Cosillah River, Kumaon, December 28th, 1843'. *Calcutta Journal of Natural History* 4 (1844): 537–39.

Batten, John Hallett. 'Letter to Henry Torrens, 8 February 1842'. *Journal of the Asiatic Society of Bengal* 11, part 1 (1842): 583–85.

Batten, John Hallett. 'Line of Perpetual Snow'. *Calcutta Journal of Natural History* 5 (1845): 383–88.

Batten, John Hallett. 'Note of a Visit to the Niti Pass of the Grand Himalayan Chain'. *Journal of the Asiatic Society of Bengal* 7, part 1 (1838): 310–16.

Bell, James. *A System of Geography, Popular and Scientific: Or a Physical, Political, and Statistical Account of the World and Its Various Divisions*. 6 vols. Glasgow: A. Fullarton, 1832.
Bell, James and John Bell. *Critical Researches in Philology and Geography*. Glasgow: James Brash, 1824.
Berghaus, Heinrich. *Dr. Heinrich Berghaus' Physikalischer Atlas*. Gotha: Justus Perthes, 1845.
Berghaus, Heinrich. *Dr. Heinrich Berghaus' Physikalischer Atlas*. 2nd ed. Gotha: Justus Perthes, 1852.
Bert, Paul. *La pression barométrique: recherches de physiologie expérimentale*. Paris: G. Masson, 1878.
'Biographical Sketch of the Late James Gilbert Gerard'. *Asiatic Journal (Third Series)* 4 (1844): 66–73.
Blanford, Henry Francis. 'On Dr. Gerard's Collection of Fossils from the Spiti Valley, in the Asiatic Society's Museum'. *Journal of the Asiatic Society of Bengal* 32 (1863): 124–38.
Buckland, William. *Reliquiae Diluvianae, or Observations on the Organic Remains Contained in Caves, Fissures, and Diluvial Gravel, and on Other Geological Phenomena, Attesting the Action of an Universal Deluge*. London: John Murray, 1823.
Burnes, Alexander. *Travels into Bokhara*. 3 vols. London: John Murray, 1834.
Clarke, C. C. *The Hundred Wonders of the World, and of the Three Kingdoms of Nature*. 15th ed. London: Richard Phillips, 1822.
Colebrooke, Henry Thomas. 'Height of the Himalaya Mountains'. *Journal of Science and the Arts* 6 (1819): 51–65.
Colebrooke, Henry Thomas. 'On the Height of the Dhawalagiri, the White Mountain of Himalaya'. *Quarterly Journal of Science, Literature and the Arts* 11 (1821): 240–46.
Colebrooke, Henry Thomas. 'On the Height of the Himálaya Mountains'. *Asiatick Researches* 12 (1816): 251–86.
Colebrooke, Henry Thomas. 'On the Limit of Constant Congelation in the Himalaya Mountains'. *Quarterly Journal of Literature, Science and the Arts* 7 (1819): 38–43.
Colebrooke, Henry Thomas. 'On the Sources of the Ganges, in the Himádri or Emodus'. *Asiatic Researches* 11 (1810): 429–46.
Cunningham, Alexander. *Ladak, Physical, Statistical, and Historical; with Notices of the Surrounding Countries*. London: W. H. Allen, 1854.
Cunningham, Joseph Davey. 'Note on the Limits of Perpetual Snow in the Himalayas'. *Journal of the Asiatic Society of Bengal* 18, part 2 (1849): 694–97.
Darwin, Charles. *On the Origin of Species by Means of Natural Selection, or the Preservation of Favoured Races in the Struggle for Life*. London: John Murray, 1859.
Everest, Robert. 'Memorandum on the Fossil Shells Discovered in the Himalayan Mountains'. *Asiatic Researches* 18, no. 2 (1833): 107–14.
Falconer, Hugh. 'On the Aptitude of the Himalayan Range for the Culture of the Tea Plant'. *Journal of the Asiatic Society of Bengal* 3 (1834): 178–88.

Falconer, Hugh. *Palaeontological Memoirs and Notes of the Late Hugh Falconer.* Edited by Charles Murchison. 2 vols. London: Hardwicke, 1868.
Fraser, James Baillie. *Journal of a Tour through Part of the Snowy Range of the Himala Mountains, and to the Sources of the Rivers Jumna and Ganges.* London: Rodwell and Martin, 1820.
'Geological Map of Captain Herbert's Himalaya Survey'. *Journal of the Asiatic Society of Bengal* 13, part 1 (1844): 170.
Gerard, Alexander. *Account of Koonawur, in the Himalaya.* Edited by George Lloyd. London: James Madden, 1841.
Gerard, Alexander. 'Journal of an Excursion through the Himalayah Mountains, from Shipke to the Frontiers of Chinese Tartary'. *Edinburgh Journal of Science* 1 (1824): 41–52, 215–25.
Gerard, Alexander. 'Narrative of a Journey from Soobathoo to Shipke, in Chinese Tartary'. *Journal of the Asiatic Society of Bengal* 11, part 1 (1842): 363–91.
Gerard, James Gilbert. 'A Letter from the Late Mr J. G. Gerard'. In *Narrative of a Journey from Caunpoor to the Boorendo Pass, in the Himalaya Mountains*, edited by George Lloyd, Vol. 1. London: J. Madden, 1840, 287–347.
Gerard, James Gilbert. 'Letter from a Correspondent in the Himalaya'. *Gleanings in Science* 1 (1829): 109–11.
Gerard, James Gilbert. 'Observations on the Spiti Valley and the Circumjacent Country within the Himalaya'. *Asiatic Researches* 18, no. 2 (1833): 238–78.
[Gerard, James Gilbert]. 'The Himalaya Country'. *Asiatic Journal (New Series)* 6, no. 2 (1831): 110–12.
Gerard, Patrick. 'Abstract of a Meteorological Journal, Kept at Kotgarh'. Edited by [James Prinsep]. *Journal of the Asiatic Society of Bengal* 2 (1833): 615–22.
Gordon, Thomas Edward. *The Roof of the World: Being the Narrative of a Journey over the High Plateau of Tibet to the Russian Frontier and the Oxus Sources on Pamir.* Edinburgh: Edmonston and Douglas, 1876.
Govan, George. 'On the Natural History and Physical Geography of the Districts of the Himalayah Mountains between the River-Beds of the Jumna and Sutluj'. *Edinburgh Journal of Science* 2, no. 3–4 (1825): 17–38, 277–87.
Graham, William Woodman. 'Travel and Ascents in the Himálaya'. *Proceedings of the Royal Geographical Society* 6, no. 8 (1884): 429–47.
Griffith, William. *Itinerary Notes of Plants Collected in the Khasyah and Bootan Mountains, 1837–8 and in Affghanistan and Neighbouring Countries, 1839–41.* Edited by John McClelland. Calcutta: J. P. Bellamy, 1848.
Griffith, William. *Journals of Travels in Assam, Burma, Bootan, Affghanistan and the Neighbouring Countries.* Edited by John McClelland. Calcutta: Bishop's College Press, 1847.
Griffith, William. 'Tables of Barometrical and Thermometrical Observations, Made in Affghanistan, Upper Scinde, and Kutch Gundava, during the Years 1839–40'. *Journal of the Asiatic Society of Bengal* 11, part 1 (1842): 49–90.

Herbert, James. 'An Account of a Tour Made to Lay Down the Course and Levels of the River Setlej'. *Asiatic Researches* 15 (1825): 339–428.

Herbert, James. 'Notice on the Occurrence of Coal, within the Indo Gangetic Tract of Mountains'. *Asiatic Researches* 16 (1828): 397–408.

Herbert, James. 'On the Mineral Productions of That Part of the Himalaya Mountains, Lying between the Satlaj and the Kali, (Gagra) Rivers'. *Asiatic Researches* 18 (1833): 227–58.

Herbert, James. 'On the Organic Remains Found in the Himmalaya'. *Gleanings in Science* 3 (1831): 265–72.

Herbert, James. 'Report upon the Mineralogical Survey of the Himalayan Mountains'. *Journal of the Asiatic Society of Bengal* 11, part 2 (1842): i–clxiii.

Herbert, James and John Hodgson. 'Description of Passes in the Himalaya'. *Asiatic Journal* 9 (1820): 589–92.

Hervey, [Mrs]. *The Adventures of a Lady in Tartary, Thibet, China and Kashmir*. 3 vols. London: Hope, 1853.

'Himalaya Mountains and Lake Manasawara'. *The London Quarterly Review* 17 (1817): 403–41.

Hodgson, Brian Houghton. *Essays on the Languages, Literature, and Religion of Nepal and Tibet, Together with Further Papers on the Geography, Ethnology, and Commerce of Those Countries*. London: Trübner, 1874.

Hodgson, Brian Houghton. 'On the Mammalia of Nepal'. *Journal of the Asiatic Society of Bengal* 1 (1832): 335–49.

Hodgson, John. 'Journal of a Survey to the Heads of the Rivers, Ganges and Jumna'. *Asiatic Researches* 14 (1822): 60–152.

[Hodgson, John]. 'Letter from the Himmalaya'. *Gleanings in Science* 2 (1830): 48–52.

Hodgson, John and James Herbert. 'An Account of Trigonometrical and Astronomical Operations for Determining the Heights and Positions of the Principal Peaks of the Himalaya Mountains'. *Asiatic Researches* 14 (1822): 187–372.

Hooker, Joseph Dalton. *Himalayan Journals: Or, Notes of a Naturalist in Bengal, the Sikkim and Nepal Himalayas, the Khasia Mountains*. 2 vols. London: John Murray, 1854.

Hooker, Joseph Dalton. *Rhododendrons of the Sikkim Himalaya*. London: Reeve, Benham and Reeve, 1849.

Humboldt, Alexander von. *Ansichten Der Natur*. 3rd ed. Stuttgart: J. G. Cotta, 1849.

Humboldt, Alexander von. *Cosmos: A Sketch of a Physical Description of the Universe*. Translated by Elise Charlotte Otté. London: Henry G. Bohn, 1848.

Humboldt, Alexander von. *Essay on the Geography of Plants*. Edited by Stephen T. Jackson. Translated by Sylvie Romanowski. Chicago: University of Chicago Press, 2013.

Humboldt, Alexander von. 'Sur la limité inférieure des neiges perpétuelles dans les montagnes de l'Himalaya et les régions équatoriales'. *Annales de chimie et de physique* 14 (1820): 5–56.

Humboldt, Alexander von. 'Sur l'elévation des montagnes de l'Inde'. *Annales de chimie et de physique* 3 (1816): 297–317.

Humboldt, Alexander von. *Views of Nature: Or Contemplations on the Sublime Phenomena of Creation.* Translated by Elise Charlotte Otté and Henry George Bohn. London: Henry G. Bohn, 1850.

Humboldt, Alexander von and Aimé Bonpland. *Essai sur la géographie des plantes.* Paris: F. Schoell, 1807.

Hutton, Thomas. 'Correction of the Erroneous Doctrine that the Snow Lies Longer and Deeper on the Southern than on the Northern Aspect of the Himalaya'. *Calcutta Journal of Natural History* 4 (1844): 275–82.

Hutton, Thomas. 'Extract of a Letter from Thomas Hutton, Soongum, 5 July 1838'. *Journal of the Asiatic Society of Bengal* 7, part 2 (1838): 667–68.

Hutton, Thomas. 'Geological Report on the Valley of the Spiti'. *Journal of the Asiatic Society of Bengal* 10, part 1 (1841): 198–229.

Hutton, Thomas. 'Journal of a Trip through Kunawur, Hungrung, and Spiti, Undertaken in the Year 1838 (Part I)'. *Journal of the Asiatic Society of Bengal* 8 (1839): 901–50.

Hutton, Thomas. 'Journal of a Trip through Kunawur, Hungrung, and Spiti, Undertaken in the Year 1838 (Part II)'. *Journal of the Asiatic Society of Bengal* 9, part 1 (1840): 489–513.

Hutton, Thomas. 'Journal of a Trip through Kunawur, Hungrung, and Spiti, Undertaken in the Year 1838 (Part III)'. *Journal of the Asiatic Society of Bengal* 9, part 1 (1840): 555–81.

Hutton, Thomas. 'Journal of a Trip to Burenda Pass in 1836'. *Journal of the Asiatic Society of Bengal* 6, part 2 (1837): 901–38.

Hutton, Thomas. 'Note on the Snow Line on the Himalaya'. *Calcutta Journal of Natural History* 5 (1845): 379–83.

Hutton, Thomas. 'Read the Following Letter from Lieut. Thomas Hutton, 37th N.I. Dated Simla, 27th August, 1837'. *Journal of the Asiatic Society of Bengal* 6, part 2 (1837): 897–98.

Hutton, Thomas. 'Remarks on the Snow Line in the Himalaya'. *Journal of the Asiatic Society of Bengal* 18, part 2 (1849): 954–66.

Hutton, Thomas. 'To the Editor of the Calcutta Journal of Natural History'. *Calcutta Journal of Natural History* 6 (1846): 56–59.

Irvine, Robert Hamilton. 'A Few Observations on the Probable Results of a Scientific Research after Metalliferous Deposits in the Sub-Himalayan Range around Darjeeling'. *Journal of the Asiatic Society of Bengal* 17, part 1 (1848): 137–44.

Jacquemont, Victor. *Letters from India.* Translated by anon. 2 vols. London: Edward Churlton, 1834.

Jacquemont, Victor. *Voyage dans l'Inde pendant les années 1828 à 1832.* 4 vols. Paris: Firmin Didot frères, 1841.

Johnston, Alexander Keith. *The Physical Atlas of Natural Phenomena.* 2nd ed. Edinburgh: William Blackwood, 1856.

Lal, Mohan. *Travels in the Panjab, Afghanistan, Turkistan, to Balk, Bokhara, and Herat; and a Visit to Great Britain and Germany.* London: W. H. Allen, 1846.

'Lieutenant White's Views in India, Edited by Emma Roberts'. *The Spectator.* 16 December 1837.

'Literary and Philosophical Intelligence'. *Asiatic Journal* 11 (1821): 375–79.

Lloyd, William and Alexander Gerard. *Narrative of a Journey from Caunpoor to the Boorendo Pass, in the Himalaya Mountains*. 2 vols. London: J. Madden, 1840.

Madden, Edward. 'Diary of an Excursion to the Shatool and Boorun Passes over the Himalaya, in September, 1845'. *Journal of the Asiatic Society of Bengal* 15 (1846): 79–135.

Madden, Edward. 'Notes of an Excursion to the Pindree Glacier, in September 1846'. *Journal of the Asiatic Society of Bengal* 16, part 1 (1847): 226–66.

Madden, Edward. 'On the Occurrence of Palms and Bambus, with Pines and Other Forms Considered Northern, at Considerable Elevations in the Himalaya'. *Transactions of the Botanical Society of Edinburgh* 4 (1853): 185–96.

Madden, Edward. 'The Turaee and Outer Mountains of Kumaoon'. *Journal of the Asiatic Society of Bengal* 17, 18 (1848): 349–450, 603–44.

Manson, James. 'Capt. Manson's Journal of a Visit to Melum and the Oonta Dhoora Pass in Juwahir'. Edited by John Hallett Batten. *Journal of the Asiatic Society of Bengal* 11, part 2 (1842): 1157–82.

[Manson, James]. 'On the Distress and Exhaustion Consequent to Exertion at Great Elevations'. *Gleanings in Science* 1 (1829): 330–31.

Markham, Clements. *A Memoir on the Indian Surveys*. London: W. H. Allen, 1871.

McClelland, John. *Some Inquiries in the Province of Kemaon in India, Relative to Geology*. Calcutta: Baptist Mission Press, 1835.

'Miscellaneous Notices'. *Gleanings in Science* 1 (1829): 373–75.

Moorcroft, William. 'A Journey to Lake Mánasaróvara in Ún-Dés, a Province of Little Tibet'. *Asiatic Researches* 12 (1816): 375–534.

Moorcroft, William and George Trebeck. *Travels in the Himalayan Provinces of Hindustan and the Panjab; in Ladakh and Kashmir; in Peshawar, Kabul, Kunduz, and Bokhara*. Edited by Horace Hayman Wilson. London: John Murray, 1841.

Montgomerie, Thomas George. 'Report of a Route-Survey Made by Pundit, from Nepal to Lhasa, and Thence through the Upper Valley of the Brahmaputra to Its Source'. *Journal of the Royal Geographical Society of London* 38 (1868): 129–219.

'Mountain Barometers'. *The Calcutta Journal of Politics and General Literature* 1, no. 26 (1823): 415–16.

'On the Most Eligible Form for the Construction of a Portable Barometer'. *Gleanings in Science* 1 (1829): 313–19.

'Passage of the Himalaya Mountains'. *Quarterly Review* 22 (1820): 415–30.

Pemberton, Robert Boileau. *Report on Bootan*. Calcutta: Bengal Military Orphan Press, 1839.

Prinsep, James. 'Table for Ascertaining the Heights of Mountains from the Boiling Point of Water'. *Journal of the Asiatic Society of Bengal* 2 (1833): 194–200.

'Proceedings of Societies – Asiatic Society, Wednesday, 7th March, 1832'. *Journal of the Asiatic Society of Bengal* 1 (1832): 116–17.

'Proceedings of the Asiatic Society, Physical Class, 8 February 1832'. *Journal of the Asiatic Society of Bengal* 1 (1832): 74–77.

'Proceedings of the Asiatic Society, Physical Class, Wednesday, 15th August, 1832'. *Journal of the Asiatic Society of Bengal* 1 (1832): 363–65.

Raper, Felix Vincent. 'Narrative of a Survey for the Purpose of Discovering the Sources of the Ganges'. *Asiatic Researches* 11 (1810): 446–563.

Raper, Henry and Robert FitzRoy. 'Hints to Travellers'. *Journal of the Royal Geographical Society of London* 24 (1854): 328–58.

Rennell, James. *Memoir of a Map of Hindoostan; or the Mogul Empire*. London: M. Brown, 1788.

'Report of the Curator, Museum of Economic Geology and Geological and Mineralogical Departments, March 1844'. *Journal of the Asiatic Society of Bengal* 13, part 1 (1844): xxv–xxviii.

Rey, M. 'Influence sur le corps humain, des ascensions sur les hautes montagnes'. *Nouvelle annales des voyages et des sciences géographiques* 19, no. 3 (1838): 227–44.

Royle, John Forbes. 'Account of the Honourable Company's Botanic Garden at Saharanpur'. *Journal of the Asiatic Society of Bengal* 1 (1832): 41–58.

Royle, John Forbes. *An Essay on the Antiquity of Hindoo Medicine*. London: W. H. Allen, 1837.

Royle, John Forbes. *Essay on the Productive Resources of India*. London: W. H. Allen, 1840.

Royle, John Forbes. 'Extract of a Report on the Medicinal Garden at Musoorea'. *Transactions of the Medical and Physical Society of Calcutta* 4 (1829): 406–22.

Royle, John Forbes. *Illustrations of the Botany and Other Branches of the Natural History of the Himalayan Mountains*. 2 vols. London: W. H. Allen, 1839.

Royle, John Forbes. 'List of Articles of Materia Medica, Obtained in the Bazars of the Western and Northern Provinces of India'. *Journal of the Asiatic Society of Bengal* 1 (1832): 458–71.

Royle, John Forbes. 'On the Identification of the Mustard Tree of Scripture'. *Journal of the Royal Asiatic Society of Great Britain and Ireland* 8 (1846): 113–37.

Royle, John Forbes. 'Report on the Progress of the Cultivation of the China Tea Plant in the Himalayas, from 1835 to 1847'. *Journal of the Royal Asiatic Society of Great Britain and Ireland* 12 (1849): 125–52.

Salter, John William and Henry Francis Blanford. *Palaeontology of Niti in the Northern Himalaya: Being Descriptions and Figures of the Palaeozoic and Secondary Fossils Collected by Colonel Richard Strachey*. Calcutta: O. T. Cutter, 1865.

Sherwill, Walter. 'Notes upon Some Atmospherical Phenomena Observed at Darjiling in the Himalayah Mountains, during the Summer of 1852'. *Journal of the Royal Asiatic Society of Bengal* 23 (1854): 49–57.

Strachey, Henry. 'Explanation of the Elevations of Places between Almorah and Gangri given in Lieut. Strachey's Map and Journal'. *Journal of the Asiatic Society of Bengal* 17, part 2 (1848): 527–31.

Strachey, Henry. 'Narrative of a Journey to Cho Lagan (Rakas Tal), Cho Mapan (Manasarowar), and the Valley of Pruang in Gnari, Hundes, in September and October 1846'. *Journal of the Asiatic Society of Bengal* 17, part 2 (1848): 98–120, 127–82, 327–51.

Strachey, Henry. 'Physical Geography of Western Tibet'. *Journal of the Royal Geographical Society* 23 (1853): 1–69.

Strachey, Richard. 'A Description of the Glaciers of the Pindur and Kuphinee Rivers in the Kumaon Himálaya'. *Journal of the Asiatic Society of Bengal* 16, part 2 (1847): 794–812.

Strachey, Richard. *Catalogue of Plants Found in Kumaon and Garhwal and the Adjoining Parts of Tibet*. Edited by John Firminger Duthie. London: Lovell, Reeve, 1906.

Strachey, Richard. 'Narrative of a Journey to the Lakes Rakas-Tal and Manasarowar, in Western Tibet, Undertaken in September, 1848'. *The Geographical Journal* 15, no. 4 (1900): 150–70, 243–64, 394–415.

Strachey, Richard. 'Note on the Motion of the Glacier of the Pindur in Kumaon'. *Journal of the Asiatic Society of Bengal* 17, part 2 (1848): 203–5.

Strachey, Richard. 'On the Geology of Part of the Himalaya Mountains and Tibet'. *Quarterly Journal of the Geological Society of London* 7, no. 1–2 (1851): 292–310.

Strachey, Richard. 'On the Physical Geography of the Provinces of Kumáon and Garhwál in the Himálaya Mountains, and of the Adjoining Parts of Tibet'. *The Journal of the Royal Geographical Society of London* 21 (1851): 57–85.

Strachey, Richard. 'On the Snow-Line in the Himalaya'. *Journal of the Asiatic Society of Bengal* 18, part 1 (1849): 287–310.

Thomson, Thomas. 'Sketch of the Climate and Vegetation of the Himalaya'. *Proceedings of the Philosophical Society of Glasgow* 3 (1855): 193–204.

Thomson, Thomas. *Western Himalaya and Tibet: A Narrative of a Journey through the Mountains of Northern India during the Years 1847–8*. London: Reeve, 1852.

Traill, George William. 'Statistical Report on the Bhotia Mehals of Kemaon'. *Asiatic Researches* 17 (1832): 1–50.

Turner, Samuel. *Account of an Embassy to the Court of the Teshoo Lama in Tibet, Containing a Narrative of a Journey through Bootan and Part of Tibet*. London: W. Bulmer, 1800.

Ullah, Mir Izzet. 'Travels beyond the Himalaya'. Translated by Horace Hayman Wilson. *Journal of the Royal Asiatic Society* 7, no. 14 (1843): 283–342.

Vigne, Godfrey Thomas. *Travels in Kashmir, Ladak, Iskardo, the Countries Adjoining the Mountain-Course of the Indus, and the Himalaya, North of the Panjab*. London: Henry Colburn, 1842.

'Volcano in the Himmalaya'. *Gleanings in Science* 1 (1829): 338–39.

Wallich, Nathaniel. *Plantae Asiaticae Rariores*. London: Treuttel and Würtz, 1830.

Webb, William. 'Extract of a Letter from Captain William Spencer Webb, 29th March, 1819'. *Quarterly Journal of Science, Literature and the Arts* 9 (1820): 61–69.

Webb, William. 'Memoir Relative to a Survey of Kumaon'. *Asiatic Researches* 13 (1820): 293–310.

Weller, Joseph Alexander. 'Extract from the Journal of Lieut. J. A. Weller . . . on a Trip to the Bulcha and Oonta Dhoora Passes'. *Journal of the Asiatic Society of Bengal* 12, part 1 (1843): 78–102.
White, George Francis. *Views in India: Chiefly among the Himalaya Mountains*. Edited by Emma Roberts. London: Fisher, Son, and Co, 1838.
Wilson, Andrew. *The Abode of Snow: Observations on a Journey from Chinese Tibet to the Indian Caucasus, through the Upper Valleys of the Himalaya*. Edinburgh: William Blackwood, 1875.
Wollaston, Francis John Hyde. 'Description of a Thermometrical Barometer for Measuring Altitudes'. *Philosophical Transactions of the Royal Society of London* 107 (1817): 183–96.
Wood, John. *Narrative of a Journey to the Source of the River Oxus*. London: John Murray, 1841.

Published Secondary Sources

Alam, Aniket. *Becoming India: Western Himalayas under British Rule*. New Delhi: Foundation Books, 2008.
Alder, Garry. *Beyond Bukhara: The Life of William Moorcroft*. London: Century, 1985.
Allen, Charles. *A Mountain in Tibet: The Search for Mount Kailas and the Sources of the Great Rivers of India*. London: Little, Brown, 1982.
Anthony, Patrick. 'Mining as the Working World of Alexander von Humboldt's Plant Geography and Vertical Cartography'. *Isis* 109, no. 1 (2018): 28–55.
Archer, Mildred. *Natural History Drawings in the Indian Office Library*. London: Her Majesty's Stationery Office, 1962.
Armitage, David, Alison Bashford and Sujit Sivasundaram. *Oceanic Histories*. Cambridge: Cambridge University Press, 2017.
Arnold, David. 'Envisioning the Tropics: Joseph Hooker in India and the Himalayas, 1848–1850'. In *Tropical Visions in an Age of Empire*, edited by Felix Driver and Luciana Martins, 211–34. Chicago: University of Chicago Press, 2005.
Arnold, David. 'Globalization and Contingent Colonialism: Towards a Transnational History of "British" India'. *Journal of Colonialism and Colonial History* 16, no. 2 (2015), https://muse.jhu.edu/article/587721.
Arnold, David. 'Hodgson, Hooker and the Himalayan Frontier, 1848–1850'. In *The Origins of Himalayan Studies, Brian Houghton Hodgson in Nepal and Darjeeling 1820–1858*, edited by David Waterhouse, 189–204. London: Routledge Curzon, 2004.
Arnold, David. 'India's Place in the Tropical World, 1770–1930'. *The Journal of Imperial and Commonwealth History* 26, no. 1 (1998): 1–21.
Arnold, David. 'Race, Place and Bodily Difference in Early Nineteenth-Century India'. *Historical Research* 77, no. 196 (2004): 254–73.
Arnold, David. *Science, Technology and Medicine in Colonial India*. Cambridge: Cambridge University Press, 2004.

Arnold, David. *The Tropics and the Traveling Gaze: India, Landscape, and Science, 1800–1856*. Seattle: University of Washington Press, 2006.

Arnold, David. *Toxic Histories: Poison and Pollution in Modern India*. Cambridge: Cambridge University Press, 2016.

Axelby, Richard. 'Calcutta Botanic Garden and the Colonial Re-Ordering of the Indian Environment'. *Archives of Natural History* 35, no. 1 (2008): 150–63.

Baber, Zaheer. 'The Plants of Empire: Botanic Gardens, Colonial Power and Botanical Knowledge'. *Journal of Contemporary Asia* 46, no. 4 (2016): 659–79.

Bailly, Jean-Christophe, Jean-Marc Besse and Gilles Palsky. *Le monde sur une feuille: les tableaux comparatifs de montagnes et de fleuves dans les atlas du XIXe siècle*. Lyon: Fage Éditions, 2014.

Ballantyne, Tony. *Orientalism and Race: Aryanism in the British Empire*. Basingstoke: Palgrave Macmillan, 2002.

Bashford, Alison and Sarah W. Tracy. 'Introduction: Modern Airs, Waters, and Places'. *Bulletin of the History of Medicine* 86, no. 4 (2012): 495–514.

Baud, Aymon, Philippe Forêt and Svetlana Gorshenina. *La Haute-Asie telle qu'ils l'ont vue: explorateurs et scientifiques de 1820 à 1940*. Geneva: Olizane, 2003.

Bayly, Christopher. *Empire & Information: Intelligence Gathering and Social Communication in India, 1780–1870*. Cambridge: Cambridge University Press, 1997.

Bergmann, Christoph. *The Himalayan Border Region: Trade, Identity and Mobility in Kumaon, India*. Dordrecht: Springer, 2016.

Bernbaum, Edwin. *Sacred Mountains of the World*. Berkeley: University of California Press, 1998.

Bewell, Alan. *Romanticism and Colonial Disease*. Baltimore: Johns Hopkins University Press, 1999.

Bigg, Charlotte, David Aubin and Philipp Felsch. 'Introduction: The Laboratory of Nature – Science in the Mountains'. *Science in Context* 22, no. 3 (2009): 311–21.

Bishop, Peter. *The Myth of Shangri-La: Tibet, Travel Writing, and the Western Creation of Sacred Landscape*. Berkeley: University of California Press, 1989.

Botting, Douglas. *Humboldt and the Cosmos*. London: Sphere Books, 1973.

Bourguet, Marie-Noëlle. 'A Portable World: The Notebooks of European Travellers (Eighteenth to Nineteenth Centuries)'. *Intellectual History Review* 20, no. 3 (2010): 377–400.

Bourguet, Marie-Noëlle. 'Landscape with Numbers: Natural History, Travel and Instruments in the Late Eighteenth and Early Nineteenth Centuries'. In *Instruments, Travel and Science: Itineraries of Precision from the Seventeenth to the Twentieth Century*, edited by Marie-Noëlle Bourguet, Christian Licoppe and H. Otto Sibum, 96–125. London: Routledge, 2002.

Bourguet, Marie-Noëlle. 'Measurable Difference: Botany, Climate, and the Gardener's Thermometer in Eighteenth-Century France'. In *Colonial Botany: Science, Commerce, and Politics in the Early Modern World*, edited by Londa Schiebinger and Claudia Swan, 270–86. Philadelphia: University of Pennsylvania Press, 2005.

Bourguet, Marie-Noëlle, Christian Licoppe and H. Otto Sibum. *Instruments, Travel and Science: Itineraries of Precision from the Seventeenth to the Twentieth Century*. London: Routledge, 2002.

Braun, Bruce. 'Producing Vertical Territory: Geology and Governmentality in Late Victorian Canada'. *Ecumene* 7, no. 1 (2000): 7–46.

Bravo, Michael. 'Precision and Curiosity in Scientific Travel: James Rennell and the Orientalist Geography of the New Imperial Age (1760–1830)'. In *Voyages and Visions: Towards a Cultural History of Travel*, edited by Jaś Elsner and Joan-Pau Rubiés, 162–83. London: Reaktion Books, 1999.

Bravo, Michael and Sverker Sörlin. *Narrating the Arctic: A Cultural History of Nordic Scientific Practices*. Canton, MA: Science History Publications, 2002.

Brescius, Moritz von. *German Science in the Age of Empire: Enterprise, Opportunity and the Schlagintweit Brothers*. Cambridge: Cambridge University Press, 2019.

Brockway, Lucile H. *Science and Colonial Expansion: The Role of the British Royal Botanic Gardens*. New York: Academic Press, 1979.

Browne, Janet. 'Biogeography and Empire'. In *Cultures of Natural History*, edited by Nicholas Jardine, James Secord and Emma Spary, 305–21. Cambridge: Cambridge University Press, 1996.

Browne, Janet. *The Secular Ark: Studies in the History of Biogeography*. New Haven: Yale University Press, 1983.

Burkill, Issac Henry. *Chapters on the History of Botany in India*. Calcutta: Botanical Survey of India, 1965.

Burnett, D. Graham. *Masters of All They Surveyed: Exploration, Geography, and a British El Dorado*. Chicago: University of Chicago Press, 2000.

Camerini, Jane R. 'Heinrich Berghaus's Map of Human Diseases'. *Medical History* 44, no. S20 (2000): 186–208.

Camerini, Jane R. 'The Physical Atlas of Heinrich Berghaus: Distribution Maps as Scientific Knowledge'. In *Non-Verbal Communication in Science Prior to 1900*, edited by Renato Mazzolini, 479–512. Florence: L. S. Olschki, 1993.

Campbell, Ian W. '"Our Friendly Rivals": Rethinking the Great Game in Ya'qub Beg's Kashgaria, 1867–77'. *Central Asian Survey* 33, no. 2 (2014): 199–214.

Cannon, Susan Faye. *Science in Culture: The Early Victorian Period*. New York: Science History Publications, 1978.

Carter, Paul. *The Road to Botany Bay: An Exploration of Landscape and History*. Minneapolis: University of Minnesota Press, 1987.

Chakrabarti, Pratik. *Inscriptions of Nature: Geology and the Naturalization of Antiquity*. Baltimore: Johns Hopkins University Press, 2020.

Chakrabarti, Pratik. 'Medical Marketplaces Beyond the West: Bazaar Medicine, Trade and the English Establishment in Eighteenth-Century India'. In *Medicine and the Market in England and Its Colonies, c. 1450–c. 1850*, edited by Mark S. R. Jenner and Patrick Wallis, 216–37. Basingstoke: Palgrave Macmillan, 2007.

Chakrabarti, Pratik. *Medicine and Empire: 1600–1960*. Basingstoke: Palgrave Macmillan, 2014.

Chakrabarti, Pratik. *Western Science in Modern India: Metropolitan Methods, Colonial Practices*. Delhi: Permanent Black, 2004.

Chakrabarti, Pratik and Joydeep Sen. '"The World Rests on the Back of a Tortoise": Science and Mythology in Indian History'. *Modern Asian Studies* 50, no. 3 (2016): 808–40.

Coen, Deborah R. *Climate in Motion: Science, Empire, and the Problem of Scale*. Chicago: University of Chicago Press, 2018.

Colley, Ann C. *Victorians in the Mountains: Sinking the Sublime*. Farnham: Ashgate, 2010.

Condos, Mark. *The Insecurity State: Punjab and the Making of Colonial Power in British India*. Cambridge: Cambridge University Press, 2017.

Cosgrove, Denis E. *Apollo's Eye: A Cartographic Genealogy of the Earth in the Western Imagination*. Baltimore: Johns Hopkins University Press, 2001.

Cosgrove, Denis E. and Veronica Della Dora. *High Places: Cultural Geographies of Mountains, Ice and Science*. London: I. B. Tauris, 2009.

Damodaran, Vinita. 'The East India Company, Famine and Ecological Conditions in Eighteenth-Century Bengal'. In *The East India Company and the Natural World*, edited by Vinita Damodaran, Anna Winterbottom and Alan Lester, 80–101. Basingstoke: Palgrave Macmillan, 2015.

Debarbieux, Bernard. 'The Various Figures of Mountains in Humboldt's Science and Rhetoric [Figures et Unité de l'idée de montagne chez Alexandre von Humboldt]'. *Cybergeo: European Journal of Geography* [En ligne] (2012). http://cybergeo.revues.org/25488.

Debarbieux, Bernard and Gilles Rudaz. *The Mountain: A Political History from the Enlightenment to the Present*. Translated by Jane Marie Todd. Chicago: University of Chicago Press, 2015.

Della Dora, Veronica. *Mountain: Nature and Culture*. London: Reaktion Books, 2016.

Dening, Greg. *Islands and Beaches: Discourse on a Silent Land: Marquesas, 1774–1880*. Honolulu: University Press of Hawaii, 1980.

Desmond, Ray. *The European Discovery of the Indian Flora*. Oxford: Oxford University Press, 1992.

Dettelbach, Michael. 'Global Physics and Aesthetic Empire: Humboldt's Physical Portraits of the Tropics'. In *Visions of Empire: Voyages, Botany and Representations of Nature*, edited by David Philip Miller and Peter Hanns Reill, 258–92. Cambridge: Cambridge University Press, 2002.

Dettelbach, Michael. 'The Face of Nature: Precise Measurement, Mapping, and Sensibility in the Work of Alexander von Humboldt'. *Studies in History and Philosophy of Biological & Biomedical Science* 30, no. 4 (1999): 473–504.

Dettelbach, Michael. 'The Stimulations of Travel: Humboldt's Physiological Construction of the Tropics'. In *Tropical Visions in an Age of Empire*, edited by Felix Driver and Luciana Martins, 43–58. Chicago: University of Chicago Press, 2005.

Douglas, Bronwen. *Science, Voyages, and Encounters in Oceania, 1511–1850*. London: Palgrave Macmillan, 2014.

Drayton, Richard. *Nature's Government: Science, Imperial Britain, and the 'Improvement' of the World*. New Haven: Yale University Press, 2000.

Driver, Felix. *Geography Militant: Cultures of Exploration and Empire*. Malden: Blackwell, 2001.

Driver, Felix. 'Hidden Histories Made Visible? Reflections on a Geographical Exhibition'. *Transactions of the Institute of British Geographers* 38, no. 3 (2013): 420–35.

Driver, Felix. 'Intermediaries and the Archive of Exploration'. In *Indigenous Intermediaries: New Perspectives on Exploration Archives*, edited by Shino Konishi, Maria Nugent and Tiffany Shellam, 11–30. Canberra: Australian National University Press, 2015.

Driver, Felix and Luciana Martins. *Tropical Visions in an Age of Empire*. Chicago: University of Chicago Press, 2005.

Edney, Matthew. *Mapping an Empire: The Geographical Construction of British India, 1765–1843*. Chicago: University of Chicago Press, 1997.

Ehrlich, Joshua. 'Anxiety, Chaos, and the Raj'. *The Historical Journal* 63, no. 3 (2020): 777–87.

Elden, Stuart. 'Secure the Volume: Vertical Geopolitics and the Depth of Power'. *Political Geography* 34 (2013): 35–51.

Eliasson, Pär. 'Swedish Natural History Travel'. In *Narrating the Arctic: A Cultural History of Nordic Scientific Practices*, edited by Michael Bravo and Sverker Sörlin, 125–54. Canton, MA: Science History Publications, 2002.

Elshakry, Marwa. 'When Science Became Western: Historiographical Reflections'. *Isis* 101, no. 1 (2010): 98–109.

Endersby, Jim. '"From Having No Herbarium." Local Knowledge versus Metropolitan Expertise: Joseph Hooker's Australasian Correspondence with William Colenso and Ronald Gunn'. *Pacific Science* 55, no. 4 (2001): 343–58.

Endersby, Jim. *Imperial Nature: Joseph Hooker and the Practices of Victorian Science*. Chicago: University of Chicago Press, 2008.

Fan, Fa-ti. *British Naturalists in Qing China*. Cambridge, MA: Harvard University Press, 2004.

Fan, Fa-ti. 'Science in Cultural Borderlands: Methodological Reflections on the Study of Science, European Imperialism, and Cultural Encounter'. *East Asian Science, Technology and Society* 1, no. 2 (2007): 213–31.

Fan, Fa-ti. 'The Global Turn in the History of Science'. *East Asian Science, Technology and Society: An International Journal* 6 (2012): 249–58.

Feldman, Theodore S. 'Applied Mathematics and the Quantification of Experimental Physics: The Example of Barometric Hypsometry'. *Historical Studies in the Physical Sciences* 15, no. 2 (1985): 127–95.

Felsch, Philipp. 'Mountains of Sublimity, Mountains of Fatigue: Towards a History of Speechlessness in the Alps'. *Science in Context* 22, no. 3 (2009): 341–64.

Finkelstein, Gabriel. 'Conquerors of the Künlün? The Schlagintweit Mission to High Asia, 1854–57'. *History of Science* 38, no. 2 (2000): 179–218.

Fleetwood, Lachlan. 'Bodies in High Places: Exploration, Altitude Sickness, and the Problem of Bodily Comparison in the Himalaya, 1800–1850'. *Itinerario* 43, no. 3 (2019): 489–515.

Fleetwood, Lachlan. '"No Former Travellers Having Attained Such a Height on the Earth's Surface": Instruments, Inscriptions, and Bodies in the Himalaya, 1800–1830'. *History of Science* 56, no. 1 (2018): 3–34.

Fleetwood, Lachlan. 'Science and War at the Limit of Empire: William Griffith with the Army of the Indus'. *Notes and Records: The Royal Society Journal of the History of Science* 75, no. 3 (2021): 285–310.

Gardner, Kyle. 'Moving Watersheds, Borderless Maps, and Imperial Geography in India's Northwestern Himalaya'. *The Historical Journal* 62, no. 1 (2019): 149–70.

Gardner, Kyle. *The Frontier Complex: Geopolitics and the Making of the India-China Border, 1846–1962*. Cambridge: Cambridge University Press, 2021.

Geer, Alexandra van der, Michael Dermitzakis and John de Vos. 'Fossil Folklore from India: The Siwalik Hills and the Mahâbhârata'. *Folklore* 119, no. 1 (2008): 71–92.

Gellner, David N., ed. *Borderland Lives in Northern South Asia*. Durham: Duke University Press, 2013.

Gilbert, Daniel. 'The First Documented Report of Mountain Sickness: The China or Headache Mountain Story'. *Respiration Physiology* 52, no. 3 (1983): 315–26.

Godlewska, Anne Marie Claire. 'From Enlightenment Vision to Modern Science? Humboldt's Visual Thinking'. In *Geography and Enlightenment*, edited by David N. Livingstone and Charles W. J. Withers, 236–76. Chicago: University of Chicago Press, 1999.

Gorshenina, Svetlana. *Explorateurs en Asie centrale: voyageurs et aventuriers de Marco Polo à Ella Maillart*. Geneva: Editions Olizane, 2003.

Grout, Andrew. 'Geology and India, 1775–1805: An Episode in Colonial Science'. *South Asia Research* 10, no. 1 (1990): 1–18.

Grout, Andrew. 'Possessing the Earth: Geological Collections, Information and Education in India, 1800–1850'. In *The Transmission of Knowledge in South Asia: Essays on Education, Religion, History, and Politics*, edited by Nigel Crook, 245–79. Delhi: Oxford University Press, 1996.

Grove, Richard. *Green Imperialism: Colonial Expansion, Tropical Island Edens and the Origins of Environmentalism, 1600–1860*. Cambridge: Cambridge University Press, 1995.

Hansen, Peter H. 'Partners: Guides and Sherpas in the Alps and Himalayas, 1850s–1950s'. In *Voyages and Visions: Towards a Cultural History of Travel*, edited by Jaś Elsner and Joan-Pau Rubiés, 210–31. London: Reaktion Books, 1999.

Hansen, Peter H. *The Summits of Modern Man: Mountaineering after the Enlightenment*. Cambridge, MA: Harvard University Press, 2013.

Hardenberg, Wilko Graf von and Martin Mahony. 'Introduction – Up, Down, Round and Round: Verticalities in the History of Science'. *Centaurus* 62, no. 4 (2020): 595–611.

Harrison, Mark. *Climates and Constitutions: Health, Race, Environment and British Imperialism in India, 1600–1850*. Delhi: Oxford University Press, 1999.

Harrison, Mark. 'The Calcutta Botanic Garden and the Wider World, 1817–46'. In *Science and Modern India: An Institutional History, c.1784–1947*, edited by Uma Das Gupta, 235–53. Delhi: Pearson Education India, 2011.

Hautdidier, Baptiste. 'The Comparative Tableau of Mountains and Rivers: Emulation and Reappraisal of a Popular 19th-Century Visualization Design'. *Environment and Planning A* 47, no. 6 (2015): 1265–82.

Heggie, Vanessa. 'Experimental Physiology, Everest and Oxygen: From the Ghastly Kitchens to the Gasping Lung'. *British Journal for the History of Science* 46, no. 1 (2013): 123–47.

Heggie, Vanessa. *Higher and Colder: A History of Extreme Physiology and Exploration*. Chicago: University of Chicago Press, 2019.

Herbert, Eugenia W. *Flora's Empire: British Gardens in India*. Philadelphia: University of Pennsylvania Press, 2011.

Hevly, Bruce. 'The Heroic Science of Glacier Motion'. *Osiris* 11 (1996): 66–86.

Hillemann, Ulrike. *Asian Empire and British Knowledge: China and the Networks of British Imperial Expansion*. Basingstoke: Palgrave Macmillan, 2009.

Hopkins, Benjamin. *The Making of Modern Afghanistan*. London: Palgrave Macmillan, 2008.

Hopkirk, Peter. *The Great Game: On Secret Service in High Asia*. London: John Murray, 1990.

Hyde, H. Montgomery. 'Dr. George Govan and the Saharanpur Botanical Gardens'. *Journal of The Royal Central Asian Society* 49, no. 1 (1962): 47–57.

Insley, Jane. 'Making Mountains out of Molehills? George Everest and Henry Barrow 1830–39'. *Indian Journal of History of Science* 30, no. 1 (1995): 47–55.

Isserman, Maurice and Stewart Weaver. *Fallen Giants: A History of Himalayan Mountaineering from the Age of Empire to the Age of Extremes*. New Haven: Yale University Press, 2008.

Jahoda, Christian. *Socio-Economic Organisation in a Border Area of Tibetan Culture: Tabo, Spiti Valley, Himachal Pradesh, India*. Vienna: Austrian Academy of Sciences Press, 2015.

Jasanoff, Maya. *Edge of Empire: Conquest and Collecting in the East, 1750–1850*. London: Harper Perennial, 2006.

Jepson, Wendy. 'Of Soil, Situation, and Salubrity: Medical Topography and Medical Officers in Early Nineteenth-Century British India'. *Historical Geography* 32 (2004): 137–55.

Jouty, Sylvain. 'Naissance de l'altitude'. *Compar(a)ison: An International Journal of Comparative Literature* 1 (1998): 17–32.

Kakalis, Christos and Emily Goetsch, eds. *Mountains, Mobilities and Movement*. London: Palgrave Macmillan, 2018.

Keay, John. *The Great Arc: The Dramatic Tale of How India Was Mapped and Everest Was Named*. London: HarperCollins, 2000.

Keay, John. *When Men and Mountains Meet: The Explorers of the Western Himalayas, 1820–75*. London: John Murray, 1977.

Keenan, Brigid. *Travels in Kashmir: A Popular History of Its People, Places and Crafts*. Gurgaon: Hachette India, 2013.

Keighren, Innes M., Charles W. J. Withers and Bill Bell, eds. *Travels into Print: Exploration, Writing, and Publishing with John Murray, 1773–1859*. Chicago: University of Chicago Press, 2015.

Kelley, Theresa M. *Clandestine Marriage: Botany and Romantic Culture*. Baltimore: Johns Hopkins University Press, 2012.

Kennedy, Dane. 'British Exploration in the Nineteenth Century: A Historiographical Survey'. *History Compass* 5, no. 6 (2007): 1879–900.

Kennedy, Dane, ed. *Reinterpreting Exploration: The West in the World*. Oxford: Oxford University Press, 2014.

Kennedy, Dane. *The Last Blank Spaces: Exploring Africa and Australia*. Cambridge, MA: Harvard University Press, 2013.

Kennedy, Dane. *The Magic Mountains: Hill Stations and the British Raj*. Berkeley: University of California Press, 1996.

Kennedy, Kenneth. *God-Apes and Fossil Men: Paleoanthropology of South Asia*. Ann Arbor: University of Michigan Press, 2000.

Kennedy, Kenneth and Russell Ciochon. 'A Canine Tooth from the Siwaliks: First Recorded Discovery of a Fossil Ape?' *Human Evolution* 14, no. 3 (1999): 231–53.

Kochhar, Rajesh. 'Natural History in India during the 18th and 19th Centuries'. *Journal of Biosciences* 38, no. 2 (2013): 201–24.

Kohler, Robert E. and Jeremy Vetter. 'The Field'. In *A Companion to the History of Science*, edited by Bernard Lightman, 282–95. Chichester: John Wiley, 2016.

Konishi, Shino, Maria Nugent and Tiffany Shellam, eds. *Indigenous Intermediaries: New Perspectives on Exploration Archives*. Canberra: Australian National University Press, 2015.

Kumar, Deepak. 'Botanical Explorations and the East India Company: Revisiting "Plant Colonialism"'. In *The East India Company and the Natural World*, edited by Vinita Damodaran, Anna Winterbottom and Alan Lester, 16–34. Basingstoke: Palgrave Macmillan, 2015.

Kumar, Deepak. *Science and the Raj, 1857–1905*. New York: Oxford University Press, 1995.

Kumar, Deepak. 'Scientific Institutions as Sites for Dissemination and Contestation: Emergence of Colonial Calcutta as a Science City'. In *Sites of Modernity: Asian Cities in the Transitory Moments of Trade, Colonialism, and Nationalism*, edited by Wasana Wongsurawat, 33–46. Berlin: Springer, 2016.

Lamb, Alastair. *British India and Tibet, 1766–1910*. London: Routledge & Kegan Paul, 1986.

Latour, Bruno. *Science in Action: How to Follow Scientists and Engineers through Society*. Cambridge, MA: Harvard University Press, 1987.

Laudan, Rachel. *From Mineralogy to Geology: The Foundations of a Science, 1650–1830*. Chicago: University of Chicago Press, 1987.

Lawrence, Christopher and Steven Shapin. *Science Incarnate: Historical Embodiments of Natural Knowledge*. Chicago: University of Chicago Press, 1998.

Leask, Nigel. *Curiosity and the Aesthetics of Travel Writing, 1770–1840: From an Antique Land*. Oxford: Oxford University Press, 2002.

Livingstone, David N. *Putting Science in Its Place: Geographies of Scientific Knowledge*. Chicago: University of Chicago Press, 2003.

Livingstone, David N. and Charles W. J. Withers. *Geographies of Nineteenth-Century Science*. Chicago: University of Chicago Press, 2011.

Lossio, Jorge. 'British Medicine in the Peruvian Andes: The Travels of Archibald Smith M.D. (1820–1870)'. *História, Ciências, Saúde – Manguinhos* 13, no. 4 (2006): 833–50.

Ludden, David. 'The Process of Empire: Frontiers and Borderlands'. In *Tributary Empires in Global History*, edited by Peter Fibiger Bang and Christopher Bayly, 132–50. Basingstoke: Palgrave Macmillan, 2011.

MacDonald, Fraser and Charles W. J. Withers, eds. *Geography, Technology and Instruments of Exploration*. Farnham: Ashgate, 2015.

Macfarlane, Robert. *Mountains of the Mind: A History of a Fascination*. London: Granta Books, 2003.

Martin, Emma. 'Translating Tibet in the Borderlands: Networks, Dictionaries, and Knowledge Production in Himalayan Hill Stations'. *Transcultural Studies* 7, no. 1 (2016): 86–120.

Martin, Peter R. and Edward Armston-Sheret. 'Off the Beaten Track? Critical Approaches to Exploration Studies'. *Geography Compass* 14, no. 1 (2020): e12476. https://doi.org/10.1111/gec3.12476.

Mas Galvañ, Cayetano. 'Los primeros contactos de los europeos con las grandes altitudes'. *Revista de Historia Moderna. Anales de la Universidad de Alicante* 29 (2011): 139–68.

Mason, Kenneth. *Abode of Snow: A History of Himalayan Exploration and Mountaineering from Earliest Times to the Ascent of Everest*. London: Rupert Hart-Davis, 1955.

Mathieu, Jon. 'Globalisation of Mountain Perception: How Much of a Western Imposition?' *Summerhill: IIAS Review* 20, no. 1 (2014): 8–17.

Mathieu, Jon. 'Long-Term History of Mountains: Southeast Asia and South America Compared'. *Environmental History* 18, no. 3 (2013): 557–75.

Mathieu, Jon. *The Third Dimension: A Comparative History of Mountains in the Modern Era*. Translated by Katherine Brun. Cambridge: White Horse Press, 2011.

Mathur, Nayanika. 'Naturalizing the Himalaya-as-Border in Uttarakhand'. In *Borderland Lives in Northern South Asia*, edited by David N. Gellner, 72–93. Durham: Duke University Press, 2013.

Mayor, Adrienne. *The First Fossil Hunters: Paleontology in Greek and Roman Times*. Princeton: Princeton University Press, 2000.

McKay, Alex. 'Fit for the Frontier: European Understandings of the Tibetan Environment in the Colonial Era'. *New Zealand Journal of Asian Studies* 9, no. 1 (2007): 118–32.

McKay, Alex. *Kailas Histories: Renunciate Traditions and the Construction of Himalayan Sacred Geography*. Leiden: Brill, 2015.

Mehra, Parshotam. *An 'Agreed' Frontier: Ladakh and India's Northernmost Borders, 1846–1947*. Delhi: Oxford University Press, 1992.

Michael, Bernardo A. *Statemaking and Territory in South Asia: Lessons from the Anglo-Gorkha War (1814–1816)*. London: Anthem Press, 2012.

Michaud, Jean. 'Editorial – Zomia and Beyond'. *Journal of Global History* 5, no. 2 (2010): 187–214.

Morrison, Alexander. 'Introduction: Killing the Cotton Canard and Getting Rid of the Great Game: Rewriting the Russian Conquest of Central Asia, 1814–1895'. *Central Asian Survey* 33, no. 2 (2014): 131–42.

Morrison, Alexander. 'Twin Imperial Disasters: The Invasions of Khiva and Afghanistan in the Russian and British Official Mind, 1839–1842'. *Modern Asian Studies* 48, no. 1 (2014): 253–300.

Mosca, Matthew W. 'Kashmiri Merchants and Qing Intelligence Networks in the Himalayas: The Ahmed Ali Case of 1830'. In *Asia Inside Out: Connected Places*, edited by Eric Tagliacozzo, Helen F. Siu and Peter C. Perdue, 219–42. Cambridge, MA: Harvard University Press, 2015.

Mosca, Matthew W. *From Frontier Policy to Foreign Policy: The Question of India and the Transformation of Geopolitics in Qing China*. Stanford: Stanford University Press, 2013.

Mueggler, Erik. *The Paper Road: Archive and Experience in the Botanical Exploration of West China and Tibet*. Berkeley: University of California Press, 2011.

Nagar, Khyati. 'Between Calcutta and Kew: The Divergent Circulation and Production of Hortus Bengalensis and Flora Indica'. In *The Circulation of Knowledge between Britain, India and China: The Early-Modern World to the Twentieth Century*, edited by Bernard Lightman, Gordon McOuat and Larry Stewart, 153–78. Leiden: Brill, 2013.

Nair, Savithri Preetha. '"Eyes and No Eyes": Siwalik Fossil Collecting and the Crafting of Indian Palaeontology (1830–1847)'. *Science in Context* 18, no. 3 (2005): 359–92.

Nair, Savithri Preetha. 'Native Collecting and Natural Knowledge (1798–1832): Raja Serfoji II of Tanjore as a "Centre of Calculation"'. *Journal of the Royal Asiatic Society* 15, no. 3 (2005): 279–302.

Naylor, Simon and Simon Schaffer. 'Nineteenth-Century Survey Sciences: Enterprises, Expeditions and Exhibitions'. *Notes and Records: The Royal Society Journal of the History of Science* 73, no. 2 (2019): 135–47.

Neuhaus, Tom. *Tibet in the Western Imagination*. New York: Palgrave Macmillan, 2012.

Nicolson, Malcolm. 'Alexander von Humboldt and the Geography of Vegetation'. In *Romanticism and the Sciences*, edited by Andrew Cunningham and Nicholas Jardine, 169–85. Cambridge: Cambridge University Press, 1990.

Nicolson, Malcolm. 'Alexander von Humboldt, Humboldtian Science and the Origins of the Study of Vegetation'. *History of Science* 25, no. 2 (1987): 167–94.

Nicolson, Marjorie Hope. *Mountain Gloom and Mountain Glory: The Development of the Aesthetics of the Infinite*. Ithaca: Cornell University Press, 1959.

Noltie, Henry. *Indian Botanical Drawings, 1793–1868, from the Royal Botanic Garden*. Edinburgh: Royal Botanic Garden Edinburgh, 1999.

Noltie, Henry. *Robert Wight and the Botanical Drawings of Rungiah and Govindoo*. 3 vols. Edinburgh: Royal Botanic Garden Edinburgh, 2007.

Oelz, Oswald and Elisabeth Simons. *Kopfwehberge: Eine Geschichte Der Höhenmedizin*. Zurich: AS Verlag, 2001.

Ortner, Sherry B. *Life and Death on Mt. Everest: Sherpas and Himalayan Mountaineering*. Princeton: Princeton University Press, 2001.

Outram, Dorinda. 'On Being Perseus: New Knowledge, Dislocation, and Enlightenment Exploration'. In *Geography and Enlightenment*, edited by David N. Livingstone and Charles W. J. Withers, 281–94. Chicago: University of Chicago Press, 1999.

Pandit, Maharaj K. *Life in the Himalaya: An Ecosystem at Risk*. Cambridge, MA: Harvard University Press, 2017.

Pang, Alex Soojung-Kim. *Empire and the Sun: Victorian Solar Eclipse Expeditions*. Stanford: Stanford University Press, 2002.

Phillimore, Reginald Henry. *Historical Records of the Survey of India*. 5 vols. Dehradun: Survey of India, 1945–68.

Pietsch, Tamson. 'Bodies at Sea: Travelling to Australia in the Age of Sail'. *Journal of Global History* 11, no. 2 (2016): 209–28.

Pont, Jean-Claude and Jan Lacki, eds. *Une cordée originale: Histoire des relations entre science et montagne*. Chêne-Bourg: Georg, 2000.

Poskett, James. *Materials of the Mind: Phrenology, Race, and the Global History of Science, 1815–1920*. Chicago: University of Chicago Press, 2019.

Pratt, Mary Louise. *Imperial Eyes: Travel Writing and Transculturation*. 2nd ed. London: Routledge, 2008.

Raj, Kapil. *Relocating Modern Science: Circulation and the Construction of Knowledge in South Asia and Europe, 1650–1900*. New York: Palgrave Macmillan, 2007.

Raj, Kapil. 'When Human Travellers Become Instruments: The Indo-British Exploration of Central Asia in the Nineteenth Century'. In *Instruments, Travel and Science: Itineraries of Precision from the Seventeenth to the Twentieth Century*, edited by Christian Licoppe, Marie-Noëlle Bourguet and H. Otto Sibum, 156–88. London: Routledge, 2002.

Raza, Rosemary. *In Their Own Words: British Women Writers and India, 1740–1857*. Oxford: Oxford University Press, 2006.

Reidy, Michael S. 'From the Oceans to the Mountains: Spatial Science in an Age of Empire'. In *Knowing Global Environments: New Historical Perspectives on the Field Sciences*, edited by Jeremy Vetter, 17–38. New Brunswick: Rutgers University Press, 2011.

Reidy, Michael S. 'Mountaineering, Masculinity, and the Male Body in Mid-Victorian Britain'. *Osiris* 30, no. 1 (2015): 158–81.

Reidy, Michael S. 'The Most Recent Orogeny: Verticality and Why Mountains Matter'. *Historical Studies in the Natural Sciences* 47, no. 4 (2017): 578–87.

Reidy, Michael S. *Tides of History: Ocean Science and Her Majesty's Navy*. Chicago: University of Chicago Press, 2009.

Roberts, Lissa. 'Situating Science in Global History: Local Exchanges and Networks of Circulation'. *Itinerario* 33, no. 1 (2009): 9–30.

Robinson, Michael. 'Science and Exploration'. In *Reinterpreting Exploration: The West in the World*, edited by Dane Kennedy, 21–37. Oxford: Oxford University Press, 2014.

Roche, Clare. 'Women Climbers 1850–1900: A Challenge to Male Hegemony?' *Sport in History* 33, no. 3 (2013): 236–59.

Rocher, Rosane and Ludo Rocher. *The Making of Western Indology: Henry Thomas Colebrooke and the East India Company*. Abingdon: Routledge, 2014.

Rogaski, Ruth. 'Knowing a Sentient Mountain: Space, Science, and the Sacred in Ascents of Mount Paektu/Changbai'. *Modern Asian Studies* 52, no. 2 (2018): 716–52.

Roller, Heather F. 'River Guides, Geographical Informants, and Colonial Field Agents in the Portuguese Amazon'. *Colonial Latin American Review* 21, no. 1 (2012): 101–26.

Romanowski, Sylvie. 'Humboldt's Pictorial Science: An Analysis of the Tableau physique des Andes et pays voisins'. In *Essay on the Geography of Plants*, edited by Stephen T. Jackson, 157–98. Chicago: University of Chicago Press, 2013.

Rudwick, Martin. *Bursting the Limits of Time: The Reconstruction of Geohistory in the Age of Revolution*. Chicago: University of Chicago Press, 2005.

Rudwick, Martin. 'Minerals, Strata and Fossils'. In *Cultures of Natural History*, edited by Nicholas Jardine, James A. Secord and Emma C. Spary, 266–86. Cambridge: Cambridge University Press, 1996.

Rudwick, Martin. *Worlds before Adam: The Reconstruction of Geohistory in the Age of Reform*. Chicago: University of Chicago Press, 2008.

Rupke, Nicholaas. 'Humboldtian Distribution Maps: The Spatial Ordering of Scientific Knowledge'. In *The Structure of Knowledge: Classifications of Science and Learning Since the Renaissance*, edited by Tore Frängsmyr, 93–116. Berkeley: Office for History of Science and Technology, University of California, 2001.

Sangwan, Satpal. 'From Gentlemen Amateurs to Professionals: Reassessing the Natural Science Tradition in Colonial India 1780–1840'. In *Nature and the Orient: The Environmental History of South and Southeast Asia*, edited by Richard Grove, Vinita Damodaran and Satpal Sangwan, 210–36. Delhi: Oxford University Press, 1998.

Sangwan, Satpal. 'Reordering the Earth: The Emergence of Geology as a Scientific Discipline in Colonial India'. *The Indian Economic & Social History Review* 31, no. 3 (1994): 291–310.

Sangwan, Satpal. 'The Strength of a Scientific Culture: Interpreting Disorder in Colonial Science'. *The Indian Economic and Social History Review* 34, no. 2 (1997): 217–50.

Schaffer, Simon. 'Easily Cracked: Scientific Instruments in States of Disrepair'. *Isis* 102, no. 4 (2011): 706–17.

Schaffer, Simon. 'The Bombay Case: Astronomers, Instrument Makers and the East India Company'. *Journal for the History of Astronomy* 43, no. 2 (2012): 151–80.

Schaffer, Simon, Lissa Roberts, Kapil Raj and James Delbourgo, eds. *The Brokered World: Go-betweens and Global Intelligence, 1770–1820*. Sagamore Beach: Science History Publications, 2009.

Schär, Bernhard C. 'On the Tropical Origins of the Alps: Science and the Colonial Imagination of Switzerland, 1700–1900'. In *Colonial Switzerland: Rethinking Colonialism from the Margins*, edited by Patricia Purtschert and Harald Fischer-Tiné, 29–49. London: Palgrave Macmillan, 2015.

Schiebinger, Londa L. *Plants and Empire: Colonial Bioprospecting in the Atlantic World*. Cambridge, MA: Harvard University Press, 2004.

Schiebinger, Londa L. and Claudia Swan. *Colonial Botany: Science, Commerce, and Politics in the Early Modern World*. Philadelphia: University of Pennsylvania Press, 2005.

Scott, James C. *The Art of Not Being Governed: An Anarchist History of Upland Southeast Asia*. New Haven: Yale University Press, 2009.

Searle, Mike. *Colliding Continents: A Geological Exploration of the Himalaya, Karakoram, and Tibet*. Oxford: Oxford University Press, 2013.

Secord, James A. 'Knowledge in Transit'. *Isis* 95, no. 4 (2004): 654–72.
Sen, Joydeep. *Astronomy in India, 1784–1876*. London: Pickering & Chatto, 2014.
Shapin, Steven and Simon Schaffer. *Leviathan and the Air-Pump: Hobbes, Boyle, and the Experimental Life*. Princeton: Princeton University Press, 1985.
Sharma, Jayeeta. 'A Space That Has Been Laboured On: Mobile Lives and Transcultural Circulation Around Darjeeling and the Eastern Himalayas'. *Transcultural Studies* 7, no. 1 (2016): 54–85.
Sharma, Jayeeta. *Empire's Garden: Assam and the Making of India*. Durham: Duke University Press, 2011.
Sharma, Jayeeta. 'Producing Himalayan Darjeeling: Mobile People and Mountain Encounters'. *Himalaya, the Journal of the Association for Nepal and Himalayan Studies* 35, no. 2 (2016): 87–101.
Shellam, Tiffany, Maria Nugent, Shino Konishi and Allison Cadzow, eds. *Brokers and Boundaries: Colonial Exploration in Indigenous Territory*. Canberra: Australian National University Press, 2016.
Shneiderman, Sara. 'Are the Central Himalayas in Zomia? Some Scholarly and Political Considerations Across Time and Space'. *Journal of Global History* 5, no. 2 (2010): 289–312.
Shneiderman, Sara. 'Himalayan Border Citizens: Sovereignty and Mobility in the Nepal–Tibetan Autonomous Region (TAR) of China Border Zone'. *Political Geography* 35 (2013): 25–36.
Simpson, Thomas. '"Clean Out of the Map": Knowing and Doubting Space at India's High Imperial Frontiers'. *History of Science* 55, no. 1 (2017): 3–36.
Simpson, Thomas. 'Forgetting Like a State in Colonial North-East India'. In *Mountstuart Elphinstone in South Asia: Pioneer of British Colonial Rule*, edited by Shah Mahmoud Hanifi, 223–48. Oxford: Oxford University Press, 2019.
Simpson, Thomas. 'Modern Mountains from the Enlightenment to the Anthropocene'. *The Historical Journal* 62, no. 2 (2019): 553–81.
Simpson, Thomas. *The Frontier in British India: Space, Science, and Power in the Nineteenth Century*. Cambridge: Cambridge University Press, 2021.
Singh, Chetan, ed. *Recognizing Diversity: Society and Culture in the Himalaya*. Oxford: Oxford University Press, 2011.
Sivaramakrishnan, Kalyanakrishnan. 'Science, Environment and Empire History: Comparative Perspectives from Forests in Colonial India'. *Environment and History* 14, no. 1 (2008): 41–65.
Sivasundaram, Sujit. 'Focus: Global Histories of Science: Introduction'. *Isis* 101, no. 1 (2010): 95–97.
Sivasundaram, Sujit. *Islanded: Britain, Sri Lanka, and the Bounds of an Indian Ocean Colony*. Chicago: University of Chicago Press, 2013.
Sivasundaram, Sujit. 'Islanded: Natural History in the British Colonization of Ceylon'. In *Geographies of Nineteenth Century Science*, edited by David N. Livingstone and Charles W. J. Withers, 123–48. Chicago: University of Chicago Press, 2011.
Sivasundaram, Sujit. 'Race, Empire and Biology before Darwin'. In *Biology and Ideology: From Descartes to Dawkins*, edited by Denis R. Alexander and Ronald L. Numbers, 114–38. Chicago: University of Chicago Press, 2010.

Sivasundaram, Sujit. 'Sciences and the Global: On Methods, Questions, and Theory'. *Isis* 101, no. 1 (2010): 146–58.

Sohrabuddin, Mohammed. 'Construction of the "Himalayas": European Naturalists and the Oriental Mountains'. In *Force of Nature: Essays on History and Politics of Environment*, edited by Sajal Nag, 87–108. Abingdon: Routledge, 2018.

Sorkhabi, Rasoul B. 'Historical Development of Himalayan Geology'. *Journal of the Geological Society of India* 49 (1997): 89–108.

Stafford, Robert. *Scientist of Empire: Sir Roderick Murchison, Scientific Exploration and Victorian Imperialism*. Cambridge: Cambridge University Press, 1989.

Stewart, Gordon. *Journeys to Empire: Enlightenment, Imperialism and the British Encounter with Tibet*. Cambridge: Cambridge University Press, 2009.

Stoler, Ann Laura. *Along the Archival Grain: Epistemic Anxieties and Colonial Common Sense*. Princeton: Princeton University Press, 2010.

Teltscher, Kate. *The High Road to China: George Bogle, the Panchen Lama and the First British Expedition to Tibet*. London: Bloomsbury, 2006.

Théodoridès, Jean. 'Humboldt and England'. *The British Journal for the History of Science* 3, no. 1 (1966): 39–55.

Thomas, Nicholas. *Islanders: The Pacific in the Age of Empire*. New Haven: Yale University Press, 2010.

Tracy, Sarah W. 'The Physiology of Extremes: Ancel Keys and the International High Altitude Expedition of 1935'. *Bulletin of the History of Medicine* 86, no. 4 (2012): 627–60.

Trautmann, Thomas. 'Foreword'. In *The Origins of Himalayan Studies: Brian Houghton Hodgson in Nepal and Darjeeling*, edited by David Waterhouse, xiii–xix. London: Routledge Curzon, 2004.

Tresch, John. 'Even the Tools Will Be Free: Humboldt's Romantic Technologies'. In *The Heavens on Earth: Observatories and Astronomy in Nineteenth Century Science and Culture*, edited by David Aubin, Charlotte Bigg and Otto Sibum, 253–85. Durham: Duke University Press, 2010.

Trouillot, Michel-Rolph. *Silencing the Past: Power and the Production of History*. Boston: Beacon Press, 1995.

Turner, Gerard L'Estrange. *Nineteenth-Century Scientific Instruments*. London: Sotheby, 1983.

Unsworth, Walt. *Hold the Heights: The Foundations of Mountaineering*. London: Hodder & Stoughton, 1993.

Van Schendel, Willem. 'Geographies of Knowing, Geographies of Ignorance: Jumping Scale in Southeast Asia'. *Environment and Planning D – Society & Space* 20, no. 6 (2002): 647–68.

Vicziany, Marika. 'Imperialism, Botany and Statistics in Early Nineteenth-Century India: The Surveys of Francis Buchanan (1762–1829)'. *Modern Asian Studies* 20, no. 4 (1986): 625–60.

Waller, Derek. *The Pundits: British Exploration of Tibet and Central Asia*. Lexington: University Press of Kentucky, 1990.

Ward, Michael. 'Mountain Medicine and Physiology: A Short History'. *The Alpine Journal* 95 (1990): 191–98.

Waterhouse, David, ed. *The Origins of Himalayan Studies: Brian Houghton Hodgson in Nepal and Darjeeling*. London: Routledge Curzon, 2004.
Weizman, Eyal. *Hollow Land: Israel's Architecture of Occupation*. London: Verso, 2007.
West, John B. *High Life: A History of High-Altitude Physiology and Medicine*. Oxford: Oxford University Press, 1998.
Wilson, Jon. *The Domination of Strangers: Modern Governance in Eastern India, 1780–1835*. Basingstoke: Palgrave Macmillan, 2008.
Winterbottom, Anna. *Hybrid Knowledge in the Early East India Company World*. Basingstoke: Palgrave Macmillan, 2016.
Withers, Charles W. J. 'Geography and "Thing Knowledge": Instrument Epistemology, Failure, and Narratives of 19th-Century Exploration'. *Transactions of the Institute of British Geographers* 44, no. 4 (2019): 676–91.
Withers, Charles W. J. 'Science, Scientific Instruments and Questions of Method in Nineteenth-Century British Geography'. *Transactions of the Institute of British Geographers* 38, no. 1 (2013): 167–79.
Withers, Charles W. J. and Innes M. Keighren. 'Travels into Print: Authoring, Editing and Narratives of Travel and Exploration, c.1815–c.1857'. *Transactions of the Institute of British Geographers* 36, no. 4 (2011): 560–73.
Wolter, John A. 'The Heights of Mountains and the Lengths of Rivers'. *The Quarterly Journal of the Library of Congress* 29, no. 3 (1972): 186–205.
Woodman, Dorothy. *Himalayan Frontiers: A Political Review of British, Chinese, Indian and Russian Rivalries*. London: The Cresset Press, 1969.
Yü, Dan Smyer. 'Introduction: Trans-Himalayas as Multistate Margins'. In *Trans-Himalayan Borderlands: Livelihoods, Territorialities, Modernities*, edited by Dan Smyer Yü and Jean Michaud, 11–41. Amsterdam: Amsterdam University Press, 2017.

Unpublished Theses

Grout, Andrew. 'Geology and India, 1770–1851: A Study in the Methods and Motivations of a Colonial Science'. PhD Dissertation, School of Oriental and African Studies, 1995.
Jones, Lowri. 'Local Knowledge and Indigenous Agency in the History of Exploration'. PhD Dissertation, Royal Holloway, University of London, 2010.
Mathur, Tapsi. 'How Professionals Became Natives: Geography and Trans-Frontier Exploration in Colonial India'. PhD Dissertation, University of Michigan, 2018.
Thomas, Adrian. 'Calcutta Botanic Garden: Knowledge Formation and the Expectations of Botany in a Colonial Context, 1833–1914'. PhD Dissertation, King's College London, 2016.

Index

www.ingramcontent.com/pod-product-compliance
Ingram Content Group UK Ltd.
Pitfield, Milton Keynes, MK11 3LW, UK
UKHW020403180125